Measurement Errors: Theory and Practice

Measurement Errors: Theory and Practice

Semyon Rabinovich

Translated by
M.E. Alferieff

American Institute of Physics **New York**

In recognition of the importance of preserving what has been written, it is a policy of the American Institute of Physics to have books published in the United States printed on acid-free paper.

Printed in the United States of America.

AIP Press
American Institute of Physics
500 Sunnyside Boulevard
Woodbury, NY 11797-2999

Library of Congress Cataloging-in-Publication Data
Rabinovich, S. G.
[Pogreshnosti izmernii. Teoriy i prakticheskie metody. English]
Measurement errors: theory and practice / Semyon Rabinovich;
translated by M. E. Alferieff.
p. cm.
Translation of: Pogreshnosti izmernii. Teoriy i prakticheskie metody.
Includes bibliographical references and index.
ISBN 1-56396-323-X
1. Mensuration. I. Title.
T50. R24 1992 92-28122
530.8-dc20 CIP

10 9 8 7 6 5 4 3 2 1

Contents

Preface

The major objective of this book is to give methods for estimating errors and uncertainties of real measurements: measurements that are performed in industry, commerce, and experimental research.

The need for such a book is due to the fact that the existing theory of measurement errors was historically developed as an abstract mathematical discipline. As a result, this theory allows estimation of uncertainties of some ideal measurements only and is not applicable to the great majority of practical cases. In particular, it is not applicable to single measurements. This situation did not bother mathematicians, while engineers, not being bold enough to assert that the mathematical theory of errors cannot satisfy their needs, solved their particular problems in one or another *ad hoc* manner.

Actually, any measurement of a physical quantity is not abstract, but involves an entirely concrete procedure that is always implemented using concrete technical devices—measuring instruments—under concrete conditions. Therefore, in order to obtain realistic estimates of measurement uncertainties, mathematical methods must be supplemented with methods that make it possible to take into account data on properties of measuring instruments, the conditions under which measurements are performed, the measurement procedure, and other features of measurements.

The importance of the methods of estimating measurement inaccuracies for practice can scarcely be exaggerated. Indeed, in another stage of planning a measurement or using a measurement result, one must know its error limits or uncertainty. Inaccuracy of a measurement determines its quality and is related to its cost. Reliability of product quality control also depends on inaccuracy of measurements. Without estimating measurement inaccuracies, one cannot compare measurement results obtained by different authors. Finally, it is now universally recognized that the precision with which any calculation using experimental data is performed must be in concordance with the accuracy of these data.

In this book the entire hierarchy of questions pertaining to measurement errors is studied, a theory of measurement inaccuracy is developed, and specific recommendations are made for solving the basic problems arising in practice. In addition, methods are presented for calculating the errors of measuring instruments. The attention devoted to the properties of measuring instruments, taking into account their relations with measurement inaccuracies, is one of the highlights of this book.

This book is a product of my professional scientific experience accumulated over many years of work in instrumentation and metrology. From 1948 to 1964 I was involved in the investigation and development of various electric measuring instruments, including calibrating potentiometers and stabilizers, extremely sensitive dc voltage and current amplifiers, automatic plotters, etc. This experience gave me a grip in understanding problems arising in real measurements. Then, in 1965 I organized and until 1980 directed a laboratory of theoretical metrology. I focused on the analysis and generalization of theoretical problems in metrology. In particular, because I discovered that there is a rift between theory and practice (as mentioned above), I concentrated on the problem of estimating measurement errors. The results achieved during these years formed the foundation of my book *Measurement Errors* (Ref. 51). Further work and new results led to the writing of this book.

The book has eleven chapters. Chapter 1 contains general information on measurements and metrology. Although introductory, the chapter includes some questions that are solved or presented anew. Partially introductory is also Chap. 2, devoted to measuring instruments. However, a large portion of it presents analysis of methods of standardization of the metrological characteristics of measuring instruments, which are important for practice and necessary for estimating measurement errors and uncertainties. Statistical analysis of errors of several batches of various measuring instruments obtained by standards laboratories is given. The analysis shows that such data are statistically unstable and hence cannot be used as the basis for obtaining a distribution function of errors of measuring instruments. This important result influenced the ways in which many problems are covered in this book.

The inaccuracy of measurements always has to be estimated based on indirect data by finding and then summing the elementary components of the inaccuracy. In Chap. 3, a general analysis of elementary errors of measurements is given. Also, the classification of elementary errors is presented and their mathematical models are introduced. Two fundamentally important methods of constructing a convolution of distribution functions are presented. These methods are necessary for summing elementary errors.

Chapter 4 contains methods of mathematical statistics as applied to idealized multiple measurements. In essence, these methods constitute the classical theory of measurement errors. They are supplemented here with some new results.

In Chap. 5, real direct measurements are considered. It is shown that single measurements should be considered as the basic form of measurements. Various methods for estimating and combining systematic and random errors are considered, and a comparative analysis of these methods is given. Special attention is paid to taking into account the errors of measuring instruments. For instance, it is shown how the uncertainty of a measurement result decreases when more accurate information on the properties of measuring instruments is used. This chapter concludes with a step-by-step procedure for estimating errors or uncertainties of direct measurements.

In Chap. 6, indirect measurements are considered, while Chap. 7 is devoted to simultaneous and combined measurements. The problems of indirect measurements are solved using the traditional Taylor's series. However, it is shown that if the second-order members in the Taylor's series are retained, an adjustment of the measurement result is appropriate. Also, a method for solving the problems of indirect measurements by bringing them to the scheme of direct measurements is proposed.

For simultaneous and combined measurements the well-known least-squares method is used. At the same time, it is noted that if one cannot assume that measurement errors have a normal distribution, this method becomes nonoptimal.

Chapter 8 contains methods for combining measurement results. Such methods are necessary in the cases where the same measurand is measured in multiple stages or in different laboratories. Along with the traditional solution, which takes into consideration only random errors, Chap. 8 includes a method taking into account systematic errors as well.

In Chapters 9 and 10, I return to considering measuring instruments. Chapter 9 gives general methods for calculating their total errors that are useful during the development of the instruments. In Chap. 10, calibration methods that tie measuring instruments to corresponding standards are considered.

Chaper 11 contains methods for accounting for dependencies between components of measurement uncertainty. One of them is the usage of correlation coeffieicient. Another one is the method of reduction described in Chaper 6. It is shown that method of reduction is a good alternative to calculations based on correlation coefficient.

The book is targeted for practical use and, to this end, includes many concrete examples, many of which illustrate typical problems arising in the practice of measurements.

This book is intended for anyone who is concerned with measurements in any field of science or technology, who designs technological processes and chooses for them instruments having appropriate accuracy, and who designs and tests new measuring devices. I also believe this book will prove useful to many university and college students. Indeed, measurements are of such fundamental importance for modern science and engineering that every engineer and every scientist doing experimental research must know the basics of the theory of measurements and especially how to estimate their accuracy.

In conclusion, remembering with pleasure the many years of collaboration, I would like to thank Dr. Abram Kagan, now Professor at the University of Maryland, for the discussions on several issues covered in the book and for composing, at my request, a review of the problems of mathematical statistics in metrology (Ref. 37).

This softcover edition has a new chapter, 11, based on my paper presented to the Measurement Science Conference that took place in Pasadena, California January 27 and 28, 1994.

Chapter 1

General information about measurements

1.1. Basic concepts and terms

The theory of measurement errors is a branch of metrology—the science of measurements. In presenting the theory we shall adhere, whenever possible, to the terminology given in the International Vocabulary of Basic and General Terms of Metrology.[1] We shall discuss the terms that are most important for this book.

A measurable quantity (briefly—quantity) is a property of phenomena, bodies, or substances that can be defined qualitatively and expressed quantitatively.

The first measurable quantities were probably length, mass, and time, i.e., quantities that people employed in everyday life, and these concepts appeared unconsciously. Later, with the development of science, measurable quantities came to be introduced consciously in order to study the corresponding laws in physics, chemistry, and biology.

Measurable quantities are also called physical quantities. The principal feature of physical quantities is that they can be measured.

The term *quantity* is used in both the general and particular sense. It is used in the general sense when referring to the general properties of objects, for example, length, mass, temperature, or electric resistance. It is used in the particular sense when referring to the properties of a specific object: the length of a given rod, the electric resistance of a given segment of wire, etc.

Measurement is the process of finding the value of a physical quantity experimentally with the help of special technical means called *measuring instruments.*

The *result of a measurement* is the value of a physical quantity expressed in the units adopted for it and obtained by means called measuring instruments.

The definitions presented above underscore three features of measurement:

(1) The result of a measurement must always be a concrete denominated number expressed in sanctioned units of measurements. The purpose of measurement is essentially to represent a property of an object by a number.

(2) A measurement is always performed with the help of some measuring instrument; measurement is impossible without measuring instruments.

(3) Measurement is always an experimental procedure.

The *true value of a measurable quantity* is the value of the measured physical quantity, which, being known, would ideally reflect, both qualitatively and quantitatively, the corresponding property of the object.

Measuring instruments are created by humans, and every measurement on the whole is an experimental procedure. Therefore results of measurements cannot be absolutely accurate. This unavoidable imperfection of measurements is expressed in their inaccuracy. Quantitatively the measurement inaccuracy is characterized by the notion of either error or uncertainty.

The *measurement error* is the deviation of the result of measurement from the true value of the measurable quantity, expressed in absolute or relative form.

If A is the true value of the measurable quantity and $\tilde{A}$ is the result of measurement, then the absolute error of measurement is $\zeta = \tilde{A} - A$.

The error expressed in absolute form is called the absolute measurement error. The error expressed in relative form is called the relative measurement error.

The absolute error is usually identified by the fact that it is expressed in the same units as the measurable quantity.

The relative error is the error expressed as a fraction of the true value of the measurable quantity $\varepsilon = (\tilde{A} - A)/A$. Relative errors are normally given as a percent and sometimes per thousand (denoted by ‰). Very small errors, which are encountered in the most precise measurements, are customarily expressed directly as fractions of the measured quantity.

Uncertainty of measurement is an interval within which a true value of a measurand lies with a given probability. Uncertainty is defined with its limits that are read out from a result of measurement in compliance with the mentioned probability. Like an error, uncertainty can be specified in absolute or relative form.

Inaccuracy of measurements characterize the imperfection of measurements. A positive characteristic of measurements is their accuracy. The accuracy of a measurement reflects how close the result is to the true value of the measured quantity.

A measurement is all the more accurate the smaller its error is. Absolute errors, however, depend in general on the value of the measured quantity and for this reason they are not a suitable quantitative characteristic of measurement accuracy. Relative errors do not have this drawback. For this reason, accuracy can be characterized quantitatively by a number equal to the inverse of the relative error expressed as a fraction of the measured quantity. For example, if the limits of error of a measurement are $\pm 2 \times 10^{-3}\% = \pm 2 \times 10^{-5}$, then the accuracy of this measurement will be 5×10^{4}. The accuracy is expressed only as a positive number, so that it is calculated based on the modulus of the limits of the measurement error.

Although it is possible to introduce in this manner the quantitative characteristic of accuracy, in practice accuracy is normally not estimated quantitatively and it is usually characterized indirectly with the help of the measurement error or the uncertainty of measurement.

Other concepts and terms will be explained as they are introduced, and they are given in the Glossary.

1.2. Metrology and the basic metrological problems

Comparison is an age-old element of human thought, and the process of making comparisons lies at the heart of measurement: Uniform quantities characterizing different objects are identified and then compared; one quantity is taken to be the unit of measurement, and all other quantities are compared with it. This is how the first measures, i.e., objects the size of whose corresponding physical quantity is taken to be unity or a known number of units, arose.

There was a time when even different cities had their own units and measures. Then it was necessary to know how measures were related. This problem gave birth to the science of measures—metrology.

But the content of metrology, as that of most sciences, is not immutable. Especially profound changes started in the second half of the 19th century, when industry and science developed rapidly and, in particular, electrical technology and instrument building began. Measurements were no longer a part of production processes and commerce, and they became a powerful means of gaining knowledge—they became a tool of science. The role of measurements has increased especially today, in connection with the rapid development of science, the assimilation of atomic energy and space, and the development of electronics and other new fields of technology.

The development of science and technology, intercourse among peoples, and international trade have prompted many countries to adopt the same units of physical quantities. The most important step in this direction was the signing of the Metric Convention [(Treaty of the Meter), 1875]. This act had enormous significance not only with regard to the unification of physical quantities and dissemination of the metric system, but also with regard to unifying measurements throughout the world. The Metric Convention and the institutions created by it—the General Conference on Weights and Measures, the International Committee and International Bureau of Weights and Measures—continue their important work even now. In 1960 the General Conference adopted the international system of units (SI).[9] Most countries now use this system.

The content of metrology also changed in accordance with the change in the problems of measurements. Metrology has become the science of measurements. The block diagrams in Fig. 1.1 show the range of questions encompassed by modern metrology. The questions are incorporated into sections and subsections, whose names give an idea of their content. The content of some of them must nonetheless be explained.

(1) The study of physical, i.e., measurable, quantities and their units [Fig. 1.1(d)]. Physical quantities are introduced in different fields of knowledge, in physics, chemistry, biology, etc. The rules for introducing and classifying them and for forming systems of units and for optimizing these systems, etc., cannot be addressed in any of these sciences and already for this reason they must be included among the problems addressed in metrology. Moreover, the size of a quantity to be used as a unit of measurement and its determination are also very important for measurement accuracy. One need only recall the fact that when the distance between two markings on a platinum–irridium rod was adopted for the meter, for the most accurate measurement of length the inaccuracy was not less than 10^{-6}. When the meter was later defined as a definite number (1 650 763.73) of wavelengths of krypton-86 radiation in vacuum, this inaccuracy was reduced to 10^{-7}–10^{-8}. Now, when the definition of the meter is based on the velocity of light in vacuum, the inaccuracy in measuring length has been reduced by another order of magnitude and it can be reduced even more.

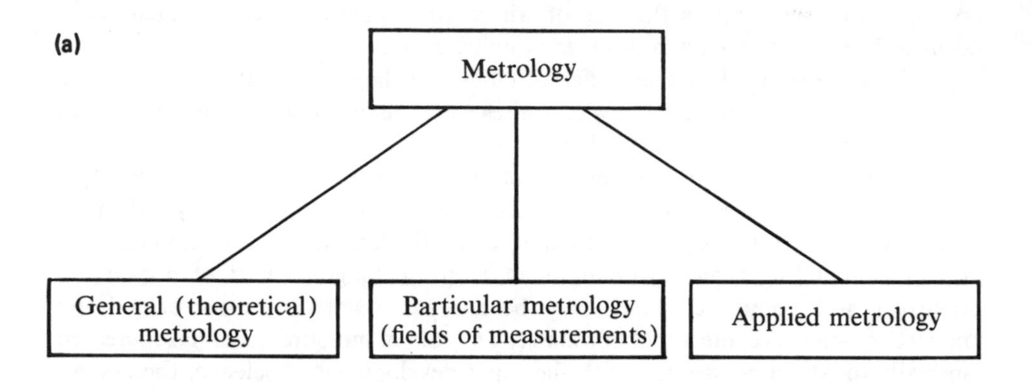

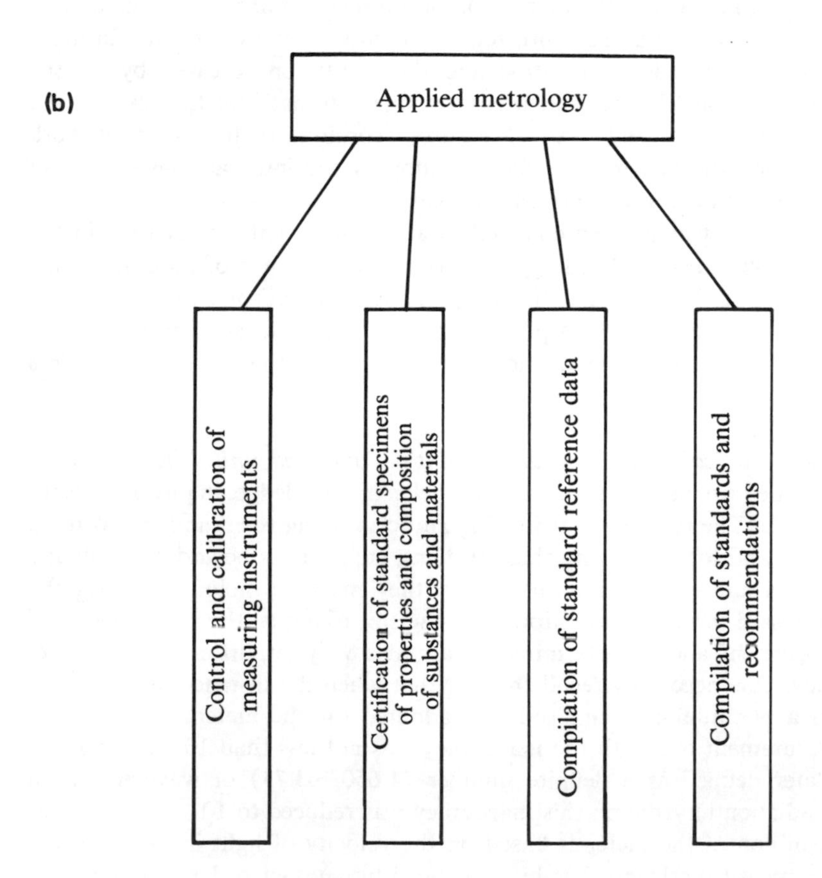

FIG. 1.1. Schematic picture of the basic problems of metrology. (a) Metrology; (b) applied metrology; (c) particular metrology; (d) general metrology.

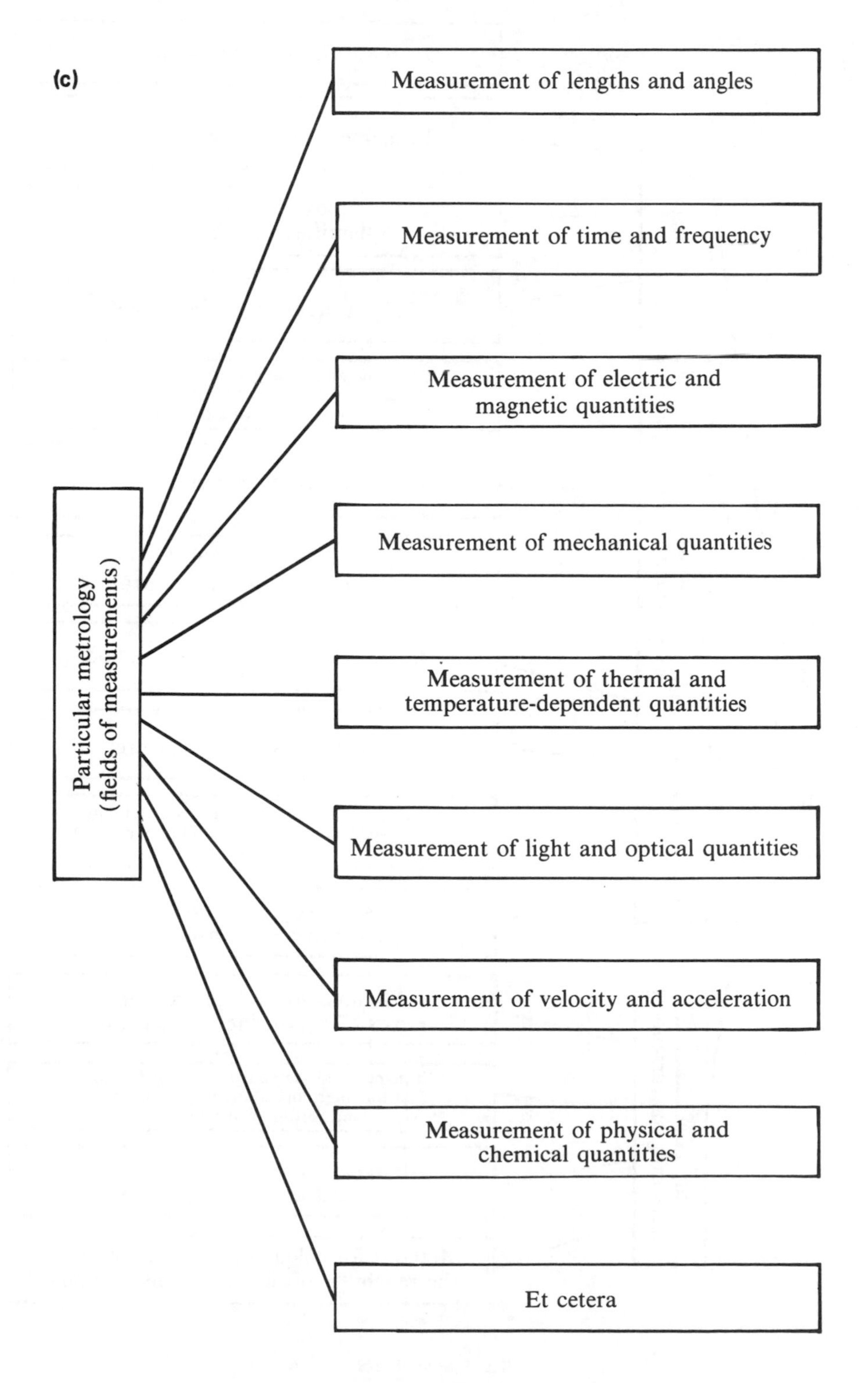

FIG. 1.1. continued.

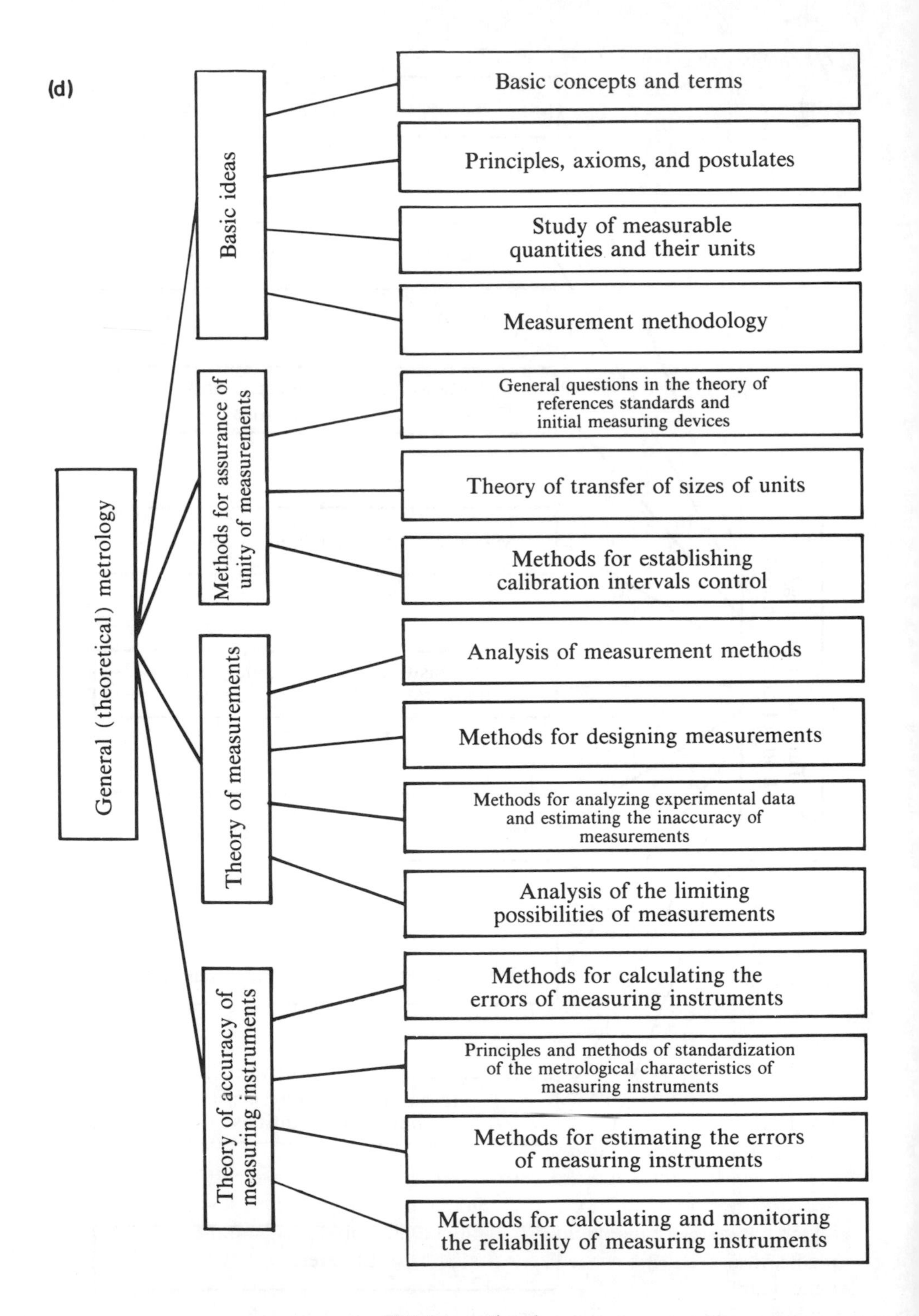

FIG. 1.1. continued.

(2) General theory of reference standards and initial measuring devices. The units of physical quantities are materialized, i.e., they are reproduced, with the help of reference standards and initial measuring devices, and for this reason these measuring devices play an exceptionally important role in the unity of measurements. The reference standard of each unit is unique and it is physically created based on the laws of specific fields of physics and technology. For this reason general metrology cannot answer the question of how a reference standard should be constructed. But metrology must determine when a reference standard must be created, and it must establish the criteria for determining when such a reference standard must be a single or group reference standard, etc. In metrology the theory and methods of comparing reference standards and monitoring their stability as well as methods for expressing errors must also be studied. Practice raises many such purely metrological questions.

(3) Theory of transfer of the size of units into measurement practice. In order that the results of all measurements be expressed in terms of established units, all means of measurement (measures, instruments, measuring transducers, measuring systems) must be calibrated with respect to reference standards. This problem cannot, however, be solved directly based on primary reference standards, i.e., reference standards that reproduce units. It is solved with the help of a system of secondary reference standards, i.e., reference standards that are calibrated with respect to the primary reference standard, and working reference standards, i.e., reference standards that are calibrated with respect to secondary reference standards. Thus the system of reference standards has a hierarchical structure. The entire procedure of calibrating reference standards and, with their help, the working measuring instruments is referred to as transfer of the sizes of units into measurement practice. The final stages of transferring the sizes of units consists of calibration of the scales of the measuring instruments, adjustment of measures, and determination of the actual values of the quantities that are reproduced by them, etc., after which all measuring instruments are checked at the time they are issued and then periodically during use.

In solving these problems there arises a series of questions. For example, how many gradations of accuracy of reference standards are required? How many secondary and working reference standards are required for each level of accuracy? How does the error increase when the size of a unit is transferred from one reference standard to another? How does this error increase from the reference standard to the working measuring instrument? What should be the relation between the accuracy of the reference standard and the measuring instrument that is calibrated (verified) with respect to it? How should complicated measurement systems be checked? Metrology should answer these questions.

The other blocks in the diagram of Fig. 1.1(d) do not require any explanations. We shall now turn to Fig. 1.1(a) and focus on the section *particular metrology*, which the fields of measurement comprise. Examples are lineal-angular measurements, measurements of mechanical quantities, measurements of electric and magnetic quantities, etc. The central problem arising in each field of measurement is the problem of creating conditions under which the measurements of the corresponding physical quantities are unified. For this purpose, in each field of measurement a system of initial measuring devices—reference standards and standard measures—is created, and methods for calibrating and checking the working measuring instruments are developed. The specific nature of each field of measurement engenders a great many problems characteristic for it. However, many problems that are common to several fields of measurement are encountered. The analysis of such problems and the development of methods for solving them are now problems of general metrology.

Applied metrology, which incorporates the metrological service and legislative metrology, is of great importance for achieving the final goals of metrology as a science. The metrological service checks and calibrates measuring instruments and certifies standards of properties and composition, i.e., it maintains the uniformity of measuring instruments employed in the country. The functions of legislative metrology are to enact laws that would guarantee uniformity of measuring instruments and unity of measurements. Thus a system of physical quantities and the sizes of their units, employed in the country, can be established only by means of legislation. The rules giving the right to manufacture measuring instruments and to check the state of these instruments when they are in use are also established by means of legislation.

We shall now define more accurately some of the expressions and terms mentioned above.

Uniformity of measuring instruments refers to the state of these instruments in which they are all carriers of the established units and their errors and other properties, which are important in order for the instruments to be used as intended, fall within the established limits.

Unity of measurements refers to a common quality of all measurements performed in a region (in a country, in a group of countries, or in the world) such that the results of measurements are expressed in terms of established units and agree with one another within the limits of estimated error.

Uniformity of measuring instruments is a necessary prerequisite for unity of measurement. But the result of a measurement depends not only on the quality of the measuring instrument employed but also on many other factors, including human factors (if measurement is not automatic). For this reason, unity of measurement in general is the limiting state that must be strived for, but which, as any ideal, is unattainable.

This is a good point at which to discuss the development of reference standards. A reference standard is always a particular measuring device: a measure, instrument, or measuring apparatus. Such measuring devices were initially employed as reference standards arbitrarily by simple volition of the institution responsible for correctness of measurements in the country. However, there is always the danger that a reference standard will be ruined. This can happen because of a natural disaster, fire, etc. An arbitrarily established reference standard, i.e., a prototype reference standard, cannot be reproduced.

Because of this scientists have for a long time strived to define units of measurement so that the reference standards embodying them could be reproducible. For this, the units of the quantities were defined based on natural phenomena. Thus the second was defined based on the period of revolution of the Earth around the Sun; the meter was defined based on the length of the Parisian meridian; etc. Historically this stage of development of metrology coincided with the creation of the metric system. Scientists hoped that these units would serve "for all time and for all peoples."

Further investigations revealed, however, that the chosen natural phenomena are not sufficiently unique or are not stable enough. Nonetheless the idea itself of defining units based on natural phenomena was not questioned. It was only necessary to seek other natural phenomena corresponding to a higher level of knowledge of nature.

It was found that the most stable or even absolutely stable phenomena are characteristic of phenomena studied in quantum physics, and that the physical constants can be employed successfully for purposes of defining units, and the corresponding effects can be employed for realizing reference standards. The meter, the second, and the volt have now been defined in this manner. It can be conjectured that in the near future, the volt, defined and reproduced based on the Josephson effect, will replace the ampere as the basic electric unit.

The physical principles for determining the most important units and the changes occurring in these definitions over time are described by D. Kamke and K. Kramer in Ref. 36. The organization of the corresponding reference standards is also described there.

The book by B. W. Petley,[46] for example, is devoted to physical constants. The last concordance of their values was performed in 1986.[20]

As one can see from the problems with which it is concerned, metrology is an applied science. However, the subject of metrology—measurement—is a tool of both fundamental sciences (physics, chemistry, and biology) and applied disciplines, and it is widely employed in all spheres of industry, in everday life, and in commerce. No other applied science has such a wide range of applications as does metrology.

We shall return once again to particular metrology. A simple list of the fields of measurement shows that the measurable quantities and therefore measurement methods and measuring instruments are extremely diverse. What then do the different fields of measurement have in common? They are united by general or theoretical metrology and primarily the general methodology of measurement and the general theory of inaccuracy of measurements. For this reason, the development of these branches of metrology is important for all fields of science and for all spheres of industry that employ measurements. The importance of these branches of metrology is also indicated by the obvious fact that a specialist in one field of measurement can easily adapt to and work in a different field of measurement.

1.3. Initial points of the theory of measurements

Measurements are so common and intuitively understandable that one would think there is no need to identify the prerequisites on which measurements are based. However, a clear understanding of the starting premises is necessary for the development of any science, and for this reason it is desirable to examine the prerequisites of the theory of measurements.

When some quantity characterizing a specific object is being measured, this object is made to interact with a measuring instrument. Thus to measure the diameter of a rod, the rod is squeezed between the jaws of a vernier caliper; to measure the voltage of an electric circuit, a voltmeter is connected to it, etc. The indication of the measuring instrument—the sliding calipers, voltmeter, etc.—gives an estimate of the measurable quantity, i.e., the result of the measurement. When necessary, the number of divisions read on the instrument scale is multiplied by a certain factor. In many cases the result of measurement is found by mathematical analysis of the indications of a instrument or several instruments. For example, the density of solid bodies, the temperature coefficients of the electric resistance of resistors, and many other physical quantities are measured in this manner.

The imperfection of measuring instruments, the inaccuracy with which the sizes of the units are transferred to them, as well as some other factors which we shall study below, result in the appearance of measurement errors. Measurement errors are in principle unavoidable, since a measurement is an experimental procedure and the true value of the measurable quantity is an abstract concept. As the measurement methods and measuring instruments improve, however, measurement errors decrease.

The introduction of measurable quantities and establishment of their units is a necessary prerequisite of measurements. Any measurement, however, is always performed on a specific object, and the general definition of the measurable quantity must be specified taking into account the properties of the object and the objective of the measurement. The true value of the measurable quantity is essentially introduced and defined in this manner. Unfortunately, this important preparatory stage of measurements is usually not formulated and not singled out.

In order to clarify this question we shall study a very simple measurement problem—the measurement of the diameter of a disk. First we shall formulate the problem. The fact that the diameter of a disk is to be measured means that the disk, i.e., the object of study, is a circle. We note that the concepts "circle" and "diameter of a circle" are mathematical, i.e., abstract, concepts. The circle is a representation or model of the given body. The diameter of the circle is the parameter of the model and is a mathematically rigorous definition of the measurable quantity. Now, in accordance with the general definition of the true value of the measurable quantity, it can be stated that the true value of the diameter of the disk is the value of the parameter of the model (diameter of the disk) that reflects, in the quantitative respect, the property of the object of interest to us; the ideal qualitative correspondence must be predetermined by the model.

Let us return to our example. The purpose of the disk permits determining the permissible measurement error and choosing an appropriate measuring instrument. Bringing the object into contact with the measuring instrument, we obtain the result of measurement. But the diameter of the circle is, by definition, invariant under rotation. For this reason, the measurement must be performed in several different directions. If the difference of the results of the measurements are less than the permissible measurement error, then any of the obtained results can be taken as the result of measurement. After the value of the measurable quantity—a concrete number, which is an estimate of the true value of the measurand has been found, the measurement can be regarded as being completed.

But it may happen that the difference of the measurements in different directions exceeds the permissible error of a given measurement. In this situation we must state that within the required measurement accuracy, our disk does not have a unique diameter, as does a circle. Therefore there is no concrete number that can be taken, with prescribed accuracy, as an estimate of the true value of the measurable quantity. Hence the adopted model does not correspond to the properties of the real object, and the measurement problem has not been correctly formulated.

If the object is a manufactured article and the model is a drawing of the article, then any disparity between them means that the article is defective. If, however, the object is a natural object, then the disparity means that the model is not applicable and it must be reexamined.

Of course, even when measurement of the diameter of the disk is assumed to be possible, in reality the diameter of the disk is not absolutely identical in different directions. But as long as this inconstancy is negligibly small, we can assume that the circle as a model corresponds to the object and therefore there exists a constant, fixed true value of the measurable quantity, and an estimate of the quantity can be found as a result of measurement. Moreover, if the measurement has been performed, we can assume that the true value of the measurable quantity lies somewhere near the obtained estimate and differs from it by not more than the measurement error.

Thus the idealization necessary for constructing a model gives rise to an unavoidable discrepancy between the parameter of the model and the real property of the object. We shall call this discrepancy the threshold discrepancy.

As we saw above, the error owing to the threshold discrepancy between the model and the object must be less than the total measurement error. If, however, this component of the error exceeds the limit of permissible measurement error, then it is impossible to make a measurement with the required accuracy. This indicates that the model is inadequate. To continue the experiment, if this is permissible for the objective of the measurement, the model must be redefined. Thus, in the example of the measurement of the diameter of a disk, a different model could be a circle circumscribing the disk.

The case studied above is very simple, but the features demonstrated for it are present in any measurement, though they are not always so easily and clearly perceived as when measuring lineal dimensions.

The foregoing considerations essentially reduce to three prerequisites:

(a) some parameter of the model of the object must correspond to a measurable property of the object;

(b) the model of the object must permit the assumption that during the time required to perform the measurement, the parameter of the object, corresponding to the property of the object being measured, is constant; and

(c) the error owing to the threshold discrepancy between the model and the object must be less than the permissible measurement error.

Generalizing all three assumptions, we formulate the following principles of metrology: a measurement with fixed accuracy can be performed only if to a measurable property of the object it is possible to associate with no less accuracy a determinate parameter of its model.

Any constant parameter is, of course, a determinate parameter. The instantaneous value of a variable (varying) quantity can also be regarded as a determinate parameter.

We note that the value of the parameter of a model of an object introduced in this manner is the true value of the measurable quantity.

The foregoing considerations are fundamental, and they can be represented in the form of postulates of the theory of measurements:[51,57]

(α) the true value of the measurable quantity exists; (β) the true value of the measurable quantity is constant; and (γ) the true value cannot be found.

The threshold discrepancy between the model and the object was employed above as a justification of the postulate (γ). There also exist, however, other unavoidable restrictions on the approximation of the true value of a measurable quantity. For example, the accuracy of measuring instruments is unavoidably limited. For this reason it is possible to formulate the fundamental statement: *the result of any measurement always contains an error.*

It was mentioned above that a necessary prerequisite of measurements is the introduction of physical quantities and their units. These questions are not directly related with the problem of estimating measurement errors and for this reason they are not studied here. These questions are investigated in a number of works. We call attention to the book by B. D. Ellis[27] and the work of K. P. Shirokov.[56]

At this point we shall discuss some examples of models which are employed for specific measurement problems.

Alternating current measurements. The object of study is an alternating current. The model of the object is a sinusoid

$$i = I_m \sin(\omega t + \varphi),$$

where t is the time and I_m, ω, and φ are the amplitude, the angular frequency, and the initial phase, and they are the parameters of the model.

Each parameter of the model corresponds to some real property of the object and can be a measurable quantity. But, in addition to these quantities (arguments), a number of other parameters that are functionally related with them are also introduced. These parameters can also be measurable quantities. Some parameters can be introduced in a manner such that by definition they are not related with the "details" of the phenomenon. An example is the effective current

$$I = \sqrt{\frac{1}{T} \int_0^T i^2 dt},$$

where $T = 2\pi/\omega$ is the period of the sinusoid.

A nonsinusoidal current is also characterized by an effective current. However, in developing measuring instruments and describing their properties the form of the current, i.e., the model of the object of study, must be taken into account.

The discrepancy between the model and the object in this case is expressed as a discrepancy between the sinusoid and the curve of the time dependence of the current strength. In this case, however, only in rare cases is it possible to discover the discrepancy between the model and the process under study by means of simple repetition of measurements of some parameters. For this reason, the correspondence between the model and the object is checked differently, for example, by measuring the form distortion factor.

The model is usually redefined by replacing one sinusoid by a sum of a certain number of sinusoids.

Measurement of the parameters of random processes. The standard model is a stationary ergodic random process on the time interval T. The constant parameters of the process are the mathematical expectation $M[X]$ and the variance $D[X]$. Suppose that we are interested in $M[X]$. The expectation $M[X]$ can be estimated, for example, with the help of the formula

$$\bar{x} = \left(\frac{\sum_{i=1}^{n} x_i}{n} \right)_T ,$$

where T is the observational time interval, x_i are the estimates of the realization of the random quantity, whose variation in time forms a random process at times $t_i \in T$, and n is the total number of realizations obtained.

Repeated measurements on other realizations of the process can give somewhat different values of $\bar{x}$. The adopted model can be regarded as corresponding to the physical phenomenon under study if the differences between the obtained estimates of the mathematical expectation of the process are not close to the permissible measurement error. If, however, the difference of the estimates of the measured quantity are close to the error or exceed it, then the model must be redefined. This is most simply done by increasing the observational interval T.

It is interesting to note that the definitions of some parameters seem, at first glance, to permit arbitrary measurement accuracy (if the errors of the measuring instrument are ignored). Examples of such parameters are the parameters of stationary random processes, the parameters of distributions of random quantities, and the average value of the quantity. One would think that in order to achieve the required accuracy in these cases it is sufficient to increase the number of observations when performing the measurements. In reality, however, the accuracy of measurement is always limited, and in particular it is limited by the correspondence between the model and the phenomenon, i.e., by the possibility of assuming that to the given phenomenon there corresponds a stationary random process or a random quantity with a known distribution, etc.

In the last few years much has been written about measurements of variable and random quantities. But these quantities, as such, do not have a true value, and for this reason they cannot be measured.

For a random quantity it is possible to measure the parameters of its distribution function, which are not random; it is possible to measure the realization of a random quantity. For a variable quantity it is possible to measure its parameters that are not variable; it is also possible to measure the instantaneous values of a variable quantity.

We shall now discuss in somewhat greater detail the measurement of instantaneous values of quantities. Suppose that we are studying an alternating current, the model of which is a sinusoid with amplitude I_m, angular frequency ω, and initial phase φ. At time t_1 to the instantaneous current there corresponds in the model the instantaneous value $i_1 = I_m \sin(\omega t_1 + \varphi)$. At a different time there will be a different instantaneous value, but at each moment it has some definite value. Thus to the measurable property of the object there always corresponds a fixed parameter of its model.

Measurement, however, requires time. The measurable quantity will change during the measurement time, and this will generate a specific error of the given measurement. The objective of the measurement permits setting a level that the measurement error as well as its component owing to the change in the measurable quantity over the measurement time must not exceed.

If this condition is satisfied, then the effect of the measurement time can be neglected, and it can be assumed that as a result we obtain an estimate of the measured instantaneous current, i.e., the current strength at a given moment in time.

In the literature, the term measurement of variable quantities usually refers to measurement of instantaneous values, and the expression *measurement of a variable quantity* is imprecise. In the case of measurement of a random quantity the writer usually has in mind the measurement of a realization of a random quantity.

Physical quantities are divided into active and passive. Active quantities are quantities that are capable of generating measurement signals without any auxiliary sources of energy, i.e., they act on the measuring instruments. Such quantities are the emf, the strength of an electric current, mechanical force, etc. Passive quantities themselves cannot act on measuring instruments, and for measurements they must be activated. Examples of passive quantities are mass, inductance, and electric resistance. Mass is usually measured based on the fact that in a gravitational field a force proportional to the mass acts on the body. Electric resistance is activated by passing an electric current through a resistor, etc.

When measuring passive physical quantities characterizing some objects, the models of the objects are constructed for the active quantities that are formed by activation of passive quantities.

1.4. Classification of measurements

In metrology there has been a long-standing tradition to distinguish direct, indirect, and combined measurements. In the last few years metrologists have begun to divide combined measurements into strictly combined measurements and simultaneous measurements.[3] This classification is connected with a definite method used for processing experimental data in order to find the result of a measurement and to estimate its errors.

In the case of direct measurements the object of study is made to interact with the measuring instrument, and the value of the measurand is read from the indications of the latter. Sometimes the instrumental readings are multiplied by some factor, corresponding corrections are made in it, etc.

In the case of indirect measurements the value of the measurable quantity is found based on a known dependence between this quantity and its arguments. The arguments are found by means of direct and sometimes indirect or simultaneous or combined measurements. For example, the density of a homogeneous solid body is found as the ratio of the mass of the body to its volume and the mass and volume of the body are measured directly.

Sometimes direct and indirect measurements are not so easily distinguished. For example, an ac wattmeter has four terminals. The voltage applied to the load is connected to one pair of terminals while the other pair of terminals is connected in series with the load. As is well known, the indications of a wattmeter are proportional to the power consumed by the load. However, the wattmeter does not respond directly to the measured power. Based on the principle of operation of the instrument, measurement of power with the help of a wattmeter would have to be regarded as indirect. In our case it is important, however, that the value of the measurable quantity can be read directly from the instrument (in this case the wattmeter). In this sense a wattmeter is in no way different from an ammeter. For this reason, in this book it is not necessary to distinguish measurement of power with the help of a wattmeter and measurement of current strength with the help of an ammeter: we shall categorize both cases as direct measurements. In other words, when referring a specific measurement to one or another category we will ignore the arrangement of the measuring instrument employed.

Simultaneous and combined measurements employ close methods for finding the measurable quantities: in both cases they are found by solving a system of equations, whose coefficients and separate terms are obtained as a result of measurements (usually direct). In both cases the method of least squares is usually employed. But the difference lies in the fact that in the case of combined measurements several quantities of the same kind are measured simultaneously while in the case of simultaneous measurements quantities of different kinds are measured simultaneously. For example, a measurement in which the electric resistance of a resistor at a temperature of $+20\,^{\circ}\mathrm{C}$ and its temperature coefficients are found based on direct measurements of the resistance and temperature performed at different temperatures is a simultaneous measurement. A measurement in which the masses of separate weights in a set are found based on the known mass of one of them and by comparing the masses of different combinations of weights from the same set is a combined measurement.

This distinction comes about because simultaneous measurements are based on known equations that reflect relations existing in nature between the properties of objects, i.e., between quantities, while combined measurements are based on equations that reflect an arbitrary combination of objects having the measurable properties. For this reason, simultaneous measurements can be regarded as a generalization of indirect measurements and combined measurements can be regarded as a generalization of direct measurements. Therefore measurements can be divided based on their physical significance only into direct and indirect measurements.

If methods for processing experimental data are to be distinguished, then it is useful to distinguish three categories of measurements: (a) direct, (b) indirect, and (c) simultaneous and combined.

Depending on the properties of the object of study, the model adopted for the object, and the definition of the measurable quantity given in the model as well as on the method of measurement and the properties of the measuring instruments, the measurements in each of the categories mentioned above are performed either with single or repeated observations. The method employed for processing the experimental data depends on the number of observations—are many measurements required or are one or two observations sufficient to obtain a measurement?

If a measurement is performed with repeated observations, then to obtain a result the observations must be analyzed statistically. These methods are not required in the case of measurements with single observations. For this reason, for us the number of observations is an important classification criterion.

We shall term measurements performed with single observations *single measurements* and measurements performed with repeated observations *multiple measurements*. An indirect measurement, in which the value of each of the arguments is found as a result of a single measurement, must be regarded as a single measurement.

Simultaneous and combined measurements can be regarded as single measurements, if the number of measurements is equal to the number of unknowns when the measurements are performed, so that each unknown is determined uniquely from the system of equations obtained.

Among simultaneous and combined measurements it is helpful to single out measurements for which the measurable quantities are related by known equations. For example, in measuring the angles of a planar triangle it is well known that the sum of all three angles is equal to 180°. This relation makes it possible to measure two angles only, and this is a single and, moreover, direct measurement. If, however, all three angles are measured, then the relation mentioned permits correlating their estimates, using, for example, the method of least squares. In the latter case this is a combined and multiple measurement.

Measurements are also divided into static and dynamic measurements. Adhering to the concept presented in Ref. 55, we shall classify as static those measurements in which, in accordance with the problem posed, the measuring instruments are employed in the static regime and as dynamic those measurements in which the measuring instruments are employed in the dynamic regime.

The static regime of a measuring instrument is a regime in which, based on the function of the instrument, the output signal can be regarded as constant. For example, for an indicating instrument the signal is constant for a time sufficient to read the instrument. A dynamic regime is a regime in which the output signal changes in time, so that to obtain a result or to estimate its accuracy this change must be taken into account.

According to these definitions, static measurements include, aside from trivial measurements of length, mass, etc., measurements of the average and effective (mean-square) values of alternating current by indicating instruments. Dynamic measurements refer to measurements of successive values of a quantity that varies in time (including stochastically). A typical example of such measurements is recording the value of a quantity as a function of time. In this case it is logical to regard the measurement as consisting not of a single measurement but rather many measurements.

Other examples of dynamic measurements are measurement of magnetic flux by the ballistic method and measurement of the high temperature of an object based on the starting section of the transfer function of a thermocouple put into contact with the object for a short time (the thermocouple would be destroyed if the contact time was long).

Static measurements also include measurements performed with the help of digital indicating instruments. According to the definition of static measurements, for this conclusion it is not important that during the measurement the state of the elements in the device itself changes. The measurement will also remain static when the indications of the instrument change from time to time, but each indication remains constant for a period of time sufficient for the indication to be read or recorded automatically.

A characteristic property of dynamic measurements is that to obtain results and estimate their accuracy in such measurements it is necessary to know a complete dynamic characteristic of the measuring instrument: a differential equation, transfer function, etc. (The dynamic characteristics of measuring instruments will be examined in Chap. 2.)

The classification of measurements as static and dynamic is justified by the difference in the methods employed to process the experimental data. At the present time, however, dynamic measurements as a branch of metrology are still in the formative stage.*

The most important characteristic of the quality of a measurement is accuracy. The material base, ensuring the accuracy of numerous measurements performed in the economy, consists of reference standards. The accuracy of any particular measurement is determined by the accuracy of the measuring instruments employed, the method of measurement employed, and sometimes by the skill of the experimenter. However, since the true value of a measurable quantity is always unknown, the errors of measurements must be estimated computationally (theoretically). This problem is solved by different methods and with different accuracy.

In connection with the accuracy of the estimation of a measurement error we shall distinguish measurements whose errors are estimated before and after the measurement. We shall refer to them as measurements with ante-measurement or *a priori* estimation of errors and measurements with post-measurement or *a posteriori* estimation of errors.

Estimates with ante-measurement estimation of errors must be performed according to an established procedure, included in the calculation of the errors. Measurements of this type include all mass measurements. For this reason, we shall call them mass measurements. Sometimes they are called technical measurements.

A posteriori estimation of errors is characteristic for measurements performed

---

*Reference 1 gives the following definition of the term *dynamic measurement*: (2.04) dynamic measurement is the determination of the instantaneous value of a quantity and, where appropriate, its variation with time. Note: the qualifier *dynamic* applies to the measurand and not to the method of measurement.

The definition itself is understandable, and the forgoing considerations are consistent with it. However the note to the definition shuffles all the cards. I think this is unfortunate: the term refers to measurement, i.e., it is organically connected with the method of measurement, while the note says that the indicator of dynamics is determined by the measurand and not by the method of measurement.

Consider, for example, the ballistic method of measurement of magnetic flux. The measurand here is simply constant, but it is measured by reading the maximum deviation of the indicator of the instrument during the motion of the indicator. It is difficult to say that this measurement is not "dynamic."

Moreover, this remark once again shows that dynamic measurements as a branch of metrology are still in the formative stage.

with an objective in mind, when it is important to know the accuracy of each result. For this reason we shall call such measurements individual measurements.

Mass measurements are very common. Their accuracy is predetermined by the use of the types (brands) of measuring instruments indicated in the procedure, the techniques for using them, as well as the stipulated conditions under which the measurements are to be performed. For this reason, the person performing the measurement is interested only in the result of measurement, and he or she knows nothing about the accuracy beforehand, i.e., whether or not it is adequate.

We shall divide individual measurements, in turn, into two groups: measurements with exact estimation of errors and measurements with approximate estimation of errors.

Measurements with exact estimation of errors are measurements in which the properties of the specific measuring instruments employed are taken into account. *Measurements with approximate estimation of errors* are measurements in which the specifications of the measuring instruments employed are taken into account.

In both cases the conditions under which the measurements are performed are taken into account. For this, the influence quantities or some of them are often measured; in other cases they are estimated.

Here we must call attention to a fact whose validity will become obvious from the further discussion. Suppose that measurements whose errors are estimated with different accuracy are performed with the help of one and the same measuring instruments. In spite of the fact that the same instruments are employed, the accuracy of the measurements in this case is different. The most accurate result will be the result obtained with exact estimation of the errors. Measurements for which the errors are estimated approximately will in most cases be more accurate than measurements whose errors are estimated beforehand. The results of measurements with ante- and post-measurement estimation of errors will be equally accurate in only separate cases.

But when measuring instruments having different accuracy are employed, this will no longer be the case. For example, measurement of voltage with the help of a potentiometer of accuracy class 0.005, performed as a mass measurement, i.e., with preestimation of errors, will be more accurate than measurement with the help of an indicating voltmeter of class 0.5, performed as an individual measurement with exact estimation of the errors.

In all of the cases studied above the objective of the measurements was to obtain an estimate of the true value of the measurable quantity, which, strictly speaking, is the problem of any measurement. However, measurements are often performed during the preliminary study of a phenomenon. We shall call such measurements *preliminary measurements*.

The purpose of preliminary measurements is to determine the conditions under which some indicator of the phenomenon can be observed repeatedly and its regular relations with other properties of the object, systems of objects, or with an external medium can be studied. Since the object of scientific investigation of the world is to establish and study regular relations between objects and phenomena, preliminary measurements are very important. Thus it is known that the first task of a scientist who is studying some phenomenon is to determine the conditions under which a given phenomenon can be observed repeatedly in other laboratories and can be checked and confirmed.

Preliminary measurements, as one can see from the concepts presented above, are required in order to construct a model of an object. For this reason, preliminary measurements are also very important for metrology.

Apart from preliminary measurements, for metrological purposes it is also possible to distinguish supplementary measurements. Supplementary measurements are measurements of influence quantities that are performed in order to determine and make corrections in the results of measurements.

There is an enormous literature on different aspects of measurements. References 17 and 44 give an idea of the wide range of questions pertaining to real measurements.

1.5. Classification of measurement errors

A measurement of a quantity whose true value is A gives an estimate $\tilde{A}$ of that quantity. The absolute measurement error ζ can be defined as the difference between $\tilde{A}$ and A: $\zeta = \tilde{A} - A$.

However, this relation cannot be used to find the error of a measurement for the simple reason that the true value of the measurable quantity is always unknown. If the true value were known, then there would be no need for a measurement.

Measurements performed for calibration of measuring instruments are an exception. In this case the value of the measurable quantity must be known with sufficient accuracy so that it can be used for this purpose instead of the true value of the quantity.

For this reason measurement errors must be estimated by using indirect data.

The necessary components of any measurement are the method of measurement and the measuring instrument; measurements are often performed with the participation of a person. The imperfection of each component of measurement contributes to the measurement error. For this reason, in the general form

$$\zeta = \zeta_m + \zeta_i + \zeta_p,$$

where ζ is the measurement error, ζ_m is the methodological error, ζ_i is the instrumental error, and ζ_p is the personal error.

Each component of the measurement error in turn can be caused by a number of factors. Thus methodological *errors* can arise as a result of an inadequate theory of the phenomena on which the measurement is based and inaccuracy of the relations that are employed to find an estimate of the measurable quantity. In particular, the error owing to the threshold discrepancy between the model of a specific object and the object itself is also a methodological error.

Instrumental measurement errors are caused by the imperfection of measuring instruments. Normally the intrinsic error of measuring instruments, i.e., the error obtained under reference conditions regarded as normal, are distinguished from additional errors, i.e., errors caused by the deviation of the influence quantities from their values under reference conditions. Properties of measuring instruments that cause the instrumental errors will be examined in detail in Chap. 2.

Personal errors: Measurements are normally performed by people. Someone reads the indications of instruments, records the moment at which an image of a filament vanishes on the screen of an optical pyrometer, etc. The individual characteristics of the person performing the measurement give rise to individual errors that are characteristic of that person. They include errors owing to incorrect reading of the tenths graduation of an instrument scale, asymmetric placement of the mark of an optical indicator between two graduation lines, etc.

Improvement of the reading and regulating mechanisms of measuring instruments has led to the fact that for modern measuring instruments the personal errors are usually insignificant; for example, for digital instruments they are virtually nonexistent.

The foregoing classification of measurement errors is based on the cause of the errors.

Another important classification of measurement errors is based on their properties. In this respect systematic and random errors are distinguished.

A measurement error is said to be *systematic* if it remains constant or changes in a regular fashion in repeated measurements of one and the same quantity. The observed and estimated systematic error is eliminated from measurements by introducing corrections. However, it is impossible to eliminate completely the systematic error in this manner. Some part of the error will remain, and then this residual error will be the systematic component of the measurement error.

In order to define a random measurement error, imagine that some quantity is measured several times. If there are differences between the results of separate measurements and these differences cannot be predicted individually and any regularities inherent to them are manifested only in a significant number of results, then the error owing to this scatter of the results is called the *random error*.

The division of measurement errors into systematic and random is very important, since these components are manifested differently and different approaches are required to estimate them.

Random errors are discovered by performing measurements of one and the same quantity repeatedly under the same conditions, whereas systematic errors can be discovered experimentally either by comparing a given result with a measurement of the same quantity performed by a different method or by using a more accurate measuring instrument. However, systematic errors are normally estimated by theoretical analysis of the measurement conditions, based on the known properties of a measured and of measuring instruments.

The quality of measurements that reflects the closeness of the results of measurements of the same quantity performed under the same conditions is called the *repeatability of measurements*. Good repeatability indicates that the random errors are small.

The quality of measurements that reflects the closeness of the results of measurements of the same quantity performed under different conditions, i.e., in different laboratories (at different locations) and using different equipment, is called the *reproducibility of measurements*. Good reproducibility indicates that both the random and systematic errors are small.

In speaking about errors we shall also distinguish gross or outlying errors and blunders. We shall call an error *gross* (outlying) if it significantly exceeds the error justified by the conditions of the measurements, the properties of the measuring instrument employed, the method of measurement, and the qualifications of the experimenter. Such measurements can arise, for example, as a result of a sharp brief change in the grid voltage (if the grid voltage in principle affects the measurements).

Outlying or gross errors in multiple measurements are discovered by statistical methods and are usually eliminated from analysis.

Blunders occur as a result of errors made by the experimenter. Examples are a slip of the pen when writing up the results of observations, an incorrect reading of the indications of an instrument, etc. Blunders are discovered by nonstatistical methods, and they must always be eliminated from the analysis.

Measurement errors are also divided into static and dynamic. Static errors were mentioned above. Dynamic errors are caused by the inertial properties of measuring instruments.

If a varying quantity is recorded with the help of a recording device, then the difference between the obtained function and the actual process of change of the recorded quantity in time (taking into account the necessary scale transformations) is the dynamic error of the given dynamic measurement. In this case it is also a function of time, and the instantaneous dynamic error can be determined for each moment in time.

We shall now study the case when the process is recorded by measuring individual instantaneous values. It is clear that if within the time of a single measurement the measurable quantity does not change significantly and the instantaneous values of the process are obtained at known times and sufficiently frequently, then the collection of points ultimately obtained gives a picture of the change of the measurement in time with a negligibly small error. Thus there will be no dynamic error here.

The inertial properties of an instrument can be such, however, that the changes in the measurable quantity over the measurement time will give rise to a definite error in the results of measurements of the instantaneous values. In this case the obtained collection of instantaneous values will not be coincident with the process of change of the measurable quantity in time, and their difference, exactly as in the case of an analog automatic-plotting instrument, will give the dynamic error. Correspondingly, it is natural to call the *instantaneous dynamic error* the error arising in the measurement of a separate instantaneous quantity as a result of the rate of change of the measurable quantity and the inertial properties of the instrument.

If some isolated instantaneous quantity, for example, the amplitude of a pulse, is measured and the measurement is performed with a special indicating device, then the difference between the shape of the pulse and the shape obtained with a calibrated instrument will give rise to a supplementary error as a result of the measurement. Based on what we have said above, one could call this error a dynamic error. However, the general term "dynamic error" is normally avoided in such situations, and such an error is given a name indicating its cause. In this example it is natural to call the error the pulse-shape error. In practice the pulse shape is characterized by a number of parameters and to each parameter there is associated a separate component of the error.

1.6. Principles of estimation of measurement errors

The problem of estimating the error of a measurement is to define the limits of uncertainty of the result. This is usually achieved by indicating the limits of the error of measurement. If random errors play the determining role, then the limits of error are determined in accordance with some probability and are called the confidence limits of the measurement error or simply the uncertainty.

Measurements are regarded metrologically to be all the better the lower their error is. However, measurements must be reproducible, since otherwise they lose their objective character and therefore become meaningless.

A measure of the nonreproducibility of a measurement permitted by the experimenter is the limits of measurement error or uncertainty estimated by the experimenter.

Correctly estimated measurement errors permit comparing the obtained result with the results obtained by other experimenters. The fact that the correctness of a given estimate is later confirmed in a more accurate measurement attests to the high skill of the experimenter.

Errors are customarily estimated based on the following considerations.

(1) The measurement is regarded as all the more accurate and, in this sense, better the smaller its relative error is:

$$\varepsilon = \frac{\zeta}{A} \approx \frac{\Delta}{\tilde{A}},$$

where $\tilde{A}$ is an estimate of the true value of the measurand A and Δ is an estimate of the limits of measurement error ζ. The relative error is studied because its value does not depend on the value of the measurable quantity. This is not true for the absolute error.

(2) The estimate of the measurement error must satisfy the inequality

$$|\zeta| \leqslant |\Delta|.$$

The meaning of this inequality is as follows. For any measurement it is, in principle, desirable that there be no error. But despite all efforts $\tilde{A} \neq A$ and correspondingly $\zeta \neq 0$, and in addition, in our case, the error can be both greater than and less than zero. In the primary estimate of any error the limits for ζ, i.e., Δ_1 and Δ_2, are established, so that $\Delta_1 \leqslant \zeta \leqslant \Delta_2$. In the calculations the value of $|\Delta_1|$ or $|\Delta_2|$, whichever is larger, is often used as the estimate for ζ. Most often $|\Delta_1| = |\Delta_2| = |\Delta|$. Thus we arrive at the inequality $|\zeta| \leqslant |\Delta|$.

If a measurement error is mainly random, its limits should be estimated with such a high probability that the above inequality is practically always satisfied.

The second assumption means that the estimate of the measurement error must be an upper estimate. This requirement should be regarded as a principle of error estimation.

It makes a great deal of sense to use the limiting values of an error as estimates of the error. First of all, humans naturally make comparisons and they establish relations of the type greater than, equal to, and less than. For this reason, limiting errors are easily perceived. Furthermore, measurement results are most easily compared when the limiting errors are known. In general, this problem can be solved if the parameters of errors, such as their standard deviations, are known, but then it is also necessary to know the probability density function. This is complicated.

The upper estimate, however, should not be exaggerated without justification. Here the words of J. DuMond are relevant:[26]

> Exaggeration of measurement error by a factor of two or three, to which some investigators resort in order to increase the reliability of their measurements, does very great harm to science. When this is done, the disagreement between the results of a given measurement and other measurements (and it could be of enormous scientific significance) is concealed because the true measurement accuracy is concealed. For whom, it is asked, does exaggeration of errors create reliability? For no one else other than the investigator himself. For him this reduces the chance that an error will later be found in his experiment or that he overrated his abilities. Such "reliability" is unworthy of an honest and courageous scientist. It is just as shameful to deliberately overestimate errors as it is to underestimate them, if, of course, the degree of overestimation is not reported honestly and clearly.

It should not be forgotten, however, that the measurement error characterizes the uncertainty of the result of a measurement and the spread and nonreproducibility of the measurement. For this reason, the estimate of error cannot be very precise, and it is not required that it be very precise. But its uncertainty, if one can say so, should be weighted toward overestimation and not underestimation. In accordance with this principle we shall still say that it is better to overestimate than underestimate an error: In the first case the quality of the measurement is reduced while in the second case the entire measurement can be made worthless. Of course, the overestimation should be kept to a minimum.

It should also be kept in mind that the correctness of estimates of measurement errors cannot be checked based on data obtained in a particular measurement. As regards a given measurement, all obtained experimental data and other reliable information, for example, corrections to the indications of instruments, are employed to find the measurement result itself, and the error must be estimated using additional information about the properties of the measuring instruments, the conditions of the measurements, and the theory. There is no point in performing a special experiment to check or estimate the measurement error. It would then be necessary to organize in parallel with the given measurement a more accurate measurement of the same measurable quantity. Then the given measurement would be meaningless: Its result would be replaced by the result of the more accurate measurement. The problem of estimating the error in a given measurement would be replaced by the problem of estimating the error of the more accurate measurement, i.e., the basic problem would remain unsolved.

The correctness of estimates of errors is nonetheless checked. It is confirmed either by the successful use of the measurement result for the purpose intended or by the fact that the measurement agrees with the results obtained by other experimenters. As in the case of measurement of physical constants, the correctness of the estimates of errors is sometimes checked with time as a result of improvements in measuring instruments and methods of measurement and increased measurement accuracy.

1.7. Presentation of results of measurements: Rules for rounding off

If $\widetilde{A}$ is the result of a measurement and Δ_U and Δ_L are the upper and lower limits of the error in the measurement, then the result of the measurement and the measurement error can be written in the form

$$\widetilde{A},\ \Delta_U,\ \Delta_L,\ \text{and}\ \widetilde{A}^{\Delta_U}_{\Delta_L}.$$

In the first case it is convenient to preserve the conventional notation, for example, $\widetilde{A} = 1.153$ cm, $\Delta_U = +0.002$ cm, and $\Delta_L = -0.001$ cm. The second case is convenient for technical documentation. For example, $1.153^{+0.002}_{-0.001}$ cm. Often $|\Delta_U| = |\Delta_L| = \Delta$. Then the result and the error are written in the form $\widetilde{A} \pm \Delta$.

If Δ is the uncertainty, then the corresponding probability must be given. For uniformity it is recommended that the probability be given in parentheses following the value of the uncertainty or as a subscript to a symbol of a measurand.

For example, if a measurement gives the value of the voltage 2.62 V and the uncertainty of this result $\delta = \pm 2\%$ was calculated for the probability 0.95, then the result will be written in the form

$$\widetilde{U} = 2.62\ \text{V},\quad \delta = \pm 2\%(0.95)$$

or in the more compact form

$$U_{0.95} = (2.62 \pm 0.05)\ \text{V}.$$

The remark regarding compactness refers to the method for indicating the value of the probability and is unrelated to the fact that the relative error is given in the first case and the absolute error is given in the second case.

If the confidence probability is not indicated in the measurement result, then the error must be assumed to have been estimated without the use of probability methods. For example,

$$U = (2.1 \pm 0.1)\ \text{V}.$$

Although an error estimate obtained without the use of probability methods can be very reliable, it cannot be associated with a probability of one or some other value. Since a probabilistic model was not employed, the probability cannot be estimated and it should not be indicated.

The form, examined above, for representing measurement errors is desirable for the final result, intended for direct practical application, for example, in quality control problems. In this case, it is usually convenient to express the error in the form of absolute errors. In many cases, however, it is desirable to know not the limiting values of the total measurement error but rather the characteristics of the random and systematic components separately. Such a representation of the error makes it easier to analyze and determine the reasons for any discrepancy between the results of measurements of one and the same quantity performed under different conditions. Such an analysis is usually necessary in the case of measurements performed for scientific purposes, for example, measurements of physical constants. It is also desirable to record the components separately in those cases when the result of a measurement is to be used for calculations together with other data that are not absolutely precise. For example, for errors of measurements of quantities measured directly in indirect measurements, recording the error in this form makes it possible to estimate more accurately the error in the result of the indirect measurement.

When the error components are recorded separately the systematic component is characterized, as a rule, by the limiting values θ_U, θ_L, and θ, if $|\theta_U| = |\theta_L| = \theta$. If these limits are calculated by probabilistic methods, then the probability employed should be indicated in parentheses immediately following the value of the error. For a random error the standard deviation $S_{\bar{x}}$, determined from the experimental data, and the number of observations n are usually indicated. Sometimes the uncertainty and the corresponding probability are given instead of $S_{\bar{x}}$.

For scientific measurements, apart from the above-indicated parameters of the error, it is helpful to describe the basic sources of error together with an estimate of their contribution to the total measurement error. For the random error it is of interest to present the form and parameters of the distribution function of the observations and how it was determined (the method employed for testing the hypothesis regarding the form of the distribution function, the significance level used in this testing, etc.).

The errors in the results of mass measurements are usually not indicated at all, since they are estimated and are known beforehand.

The number of significant figures employed in the number expressing the result of a measurement must correspond to the accuracy of the measurement. This means that the limiting error of a measurement can be equal to 1 or 2 units in the last figure of the number expressing the result of the measurement. In any case, this error should not exceed 5 units in the last figure.

Since measurement errors determine only the vagueness of the results, they need not be known precisely. For this reason, in its final form the measurement error is customarily expressed by a number with one or two significant figures. Two figures are retained for the most precise measurements, and also if the most significant digit of the number expressing the error is equal to or less than 3.

It should be noted, however, that in calculations and intermediate computations, depending on the computational operations performed, one or two significant figures more than suggested by the result should be retained in order that the roundoff error not distort the results too much.

The numerical value of the result of a measurement must be represented in a manner so that the last decimal digit is of the same rank as its error. There is no point in including a larger number of digits, since this will not reduce the uncertainty characterized by the error. But a smaller number of digits, which can be obtained by further rounding off the number, would increase the uncertainty and would make the result less accurate, and thereby it would make pointless the measures employed in the measurement.

When analyzing the results of observations and recording the results of measurements the rounding off should be done according to the following rules:

(1) The last digit retained is not changed if the adjacent digit being discarded is less than 5. Extra digits in integers are replaced by 0's, while extra digits in decimal fractions are dropped.

Examples. The numerical value of the result of a measurement 85.6342 with an error in the limits ± 0.04 should be rounded off to 85.63. If the error limits are ± 0.012, the same number should be rounded off to 85.634.

Retaining four significant figures, the number 165 245 should be rounded off to 165 200 and the number 165.245 should be rounded off to 165.2.

(2) The last digit retained is increased by 1 if the adjacent digit being discarded is greater than 5 or if it is 5 and there are digits other than 0 to its right.

Examples. If three significant digits are retained, the number 18 598 is rounded off to 18 600 and the number 152.56 is rounded off to 153.

(3) If the digit being discarded is equal to 5 and the digits to its right are unknown or are equal to 0, then the last retained digit is not changed if it is even and it is increased by 1 if it is odd.

Examples. If two significant digits are retained, the number 10.5 is rounded off to 10 and the number 11.5 is rounded off to 12.

(4) If the decimal fraction in the numerical value of the result of a measurement terminates in 0's, then the 0's are dropped only up to the digit that corresponds to the rank of the numerical value of the error.

The foregoing rules were established by convention, and for calculations performed by humans they are entirely satisfactory. In the case of calculations performed with the help of computers, however, rounding off depending on the evenness or oddness of the last retained digit [rule (3)] is inconvenient, since it complicates the algorithm. It has been suggested that this rule be dropped and the last retained figure not be changed, irrespective of whether it is even or odd. This suggestion, however, has not been adopted. The main objection is that a large number of such roundoffs of intermediate results can significantly distort the final result.

If the rules presented above are used, then the number of significant figures in the numerical value of the result of a measurement makes it possible to judge approximately the accuracy of a measurement. For this, it should be noted that the limiting error, owing to roundoff, is equal to one-half the last digit in the numerical value of the result of the measurement, and the measurement error can reach two units in the next to last and several units in the last digit.

We shall now estimate the relative roundoff error. Assume, for example, that the result of a measurement is expressed by a number with two significant figures. Then the minimum number will be equal to 10 and the maximum number will be equal to 99. Therefore the relative roundoff error will be $0.5\% \leqslant \varepsilon_2 \leqslant 5\%$.

If the result of a measurement is expressed by a number with three significant figures, this error will fall in the range $0.05\% \leqslant \varepsilon_3 \leqslant 0.5\%$, and so on.

The error limits obtained above show the effect of roundoff on the measurement error. In addition, these data permit focusing on the minimum number of significant figures necessary to record the result of a measurement with the prescribed accuracy.

When analyzing the results of observations and, especially, when estimating the errors it is useful to employ methods and formulas of approximate calculations.

1.8. Basic conventional notations

We shall employ latin letters for the measurands. Greek letters will be employed for errors and uncertainties.

TABLE 1.1. Designations of measurement errors.

Name of error	Designation	
	Error	Limits of error and uncertainty
Systematic	ϑ	θ
Random	ψ	Ψ
Total absolute	ζ	Δ
Total relative	ϵ	δ

We shall distinguish estimates of quantities from the true values by adding a tilde to the corresponding symbol. For example, $\tilde{A}$ is the estimate of the true value of A.

We shall denote the arithmetic mean with the help of an overbar on the corresponding symbol. For example, $\bar{x}$ is the arithmetic mean of the obtained values of x_i $(i = 1,...,n)$.

In addition, we shall use the following mathematical symbols: $M[X]$ is the mathematical expectation and $D[X]$ is the variance of a random quantity X.

Of the notation used for specific concepts, we present the following: p is the probability at which an event first occurs, α is the confidence probability, q is the significance level, σ^2 is the variance of a random quantity, σ is the rms or standard deviation, and S^2 and S are the estimations of σ^2 and σ.

The notations employed for errors, their limits, and uncertainties are presented in Table 1.1.

Chapter 2

Measuring instruments and their properties

2.1. Types of measuring instruments

Measuring instruments are the technical objects that are specially developed for the purpose of measuring specific physical quantities. A general property of measuring instruments is that their accuracy is standardized.

Measuring instruments are divided into material measures, measuring transducers, indicating instruments, measuring setups, and measuring systems.

A *material measure* is a measuring instrument that reproduces one or more known values of a given physical quantity. Examples of measures are balance weights, measuring resistors, and measuring capacitors.

Single-valued measures, multiple-valued measures, and collections of measures are distinguished. Examples of multiple-valued measures are graduated rulers, measuring tapes, resistance boxes, etc.

In addition to multiple-valued measures, which reproduce discrete values of quantities, there exist multiple-valued measures that continuously reproduce quantities in some range, for example, a measuring capacitor with variable capacitance. Continuous measures are usually less accurate than discrete measures.

When measures are used to perform measurements, the measurands are compared with the known quantities reproduced by the measures. The comparison is made by different methods, but so-called comparators are a specific means that are used to compare quantities. The simplest comparator is the standard equal-armed pan balance.

A *comparator* is a measuring device that makes it possible to compare similar physical quantities and has a known sensitivity.

In some cases quantities are compared without comparators, by experimenters themselves, with the help of their viewing or listening perceptions.

Thus when measuring the length of a body with the help of a ruler, the ruler is placed on the body and the observer fixes visually the graduations of the ruler (or fractions of a graduation) at the corresponding points of the body.

A *measuring transducer* is a measuring instrument that converts the measurement signals into a form suitable for transmission, processing, or storage. The measurement information at the output of a measuring transducer cannot, as a rule, be directly observed by the observer.

It is necessary to distinguish measuring transducers and the transforming elements of a complicated instrument. The former are measuring instruments and as such they have standard metrological properties (see below). The latter, on the other hand, do not have an independent metrological significance and are not used separately from the instrument of which they are a part.

Measuring transducers are very diverse. Examples are thermocouples, resistance thermometers, measuring shunts, the measuring electrodes of *p*H meters, etc. Measuring current or voltage transformers and measuring amplifiers are also measuring transducers. But this group of transducers is characterized by the fact that the signals at their inputs and outputs are a physical quantity of the same form, and only the dimension of the quantity changes. For this reason these measuring transducers are called scaling measuring transducers.

Measuring transducers that convert an analog quantity at the input (the measurand) into a discrete signal at the output are called analog-to-digital converters. Such converters are manufactured in the form of autonomous, i.e., independent measuring instruments, and in the form of units built into other instruments, in particular, in the form of integrated microcircuits. Analog-to-digital converters are a necessary component of digital devices, but they are also employed in monitoring, regulating, and control systems.

An *indicating instrument* is a measuring instrument that is used to convert measurement signals into a form that can be directly perceived by the observer.

Based on the design of the input circuits, indicating instruments are just as diverse as measuring transducers, and it is difficult to survey all of them. Moreover, such a review and even classification are more important for designing instruments than for describing their general properties.

A common feature of all indicating instruments is that they all have readout devices. If these are implemented in the form of a scale and an indicating needle, then the indications of the instrument are a continuous function of the measurand. Such instruments are called analog instruments. If the indications of instruments are in a digital form, then such instruments are called digital instruments.

The definition of digital instruments presented above formally includes both automatic digital voltmeters, bridges, and similar instruments and induction meters for measuring electrical energy. In these instruments, however, the measuring transformations are performed in a discrete form, and in the case of induction meters, all measuring transformations of signals occur in an analog form and only the output signal assumes a discrete form. The conversions of measurement information into a discrete form have a number of specific features. For example, only instruments in which the measurement conversions occur in a discrete form are usually considered to be digital instruments.

The indications of digital instruments are easily recorded and are convenient for entering into a computer. In addition, their design usually makes it possible to obtain significantly higher accuracy than analog instruments. Moreover, when digital instruments are employed there is no reading error. However analog instruments are much simpler than digital instruments, and when analog instruments are used it is easier to judge trends in the variation of the measurands.

In addition to analog and digital instruments there also exist analog-discrete measuring instruments. In these instruments the measuring conversions are performed in a discrete form, but the readout device is an analog unit. The readout device has a scale and a glowing strip, whose length changes discretely, which plays the role of the indicator. Sometimes the indicator is a glowing dot that moves along a scale.

Analog-discrete instruments combine the advantages of both analog and digital instruments. Induction meters for measuring electric energy are examples of such hybrid instruments.

In many cases measuring instruments are designed so that their indications are recorded. Such instruments are said to be *recording instruments.* Data can be recorded in the form of a continuous record of the variation of the measurand in time or in the form of a series of points of this dependence.

Instruments of the first type are called automatic-plotting instruments and instruments of the second type are called printing instruments. Printing instruments can record the values of a measurand in digital form. Printing instruments give a discrete series of values of the measurand in some interval of time. The continuous record provided by automatic-plotting instruments can be regarded as an infinite series of values of the measurand.

Sometimes measuring instruments are equipped with induction, photo-optical, or contact devices and relays for purposes of control or regulation. Such instruments are called regulating instruments. Designers strive to design regulating units so as not to reduce the accuracy of the measuring instrument. However, this is rarely possible.

Measuring instruments also customarily include comparators, mentioned above, for comparing measures, and null indicators, for example, galvanometers. This can be explained by the fact that a comparator with a collection of measures becomes a comparison measuring instrument while a galvanometer can be used as a highly sensitive indicating instrument.

A *measurement setup* is a collection of functionally and structurally integrated measuring instruments and auxiliary devices that provides efficient organization of the measurements. An example is the potentiometric setup for electric measuring instrument calibration.

A *measuring system* is a collection of functionally unified measuring, computing, and auxiliary means for obtaining measurement information and for converting and processing it in order to provide the user with information in the required form, introducing it into the control system, or for performing logical functions automatically. Modern measuring systems include microprocessors and even entire computers, and apart from processing and providing output of the measurement information, they can control the measurement process itself.

Finally, systems all of whose units must, in accordance with the purpose of the system, operate under the same conditions can be distinguished from systems whose units operate under different conditions. We shall call the former *uniform measuring systems* and the latter *nonuniform measuring systems.* This classification makes it easier to study questions concerning the metrological support of measuring systems and the calculation of the errors of such systems.

2.2. The concept of an ideal instrument: Metrological characteristics of measuring instruments

Any technical object can be described by a collection of characteristics. Measuring instruments are not an exception in this respect. We shall divide all characteristics of measuring instruments into two groups: metrological, necessary for using a measuring instrument in the manner intended, and secondary. We shall include in the latter such characteristics as mass, dimensions, degree of protection from moisture and dust, etc. We shall not discuss characteristics of the secondary group, though sometimes they determine the selection and application of an instrument, since they are not directly related with the measurement accuracy.

By metrological characteristics of a measuring instrument we mean the characteristics that make it possible to judge the suitability of the instrument for performing measurements in a known range with known accuracy, to obtain a value of the measurand, and to estimate its inaccuracy.

In order to sort the metrological characteristics, it is helpful to introduce the concept of an ideal instrument. The ideal of any technical object is its design, i.e., its model. For measuring instruments this is not sufficient, since such a device must contain also an "impression" of the corresponding unit of measurement. The impression of the unit cannot be prepared; it must be obtained from a standard. We shall give several examples.

Gauge block. The ideal is a completely regular parallelepiped, one edge of which is determined exactly in terms of the established units of length.

Measure of constant voltage. The ideal is a source of constant voltage whose value is known exactly and which is free of any noise at the output.

Measuring transformer. The ideal is a voltage or current transformer whose conversion factor is known exactly and which does not have any losses and parasitic noise in the input and output circuits.

Integrating analog-to-digital converter of voltage. The ideal is an instrument whose output voltage U_0 is related to the input voltage U_x by the dependence

$$U_0=K_1 \int_{t_1}^{t_2} U_x dt,$$

where $K_1 = \text{const}$ and $\Delta t = t_2 - t_1 = \text{const}$.

As a result we obtain $U_0 = K_2 U_x$. The voltage U_0 can be quantized and a code, reflecting the voltage at the input without any distortions (with the exception of the quantization error, which can be made to be quite small), is obtained at the output of the converter.

Moving-coil ammeter. In this instrument a constant current flows through a moving coil and forms with the help of a permanent magnet a mechanical moment, which twists a spring to a point of balance. In the process an indicating needle is deflected by an amount along a scale that is proportional to the current strength.

Each value of the current strength corresponds to a definite indication, which is fixed by calibrating the instrument.

Thus the ideal instrument performs (theoretically) a series of single-valued transformations and after calibration acquires a precise scale in the units of the measurand.

An ideal representation of each type of measuring instruments is formed for a specific model of the objects—carriers of the corresponding physical quantity.

We shall call the metrological characteristics established for ideal measuring instruments of a specific type the *nominal metrological characteristics*. An example of such a characteristic is the nominal value of the measure (10 Ω, 1 kG, etc.), the measurement range of the instrument (0–300 V, 0–1200 °C, etc.), the conversion range of the transducer, the value of the scale factor of the instrument scale, etc.

The relation between the input and output signals of instruments and transducers is determined by the transfer function. For instruments it is fixed by the scale, while for measuring transducers it is determined by a graph or an equation. Either the graph or the equation represents the nominal metrological characteristic if the graph or equation was determined (indicated) before these measuring instruments were developed.

The real characteristics of measuring instruments differ from the nominal characteristics owing to fabrication errors and changes occurring in the corresponding properties in time.

An ideal measuring instrument (transducer) would react only to the measured physical quantity or to the parameter of the input signal of interest (the informative parameter of the input signal), and its indication would not depend on the external conditions, the power supply regime, and the uninformative parameters of the input signal. For a real measuring instrument, as for other types of measuring instruments, these undesirable phenomena do occur.

The quantities characterizing the external conditions and the uninformative parameters of the input signal are called influence quantities.

For some types of measuring instruments the dependence of the output signal, the indications, or the error owing to one or another influence quantity can be represented as a functional dependence, called the influence function. The influence function can be expressed in the form of an equation (for example, the temperature dependence of the emf of standard cells) or a graph. In the case of a linear dependence it is sufficient to give the coefficient of proportionality between the output quantity and the influence quantity. We shall call this coefficient the influence factor.

Influence factors and functions make it possible to take into account the conditions under which measuring instruments are used by introducing the corresponding corrections. The imperfection of measuring instruments is also manifested by the fact that when one and the same quantity is measured repeatedly under identical conditions, the results can differ somewhat from one another. In this case it is said that the indications are nonrepeatable.

The inaccuracy of a measuring instruments is usually characterized by its error. We shall explain this concept for the example of a indication measuring instrument. Let the true value of a quantity at the input of the instrument be A_t. The instrument indicates the value A_r. The absolute error of the instrument will be

$$\zeta = A_r - A_t.$$

The nonrepeatability of the indications of the instrument is manifested by the fact that when A_t is measured repeatedly the indications of the instrument will be somewhat different. For this reason, one can talk about a random component of instrument error. This component is referred to as the repeatability error of a measuring instrument.

The random component of instrument error is normally caused by friction in the supports of a movable part of the instrument and hysteresis phenomena, and its limits are quite sharp. The limits can be found experimentally if the quantity measured by the instrument varies continuously. The strength of the electric current, the voltage, and other quantities can be varied continuously. Correspondingly, the indications of ammeters and voltmeters can vary continuously. The indications of weighing balances and a number of other instruments cannot be varied continuously.

For instruments whose indications can vary continuously, the limits of the random error are found by continuously driving the indicator of the instrument up to the same scale marker first from below and then from above (or vice versa) a marker. We will call *the dead zone* the absolute value of the difference of the values of the measurand that are obtained in such a test and that correspond to a given scale marker of the instrument.

The dead zone is the length of the range of possible values of the random component of instrument error and one half of this length is the limiting value of the random error.

Figure 2.1 shows graphs of the "input–output" in the presence of (a) only friction, (b) only hysteresis, and (c) friction together with hysteresis. These are examples of processes that reveal dead zones.

The random error of weighing scales is usually characterized by the standard deviation.[11] This characteristic of an instrument is calculated from the changes produced in the indications of the scales by a load with a known mass; the test is performed at several scale markers, including the limits of the measurement range. One method for performing the tests and the computational formula for calculating the standard deviation of weighing scales are presented in Ref. 11.

Measuring instruments are created in order to introduce certainty into the phenomena studied and to establish regular relations between the phenomena, and the uncertainty created by the non-single-valuedness of instrument indications interferes with using an instrument in the manner intended. For this reason, the first problem that must be solved when developing a new measuring device is to make its random error insignificant, i.e., either negligibly small compared with other errors or falling within prescribed limits as the limits of admissable errors for measuring devices of the given type.

If the random error is insignificant and the elements determining instrument accuracy are stable, then by calibration the measuring device can always be "tied" to a corresponding standard and the potential accuracy of the instrument can be realized.

The value of a scale division or the value of a significant figure is the value of the

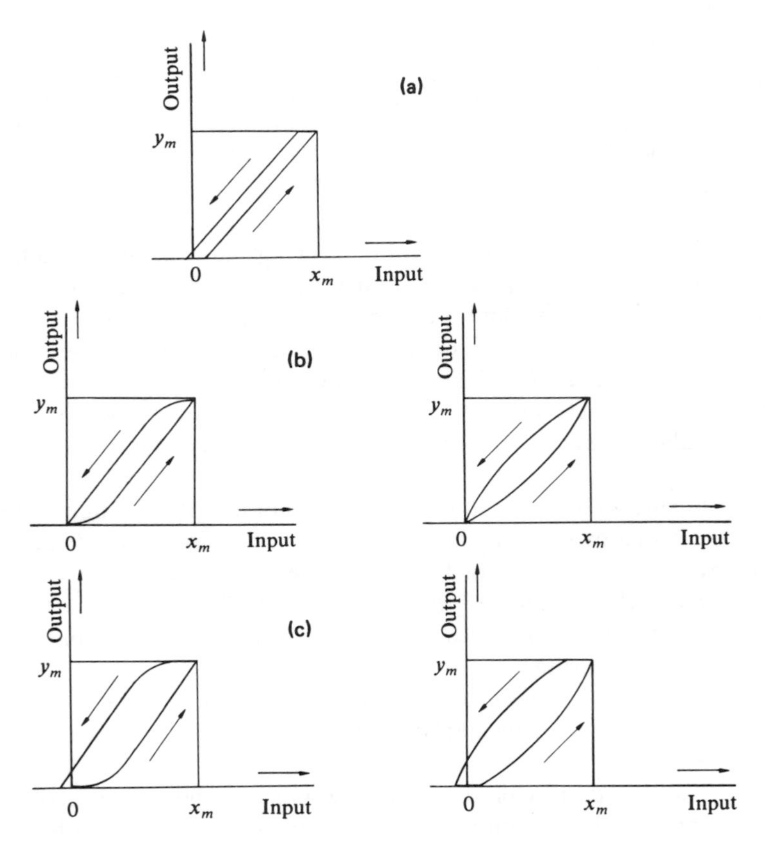

FIG. 2.1. Dependences between the input and output of the instrument with a dead zone. (a) In the presence of a friction; (b) in the presence of hysteresis (two types, for example); (c) in the presence of a friction and hysteresis (two types mentioned above).

measurand corresponding to the interval between two neighboring markers on the instrument scale or one figure of some digit of a digital readout device.

The *sensitivity* is the ratio of the change in the measurand at the output of the measuring instruments to the input value of the quantity that causes the output value to change. The sensitivity can be a nominal metrological characteristic and an actual characteristic of a real instrument.

The *discrimination threshold* is the minimum change in the input signal that causes an appreciable change in the output signal.

The *resolution* is the smallest interval between two distinguishable neighboring discrete values of the output signal.

Instability (of a measuring instrument) is a general term that expresses the change in any property of the measuring instrument in time.

Drift is the change occurring in the output signal (always in the same direction) over a period of time that is significantly longer than the measurement time when using a given measuring instrument.

The drift and also the instability do not depend on the input signal or the load, but they can depend on the external conditions. The drift is usually determined in the absence of a signal at the input.

The metrological characteristics of measuring instruments should also include their dynamic characteristics. These characteristics reflect the inertial properties of measuring instruments. It is necessary to know them in order to correctly choose and use many types of measuring instruments. The dynamical characteristics are examined below in Sec. 2.5.

The properties of measuring instruments can normally be described based on the characteristics enumerated above. For specific types of measuring instruments, however, additional characteristics are often required. Thus, for the gauge rods the so-called flatness and polishability are important. For voltmeters the input resistance is important. We shall not study such characteristics, since they refer only to individual types of measuring instruments.

2.3. Standardization of the metrological characteristics of measuring instruments

Measuring instruments can only be used as intended when their metrological properties are known. In principle, the metrological properties can be established by two methods. One method is to find the actual characteristics of a specific instrument. In the second method, the nominal metrological characteristics and the permissable deviations of the real charactersitics from the nominal characteristics are given.

The first method is laborious and for this reason it is used primarily for the most accurate and stable measuring instruments. For this reason the second method is the main method. The nominal characteristics and the permissable deviations from them are given in the technical documentation when measuring instruments are designed. This predetermines the properties of measuring instruments and ensures that they are interchangeable.

In the process of using measuring instruments, checks are made to determine whether or not the real properties of the devices deviate from the established standards. If one of the real properties deviates from its nominal value by an amount greater than demonstrated by the standards, then the measuring instrument is adjusted, remade, or discarded and no longer used.

Thus the choice of the nominal characteristics of measuring instruments and the designation of permissable deviations of the real characteristics from them—standardization of the metrological characteristics of measuring instruments—are of great importance for measurement practice. We shall examine the practice of standardization of the metrological characteristics of measuring instruments that has evolved.

Both the production of measuring instruments and the standardization of their characteristics initially arose spontaneously in each country. Later, standards, which gave order to this standardization, were developed in all countries in which instrument building was highly developed. The recommendations developed at this time by international organizations, primarily Publication 51 of the International Electrotechnical Commission (IEC), were of great importance for the preparation of national standards.[7] We should also mention the International Organization for

Standardization (ISO) and the International Organization of Legal Metrology (OIML). The terminological documents are also of great value for this work.[1,3,6]

We shall now return to the gist of the problem. The significance of nominal metrological characteristics, such as the upper limits of measurement ranges, the nominal values of the measures, the scale factors of instruments, etc., is chosen from standardized series of values of these characteristics. There is nothing special here. Another task is to standardize the accuracy characteristics, errors, and stability.

In spite of the efforts of designers, the real characteristics of measuring instruments depend to some extent on the external conditions. For this reason, some narrow ranges of values of all influence quantities are fixed first and in this manner the conditions under which measuring instruments are to be calibrated and checked are determined. These conditions are called reference conditions. The error of measuring instruments under reference conditions is called the *intrinsic error.*

In the standard in Ref. 6 this question is solved less formally: the conditions under which the characteristics of measuring instruments depend negligibly on the possible variations of influence quantities are called *reference* conditions. In other words, these are conditions under which the metrological characteristics are practically constant.

This definition of reference conditions seems very attractive. I stated the identical idea in Ref. 51. However, I also stated there my doubts in the possibility of implementing the idea. We shall return to this question in the next section.

Thus the reference conditions of measuring instruments are prescribed and the intrinsic errors of measuring instruments are determined.

In addition to the reference conditions, the normal operating conditions of measuring instruments are also established, i.e., the conditions under which the characteristics of measuring instruments remain within certain limits and the measuring instruments can be employed as intended. Understandably, errors in the normal operating conditions are larger than errors in the reference conditions (i.e., these errors are larger than the intrinsic errors).

When any influence quantity exceeds the normal value or range for a reference condition, the error of the measuring instrument changes. This change is characterized and standardized by indicating the limit of the permissable additional error, by indicating the highest permissable value of the influence factor of the corresponding influence quantity, or by indicating the limit of the permissable error under the normal operating conditions.

The errors of measuring instruments are expressed not only in the form of absolute and relative errors, adopted for estimating measurement errors, but also in the form of *fiducial errors.* The fiducial error is the ratio of the absolute error of the measuring instrument to some standardizing value—fiducial value. The latter value is established by standards on separate types of measuring instruments. For indicating instruments, for example, the fiducial value is established depending on the characteristic features and character of the scale. The fiducial errors make it possible to compare the accuracy of measuring instruments that have different measurement limits. For example, the accuracy of an ammeter with a measurement limit of 1 A can be compared with that of an ammeter with a measurement limit of 100 A.

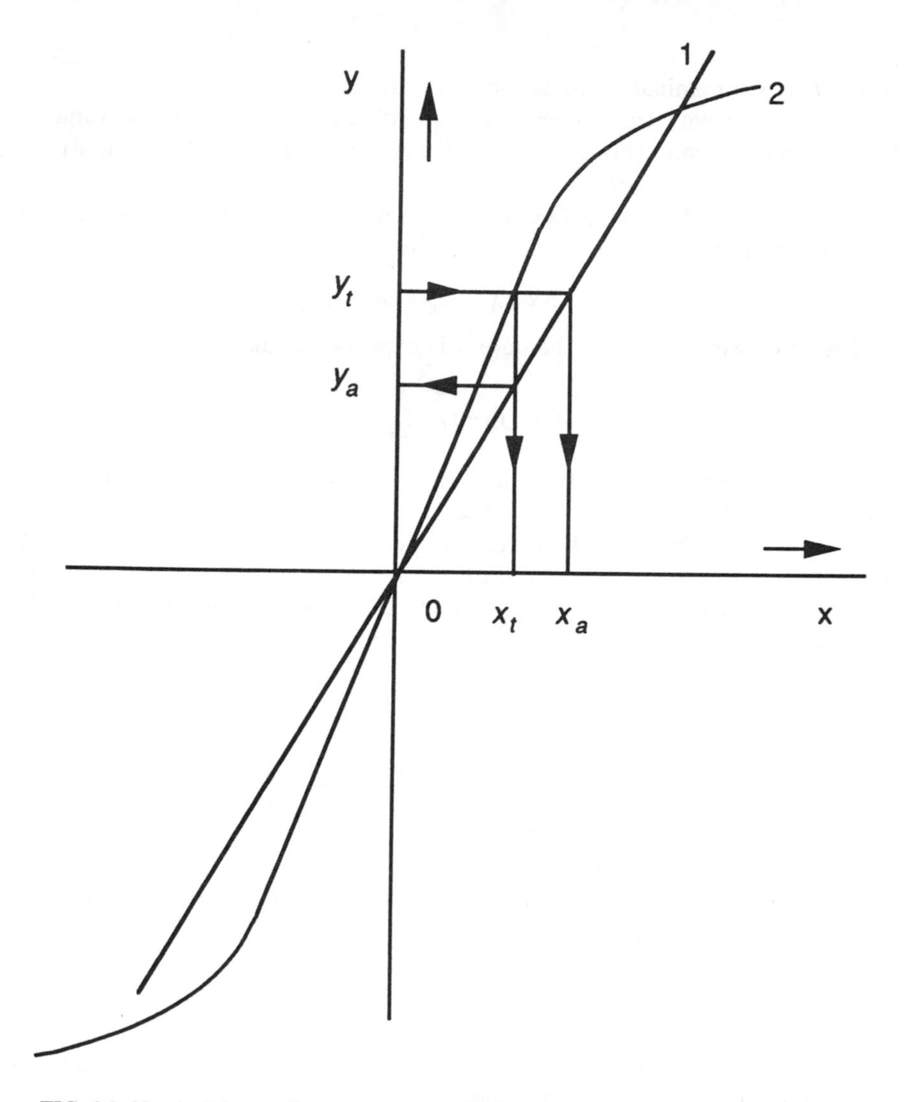

FIG. 2.2. Nominal (curve 1) and real (curve 2) functions of a measuring transducer.

In addition, cases when the error of an indicating instrument is expressed in terms of the scale graduations (normally, in fractions of a graduation) are also encountered.

For measuring transducers the errors can be represented by the *errors* relative to the input or output.

Figure 2.2 shows the nominal and, let us assume, the real transfer functions of some transducer. The nominal dependence, as done in practice whenever possible, is assumed to be linear. We shall investigate the relationship between the errors of the transducer that are scaled to the input and the output.

We denote the input quantity by x and the output quantity by y. They are related by the relation

$$x = Ky,$$

where K is the nominal transduction constant.

At the point with true values of the quantities x_t and y_t, the true value of the transduction constant will be $K_t = x_t/y_t$. Calculations based on the nominal constant K, however, give an error.

Let $x_a = Ky_t$ and $y_a = x_t/K$ be determined based on y_t and x_t (see Fig. 2.2). Then the absolute transducer error with respect to the input will be

$$\Delta x = x_a - x_t = (K - K_t)y_t.$$

The error with respect to the output is expressed analogously:

$$\Delta y = y_a - y_t = \left(\frac{1}{K} - \frac{1}{K_t}\right) x_t.$$

We note, first of all, that Δx and Δy always have different signs: if $(K - K_t) > 0$, then $(1/K - 1/K_t) < 0$.

But this is not the only difference. The quantities x and y can also have different dimensions, i.e., they can be physically different quantities, so that the absolute input and output errors are not comparable. For this reason, we shall study the relative errors:

$$\varepsilon_x = \frac{\Delta x}{x_t} = (K - K_t)\frac{y_t}{x_t} = \frac{K - K_t}{K_t},$$

$$\varepsilon_y = \frac{\Delta y}{y_t} = \frac{(K_t - K)}{KK_t}\frac{x_t}{y_t} = \frac{K_t - K}{K}.$$

Since $K_t \neq K$, we have $|\varepsilon_x| \neq |\varepsilon_y|$.

We denote the relative error in the transduction constant at the point (x_t, y_t) as ε_k, where $\varepsilon_k = (K - K_t)/K_t$. Then

$$\frac{\varepsilon_x}{\varepsilon_y} = -(1 + \varepsilon_k).$$

However, $\varepsilon_k \ll 1$, and in practice relative errors with respect to the input and output can be regarded as equal in magnitude.

We must stop to consider how the error of measures is determined: The error of measures is the difference between the nominal value of the measure and the true value of the quantity reproduced by the measure. Indeed, in the case of indicating instruments, the nominal value of measures is the analog of the indication of the instrument, and the definition given becomes obvious.

It is also interesting that measures which reproduce passive quantities, for example, mass, electric resistance, etc., have only systematic errors. The error of measures of active quantities (electric voltage, electric current, etc.) can have both systematic and random components. Multiple-valued measures of passive quantities can have random errors owing to switching elements.

So, when the errors of measuring instruments are normalized, the limits of the intrinsic and all additional errors are prescribed. At the same time the reference and normal operating conditions are indicated.

Of all forms enumerated above for expressing the errors of measuring instruments, the best is the relative error, since in this case the indication of the permissible limit of error gives the best idea of the level of measurement accuracy that can be achieved using the given measuring instrument. The relative error, however, usually changes significantly over the measurement range of the instrument, and for this reason it is difficult to use for standardization.

The absolute error is frequently more convenient than the relative error. In the case of an instrument with a scale, the limit of the permissible absolute error can be standardized using the same numerical value for the entire scale of the instrument. But then it is difficult to compare the accuracies of instruments having different measurement ranges. This difficulty disappears when the reduced errors are standardized.

In the following discussion we shall follow primarily Refs. 7 and 8.

The limit of the permissible absolute error Δ can be expressed by a single value (neglecting the sign)

$$\Delta = \pm a,$$

in the form of the linear dependence

$$\Delta = \pm (a + bx), \tag{2.1}$$

where x is the nominal value of the measure, the indication of a measuring instrument, or the signal at the input of a measuring transducer, and a and b are constants, or by a different equation,

$$\Delta = f(x).$$

When the latter dependence is complicated, it is given in the form of a table or graph.

The fiducial error γ (in percent) is defined by the formula

$$\gamma = 100\Delta / x_N,$$

where x_N is the fiducial value.

The fiducial value is assumed to be equal to the following:

(i) The value at the end of the instrument scale, if the zero marker falls on the edge or off the scale.

(ii) The span that is a sum of the end values of the instrument scale (neglecting the signs), if the zero marker falls within the scale.

(iii) The nominal value of the measurand, if it has been established.

(iv) The length of the scale, if the scale graduations become sharply narrower. In this case the error and the length of the scale are expressed in the same units.

For instruments having a scale that is calibrated in units of a quantity for which a scale with a conventional 0 is adopted (for example, in °C), the *fiducial value* is assumed to be equal to the difference of the final and starting values of the scale (the measurement range or span).

According to Recommendation 34 of OIML[8] for measuring instruments with a 0 marker within the scale, the fiducial value is taken to be equal to the larger (neglecting the sign) of the end values of the indication range of the instrument. According to Publication 51 of IEC,[7] for electrical measuring instruments it can be set equal to the sum of the end values of the scale.

A progressive and correct solution is the one recommended by OIML. Indeed, consider, for example, an ammeter with a scale 100–0–100 A and with a permissible absolute error of 1 A. In this case the fiducial error of the instrument will be 1% according to OIML and 0.5% according to IEC. But when using this instrument the possibility of performing a measurement with an error of up to 0.5% cannot be guaranteed for any point of the scale. An error not exceeding 1%, however, can be guaranteed when measuring a current of 100 A under reference conditions. The tendency to choose a fiducial value such that the fiducial error would be close to the relative error of the instrument was observed in the process of improving IEC Publication 51. Thus in the previous edition of this publication the fiducial value for instruments without a 0 marker on the scale was taken to be equal to the difference of the end values of the range of the scale and now it is taken to be equal to the larger of these values (neglecting the sign). Consider, for example, a frequency meter with a scale 45–50–55 Hz and a limit of permissible absolute error of 0.1 Hz. Previously the fiducial error of the frequency meter was assumed to be equal to 1%, and now it is equal to 0.2%. But when measuring a 50-Hz frequency its relative error indeed will not exceed 0.2% (under reference conditions), and the 1% error has no relation to the error of frequency measurement with this meter, so that the new edition is more correct.

The next step in this direction was made in Recommendation 34 of OIML. One must hope that in the future IEC will take into account the recommendation of OIML, and the stipulation mentioned regarding electrical measurement instruments in the recommendation of OIML will disappear.

The fiducial error is expressed in percent, but it is not a relative error. Since its limit is equal to that of the permissible reduced error, the limit of the permissible relative error for each value of the measurand must be calculated according to the formula

$$\delta=\gamma\frac{x_N}{x}.$$

The limit of permissible relative error δ is usually expressed in percent according to the formula

$$\delta=\frac{100\Delta}{x}=\pm c.$$

If the limit of the absolute error Δ is determined by formula (2.1), then the last expression is possible for $a\approx 0$.

For digital instruments the errors are very often standardized in the conventional form $\pm(b+q)$, where b is the relative error in percent and q is some number of figures of the least significant digit of the digital readout device. For example, an instrument with a measurement range of 0–300 mV is assigned the limits of permissible error $\pm(0.5\%+2)$. The indicator of the instrument has four digits, so that the figure 2 of the least significant digit corresponds to 0.2 mV. Now the limit of the relative error of the instrument when measuring, for example, a voltage of 300 mV can be calculated as follows:

$$\delta = \pm\left(0.5 + \frac{0.2 \times 100}{300}\right) = \pm 0.57\%.$$

Thus to estimate the limit of permissible error of an instrument some calculations must be performed. For this reason, though the conventional form gives a clear representation of the components of instrument error, it is inconvenient to use.

A more convenient form is given in Recommendation 34 of OIML: the limit of permissible relative error is expressed by the formula

$$\delta = \pm\left[c + d\left(\frac{x_e}{x} - 1\right)\right], \tag{2.2}$$

where x_e is the end value of the measurement range of the instrument or the input signal of a transducer and c and d are relative quantities.

With the adopted form of the formula (2.2), the first term on the right-hand side is the relative error of the instrument at $x = x_e$. The second term in this expression characterizes the increase of the relative error as the indications of the instrument decrease.

Formula (2.2) can be obtained from $\pm(b + q)$ as follows. To the figure q there corresponds the measurand qD, where D is the value of one figure in the same digit as the figure q, in units of the measurand. In the relative form it is equal to qD/x. Now the sum of the terms b and qD/x has the following physical meaning: It is the limit of permissible relative error of the instrument.

So,

$$\delta = \left(b + \frac{qD}{x}\right).$$

With the help of identity transformations we obtain

$$\delta = b + \frac{qD}{x} + \frac{qD}{x_e} - \frac{qD}{x_e} = \left(b + \frac{qD}{x_e}\right) + \frac{qD}{x_e}\left(\frac{x_e}{x} - 1\right).$$

Writing

$$c = b + \frac{qD}{x_e}, \quad d = \frac{qD}{x_e},$$

we obtain formula (2.2).

In application to the example of a digital millivoltmeter studied above we have

$$\delta = \pm\left[0.57 + 0.07\left(\frac{x_e}{x} - 1\right)\right].$$

It is clear that the last expression is more convenient to use and in general it is more informative than the conventional expression.

Note that for standardization the error limits are established for the total instrument error and not for the separate components. If, however, the instrument has an appreciable random component, then a permissible limit is established separately for it also. For example, aside from the limit of the permissible intrinsic error, the limit of the permissible dead zone or hysteresis is also established. Sometimes, however, the limits are nonetheless set separately for the systematic and random components. For example, the error of reference standards is customarily given in this manner in the C.I.S.

Additional errors of measuring instruments are normalized by prescribing the limits for each of the additional errors separately. The intervals of variation of the corresponding influence quantities are indicated simultaneously with the limits of the additional errors. The collection of ranges provided for all influence quantities determines the normal operating conditions of the measuring instrument. The limit of permissible additional error is often represented in proportion to the value of the influence quantity or its deviation from the limits of the interval determining the standard values of these quantities. In this case the corresponding coefficients are standardized. We shall call it the *influence coefficient*.

In the case of measuring instruments the term *variation of indications* is used as well as the term *additional error*. The term variation of indications is used, in particular, for electric measuring instruments.[7]

The additional errors arising when the influence quantities are fixed are systematic errors. For different models of instruments, however, they can have different values and, what is more, different signs. For this reason, in the overwhelming majority of standards, the limits of additional errors are set both positive and negative with equal numerical values. For example, the change in the indications of an electric measuring instrument caused by a change in the temperature of the surrounding medium should not exceed the limits ±0.5% for each 10 °C change in temperature under normal operating conditions (the numbers here are arbitrary).

If, however, the properties of a measuring device are sufficiently uniform, it is best to standardize the influence function, i.e., to indicate the dependence of the indications of the instruments or output signals of the transducers on the influence quantities and the limits of permissible deviations from each such dependence. If the influence function can be standardized, then it is possible to introduce corrections to the indications of the instruments and thereby to use the capabilities of the instruments more fully.

It should be emphasized that the properties of only the measuring instruments themselves are standardized with the help of the norms of the additional errors. The actual additional error that can arise in a measurement will depend not only on the properties of the measuring instrument itself but also on the value of the corresponding influence quantity.

The errors owing to deviations of the noninformative parameters of the input signal from the standard values should also be included among the additional errors of a measuring instrument. For example, for a voltmeter in an electromagnetic system the frequency of the alternating current is one of the noninformative parameters of the signal. According to Sec. 1.3 these errors are caused by the fact that one or more parameters of the model do not correspond to the properties of the real object. Since these errors are very characteristic for the measuring instruments in

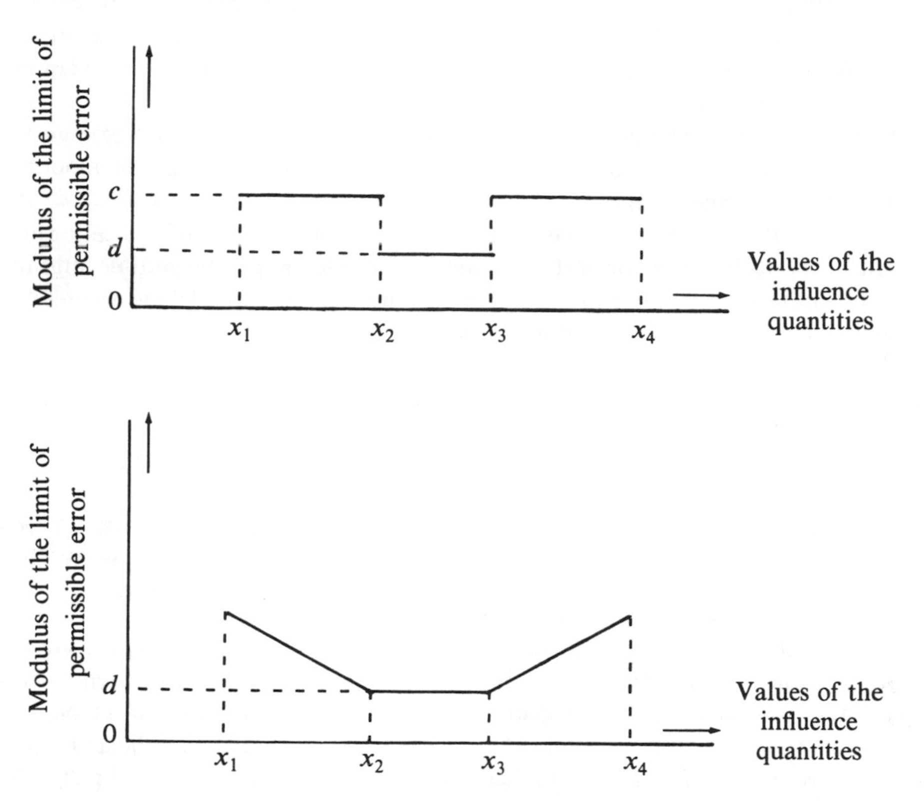

FIG. 2.3. Variants of standardization of the limits of additional errors of measuring instruments. The interval $(x_3 - x_2)$ corresponds to *reference conditions*; the interval $(x_4 - x_1)$ corresponds to the *normal operating conditions*; d is the limit of permissible intrinsic error; c is the limit of permissible error in the *normal operating conditions*; and, $(c - d)$ is the limit of permissible additional error.

which they are observed, they are usually given the name of the corresponding parameter of the model. Thus in the foregoing example the model of the signal is a sinusoidal voltage with a fixed parameter (frequency). The corresponding error is called the frequency error.

The basic cases of standardization of additional errors are shown in Fig. 2.3.

Stability of measuring instruments. Stability, like accuracy, is a positive quality of a measuring instrument. Just as the accuracy is characterized by inaccuracy (error, uncertainty), stability is characterized by instability.

Instability is normalized by the value of the limits of permissible variations of the error over a definite period of time or by prescribing different error limits to different "lifetimes" of the instrument after it is calibrated. In addition, limits are sometimes prescribed for the drift of the indications of the instrument; these limits, naturally, are indicated together with the time. It is desirable to normalize drift for automatic-plotting instruments. But normalization of the drift is also helpful for other types of measurement instruments, since it makes it possible to judge how often the indications or the 0 of the instruments must be corrected.

To correct the indications of electric measuring instruments, standard cells or electronic voltage stabilizers are often built into them. For example, weak sources of stable radioactivity are built into meters for measuring the parameters of radioactive radiations, etc.

It is significant that separate standardization of the drift does not change the standards for the instrumental error, i.e., it gives additional information about the properties of the instruments.

The second method of standardization of instability consists of indicating different standards for the error of the instrument for different periods of time after the instrument is calibrated. For example, a table with the following data is provided in the specifications of some digital instrument:

Time after calibration	24 hour	3 month	1 year	2 years
Temperature	23 ± 1 °C	23 ± 5 °C	23 ± 5 °C	23 ± 5 °C
Limit of error	0.01% + 1 digit	0.015% + 1 digit	0.02% + 1 digit	0.03% + 2 digits

The limits of error are presented here in the conventional form.

The first method for standardizing instability of instruments is widely used in the C.I.S. and the second method is widely used in the U.S. The second method reveals more fully the capabilities of instruments. For example, the limits of error of a digital instrument manufactured in the C.I.S. with the parameters indicated in the table above would have to be checked once per year ± (0.02% + 1 digit). The maximum instrument accuracy that can be realized in a short period of time after calibration, though in a more restricted temperature regime, would remain unknown.

Standardization predetermines the properties of measuring instruments and is closely related with the concept of accuracy classes of measuring instruments.

Accuracy classes were initially introduced for indicating electric measuring instruments.[7] Later this concept was also extended to all other types of measuring instruments.[8] Unification of the accuracy requirements of measuring instruments, the methods for determining them, and the notation in general are certainly useful to both the manufacturers of measuring instruments and to users, since it makes it possible to limit, without harming the manufacturers or the users, the list of instruments, and it makes it easier to use and check the instruments. We shall discuss this concept in greater detail.

In Ref. 1 the following definition is given for the term accuracy class (the following definition is very close to that given in Ref. 7): The accuracy class is a class of measuring instruments which meet certain metrological requirements that are intended to keep errors within specified limits.

Every accuracy class has conventional notation, established by agreement—the class index—that is presented in Refs. 7 and 8.

On the whole, the accuracy class is a generalized characteristic that determines the limits for all errors and standards for all other characteristics of measuring instruments that affect the accuracy of measurements performed with their help.

TABLE 2.1. Designations of accuracy classes.

Form of the expression for the error	Limit of permissible error (examples)	Designation of the accuracy class (for the given example)
Fiducial error, if the fiducial value is expressed in units of the measurand	$\gamma = \pm 1.5\%$	1.5
Fiducial error, if the fiducial value is set equal to the scale length	$\gamma = \pm 0.5\%$	0.5 (inside a check mark)
Relative error, constant	$\delta = \pm 0.5\%$	0.5 (circled)
Relative error, increasing as the measurand decreases	$\delta = \pm \left[0.02 + 0.01\left(\frac{x_e}{x} - 1\right)\right]$	0.02/0.01

For measuring instruments whose permissible limits of intrinsic error are expressed in the form of relative or fiducial errors, the following series of numbers, which determine the limits of permissible intrinsic errors and are used for denoting the accuracy classes, was established in Ref. 8:

$$1,\ 1.5,\ 1.6,\ 2,\ 2.5,\ 3,\ 4,\ 5,\ \text{and}\ 6 \times 10^n,$$

where $n = +1, 0, -1, -2, \ldots$; the numbers 1.6 and 3 can be used, but are not recommended. For any one value of n not more than five numbers of this series are allowed. The limit of permissible intrinsic error for each type of measuring instrument is set equal to one of the numbers in the indicated series.

Conventional designations of accuracy classes, employed in documentation accompanying measuring instruments, as well as the designations imposed on them, have been developed using the numbers in the indicated series. Of course, this refers to measuring instruments whose errors are standardized in the form of relative and fiducial errors. Table 2.1 gives examples of the adopted designations of accuracy classes of these measuring instruments.

In those cases when the limits of permissible errors are expressed in the form of absolute errors, the accuracy classes are designated by latin capital letters or roman numerals.

If formula (2.2) is used to determine the limit of permissible error, then both numbers c and d are introduced into the designation of the accuracy class. These numbers are selected from the series presented above, and in calculating the limits of permissible error for a specific value of x, the result is rounded off so that it would be expressed by not more than two significant figures; the roundoff error should not exceed 5% of the computed value.

The limits of all additional errors and other metrological characteristics of measuring instruments must be related with their accuracy class. In general, it is impossible to establish these relations for all types of measuring instruments simultaneously—measuring instruments are too diverse. For this reason, these relations must be given in the specifications together with the characteristics of specific types of measuring instruments, which the designers formulate.

For the fiducial value not recommended in Ref. 8 but demonstrated in Ref. 7 and equal to the sum of the end values of the measurement range of the instrument (neglecting the sign) the special designation is used. For example, if the limits of a fiducial error are ±1.5%, the designation of accuracy class in this case is |1.5|.

2.4. Some suggestions for changing methods of standardization of errors of measuring instruments and their analysis

Standardization, i.e., establishment of standards, is basically a volitional act. For this reason, in principle, different suggestions can be made for solving this question, and in the last few years several new methods for expressing the errors of measuring instruments and for standardizing them have indeed been proposed.

In order to evaluate these suggestions it is necessary to determine how well they solve problems for whose sake the properties of measuring instruments are standardized. From what we have said above it can be concluded that the purpose of standardization of errors of measuring instruments is to solve the following problems:

(1) To ensure that the entire collection of measuring instruments of the same type have the required accuracy and to ensure that they are uniform and interchangeable.

(2) To make sure that it is possible to evaluate the instrumental measurement errors according to established standards for metrological properties of measuring instruments.

(3) To ensure that measuring instruments can be compared with one another according to accuracy.

The first problem is ultimately solved by monitoring new measuring instruments during the manufacturing process and checking periodically the units that are in use. Since measuring instruments are employed individually, the standards must be established so that it is possible to check that each sample measuring instrument satisfies these standards.

To solve the second problem successfully it is desirable to know accurately the properties of measuring instruments. For this reason, the established standards must be as close as possible to the real properties of the measuring instruments. The degree of detail with which the errors of measuring instruments can be described is limited by the instability of the instruments, by the change in their errors in time, as well as by the degree of nonuniformity of the measuring instruments introduced by their construction and manufacturing technology. In addition, the calibration process must be simple. Complicated methods for describing and standardizing the errors of measuring instruments, which lead to laborious and prolonged checks, are nonviable.

Having made these preliminary remarks, we shall now examine the most interesting suggestions.

(1) The calculation of the errors of measuring instruments under real conditions involves summation of the errors and presents a number of difficulties. For this reason, it has been repeatedly suggested that the normal operating conditions be extended so as to absorb all possible values of the influence quantities. One would

think that in so doing the additional errors of measuring instruments would vanish and only the intrinsic error would remain, and all difficulties would be simply resolved.

The actual properties of measuring instruments, however, do not depend on the method by which they are standardized, and they remain unchanged. Suppose that in the usual method of standardization we have the following:

Δ_0, the limit of permissible intrinsic error and

Δ_i, the limit of permissible additional error, caused by the change in the ith influence quantity from the standard value to the limit of the range of the given influence quantity ($i = 1,...,n$) for normal operating conditions.

By transferring to a new method of standardization, the manufacturer of the instruments can adopt as the limit of permissible error of the measuring instrument only the arithmetic sum

$$\Delta = \sum_{i=1}^{n} \Delta_i.$$

The manufacturer cannot proceed otherwise, since he or she must guarantee that the errors of a given measuring instrument will be less than Δ for any combination of limiting values of the influence quantities. What then can this suggestion give?

From the standpoint of evaluating the measurement errors, it can significantly simplify the procedure. But in exchange, the error is significantly overestimated as a result of the fact that under the actual operating conditions of the measuring instrument in the overwhelming majority of the cases, the influence quantities do not all reach their limiting values simultaneously and in the most unfavorable combination. For this reason, even the arithmetic sum of the errors occurring in a specific measurement will be less than Δ and closer to the real value.

With respect to uniformity and interchangeability of measuring instruments of the same type, the suggestion worsens the existing situation, since the same value of Δ can be obtained for different values of the components. Thus to adopt this suggestion means taking a step backward compared with the present situation.

The foregoing analysis also shows that the definition of reference conditions given in Ref. 6 gives the most complete disclosure of the properties of a measuring instrument. In the overwhelming majority of the cases, however, before a measuring instrument can be developed it is necessary to establish the technical requirements that it must meet. In the process, the reference conditions and the permissible limits of the intrinsic error are determined. During the design process the investigators and designers strive to satisfy these requirements within some margin. Normally this is possible to do. This essentially means that the reference conditions established earlier can be defined more stringently. But if this path is followed, then the reference conditions would have to be redetermined after the sample measuring instruments have been built, and it would be found that they are very diverse for different types of measuring instruments. This would create great difficulties for technical monitoring services and calibrating laboratories. For this reason, on the whole, the reference conditions are best determined by agreement between specialists, and these conditions should be unified as much as possible for different types of measuring instruments. When developing measuring instruments, however, it should be kept in mind that if the intrinsic error is appreciably correlated with one or another influence quantity, then the real properties of the measuring instruments are not completely disclosed by the prescribed standards.

(2) It has been suggested that the integral accuracy index I, calculated according to the formula

$$I=\sqrt{\sum_{i=0}^{n} \varepsilon_i^2},$$

where ε_i is the limit of additional error determined by the ith influence quantity and ε_0 is the limit of intrinsic error, be standardized.

It is clear that one and the same value of I can be obtained for different values of the components. For example, one instrument can have a large temperature error and a small frequency error, while the opposite could be true for a different instrument, etc. Ultimately, replacing one instrument by another (of the same type) results in a large error, and this error cannot be estimated beforehand. Therefore, it becomes more difficult to estimate the measurement errors. In addition, uniformity of measuring instruments is not achieved. The conclusion is obvious: The suggestion is not acceptable.

(3) Another suggestion was to characterize the accuracy of instruments by the weighted mean of the permissible relative error, determined according to the formula

$$\delta_c=\int_{x_i}^{x_f} [\varepsilon(x)f(x)]dx,$$

where x_i and x_f are the initial and final (upper) values of the instrument scale, $\varepsilon(x)$ is the relative error of the instrument, and $f(x)$ is the probability distribution of the indications of the instrument.

This suggestion has the drawback that the probability distribution of the indications of instruments is, in general, unknown. More importantly, however, this weighted-mean characteristic, as any other average characteristic, is completely unsuitable for standardizing the properties of measuring instruments, since uniformity of measuring instruments cannot be achieved in this manner. For example, an instrument that has one or two very significant error components and for which other errors are small can have the same weighted-mean error as an instrument all of whose errors are approximately the same.

In addition, when using an instrument whose errors are standardized as weighted means, experimenters cannot estimate the error of a specific result they have obtained, since in this method of standardization the error of the instrument with a fixed indication can in principle be virtually arbitrarily large.

Thus none of the goals of standardization is achieved with this method of standardization of errors of measuring instruments and this method cannot be used.

(4) It has been repeatedly suggested that the additional error due to the simultaneous action of all influence quantities be standardized. It can be conjectured that some of them will mutually compensate one another so that it will be possible to use the instruments more fully or, vice versa, the error will be larger than in the case when each influence quantity acts separately.

In practice, however, normally not all influence quantities assume their worst (for us) values simultaneously, and it is impossible to take into account only some of the influence quantities by standardizing in this manner. Instead of lowering the estimate of the measurement error or increasing its accuracy, the measurement error will increase and it will not be estimated as accurately. In addition, the testing equipment would become much more complicated.

In reality, some of the additional errors can be correlated with one another, and it would be correct to determine these cases and to standardize the correlation of the corresponding errors. However, I have never encountered in practice such cases of standardization of additional errors. There does not appear to be any need to do so. However, this question deserves a detailed study.

(5) In the former USSR, a bold nationwide experiment was undertaken in 1972: A standard that decisively changed the practice of standardization of errors of measuring instruments was adopted (in the USSR standards were required for industry and for anyone who had anything to do with the object of standardization). This standard greatly complicates the standardization of errors of measuring instruments and contains wholly unrealistic requirements. They include, for example, the requirement that the mathematical expectation and the rms deviation of the systematic component of the errors of measuring instruments of each type be standardized. These characteristics must be estimated according to the formulas

$$\bar{x}=\frac{\sum_{i=1}^{m} x_i}{m}, \quad s=\sqrt{\frac{\sum_{i=1}^{m} (x_i-\bar{x})^2}{m-1}},$$

where m is the number of instruments in a batch and x_i is the systematic error of the ith instrument.

Estimates of these parameters can be calculated by checking instruments in a batch. Let us assume that they satisfy the standards. Does this mean that the systematic error is sufficiently small for all instruments in the batch? Obviously not. In exactly the same way, nothing can be said about a separate instrument, if these estimates do not satisfy the standards. According to these standards an instrument cannot be rejected, and it cannot be judged satisfactory in case of verification. Of course, once estimates have been found for a batch of instruments, then the entire batch of instruments can either be discarded or accepted. But this is absurd: A bad batch can contain several good instruments and they should not be discarded. Conversely, it is absurd to pass as satisfactory some unsatisfactory instruments simply because they are contained in the batch that has been found to satisfy the standards. Every instrument is used individually, and the standards must make it possible to determine whether the instrument is good or bad. Parameters pertaining to batches do not meet this requirement, and for this reason they cannot be used as standards for measuring instruments.

I was one of the individuals who spoke against the adoption of this standard, but the standard was adopted (GOST 8.009-72). In 1984 a new edition of this standard was published. Now the characteristics that we studied above are no longer obligatory, and this eases somewhat the situation of instrument manufacturers: They do not have to standardize these characteristics, and this will not be a violation of the standard.

In conclusion we shall formulate the basic rules for standardization of errors of measuring instruments:

(i) all properties of a measuring instrument that affect the accuracy of the results of measurements must be standardized;

(ii) every property that is to be standardized should be standardized separately;

(iii) methods of standardization must make it possible to check experimentally, and as simply as possible, how well each sample of a measuring instrument corresponds to the established standards, and

(iv) the standardization must be performed so that measuring instruments can be chosen based on the established standards and so that the measurement error can be estimated.

In some cases exceptions must be made to these rules. Such an exception is necessary for strip strain gauges that can be glued on an object only once. For this reason, the strain gauges that are checked can no longer be used for measurements, while the gauges that are used for measurements usually cannot be checked or calibrated. In this case, it is necessary to resort to regulation of the properties of a collection of strain gauges, such as, for example, the standard deviation of the sensitivity and mathematical expectation of the sensitivity. The sensitivity of a separate strain gauge, which is essentially not a random quantity, is a random quantity in application to a collection of strain gauges. Once the sensitivity x_i of every strain gauge chosen at random from a batch (sample) has been determined, it is possible to construct a statistical tolerance interval, i.e., the interval into which the sensitivity of a prescribed fraction p of the entire collection of strain gauges will fall with a chosen probability a. Since $a \neq 1$ and $p \neq 1$, there is a probability that the sensitivity of any given strain gauge falls outside these tolerance limits. For this reason the user must take special measures that exclude such a case. In particular, several strain gauges, rather than one, should be used.

2.5. Dynamic characteristics of measuring instruments and their standardization

The dynamic characteristics of measuring instruments reflect the relation between the change in the output signal and one or another action that produces this change. The most important action is a change in the input signal. In this case the dynamic characteristic is called the dynamic characteristic for the input signal. Dynamic characteristics for one or another influence quantity and for a load (for measuring instruments whose output signal is an electric current or voltage) are also studied.

Complete and partial dynamic characteristics are distinguished.[32]

The complete dynamic characteristics determine uniquely the change in time of the output signal owing to a change in the input signal or other action. Examples of such characteristics are a differential equation, transfer function, amplitude– and phase–frequency response, the transient response, and the impulse characteristic. All these characteristics are essentially equivalent, but the differential equation is still the source characteristic.

A partial dynamic characteristic is a parameter of the full dynamic characteristic or a functional of it. Examples are the response time of the indications of an instrument and the transmission band of a measuring amplifier.

Measuring instruments can most often be regarded as inertial systems of first or second order. If $x(t)$ is the signal at the input of a measuring instrument and $y(t)$ is the corresponding signal at the output, then the relation between them can be expressed with the help of first-order [Eq. (2.3)] or second-order [Eq. (2.4)] differential equations, respectively, which reflect the dynamic properties of the measuring instrument:

$$Ty'(t)+y(t)=Kx(t), \tag{2.3}$$

$$\frac{1}{\omega_0^2}y''(t)+\frac{2\beta}{\omega_0}y'(t)+y(t)=Kx(t). \tag{2.4}$$

The parameters of these equations have specific names: T is the time constant of a first-order device, K is the transduction coefficient in the static state, ω_0 is the angular frequency of free oscillations, and β is the damping ratio.

Equations (2.3) and (2.4) reflect the properties of real devices and for this reason they have zero initial conditions: for $t \leqslant 0$, $x(t)=0$ and $y(t)=0$, $y'(t)=0$, and $y''(t)=0$.

For definiteness, in what follows we shall study the second-order equation and we shall assume that it describes a moving-coil galvanometer. Then $\omega_0 = 2\pi f_0$, where f_0 is the frequency of free oscillations of the moving part of the galvanometer.

In order to obtain transfer functions from differential equations it is first necessary to transfer from signals in the time domain to their Laplace transforms, and then to form their ratio. Thus

$$\mathscr{L}[x(t)]=x(s), \quad \mathscr{L}[y(t)]=y(s), \quad \mathscr{L}[y'(t)]=sy(s),$$

$$\mathscr{L}[y''(t)]=s^2y(s) \quad \text{where } s \text{ is the Laplace operator.}$$

For the first-order system we obtain

$$W(s)=\frac{y(s)}{x(s)}=\frac{K}{1+sT}$$

and for the second-order system we obtain

$$W(s)=\frac{y(s)}{x(s)}=\frac{K}{\dfrac{1}{\omega_0^2}s^2+\dfrac{2\beta}{\omega_0}s+1}. \tag{2.5}$$

If in the transfer function the operator s is replaced by the complex frequency $j\omega$ ($s=j\omega$), then we obtain the complex frequency response. We shall study the relation between the named characteristics for the example of a second-order system. From Eqs. (2.4) and (2.5) we obtain

$$W(j\omega)=\frac{K}{(1-\omega^2/\omega_0^2)+j2\beta\omega/\omega_0}, \tag{2.6}$$

where $\omega = 2\pi f$ is the running angular frequency.

The complex frequency response is often represented in terms of its real and imaginary parts,

$$W(j\omega)=P(\omega)+jQ(\omega).$$

In our case,

$$P(\omega)=\frac{K\left(1-\frac{\omega^2}{\omega_0^2}\right)}{\left(1-\frac{\omega^2}{\omega_0^2}\right)^2+4\beta^2\frac{\omega^2}{\omega_0^2}},$$

$$Q(\omega)=-\frac{2\beta\frac{\omega}{\omega_0}K}{\left(1-\frac{\omega^2}{\omega_0^2}\right)^2+4\beta^2\frac{\omega^2}{\omega_0^2}}.$$

The complex frequency response can also be represented in the form

$$W(j\omega)=A(\omega)e^{j\varphi(\omega)},$$

where $A(\omega)$ is the amplitude–frequency response and $\varphi(\omega)$ is the frequency response of phase. In the case at hand

$$A(\omega)=\sqrt{P^2(\omega)+Q^2(\omega)}=\frac{K}{\sqrt{\left(1-\frac{\omega^2}{\omega_0^2}\right)^2+4\beta^2\frac{\omega^2}{\omega_0^2}}},$$

$$\varphi(\omega)=\arctan\frac{Q(\omega)}{P(\omega)}=-\arctan\frac{2\beta\frac{\omega}{\omega_0}}{1-\frac{\omega^2}{\omega_0^2}}. \tag{2.7}$$

Equations (2.7) have a well-known graphical interpretation.

The transient response is the function $h(t)$ representing the output signal produced by a unit step function $1(t)$ at the input. We recall that the unit step function is a function $x(t)$ satisfying the following conditions: $x(t)=0$ for $t<0$ and $x(t)=1$ for $t\geqslant 0$. Since the input is not periodic, $h(t)$ is calculated using Eq. (2.3) or (2.4). Omitting the simple but, unfortunately, complicated calculations, we arrive at the final form of the transient response of the instrument under study:

$$h(t)=e^{-\beta\tau}\frac{1}{\sqrt{1-\beta^2}}\sin\left(\tau\sqrt{1-\beta^2}+\arctan\frac{\sqrt{1-\beta^2}}{\beta}\right)\quad \text{for } \beta<1,$$

$$h(t)=e^{-\tau}(\tau+1)\quad \text{for } \beta=1,$$

$$h(t)=e^{-\beta\tau}\frac{1}{\sqrt{\beta^2-1}}\sinh\left(\tau\sqrt{\beta^2-1}+\operatorname{arctanh}\frac{\sqrt{\beta^2-1}}{\beta}\right)\quad \text{for } \beta>1. \tag{2.8}$$

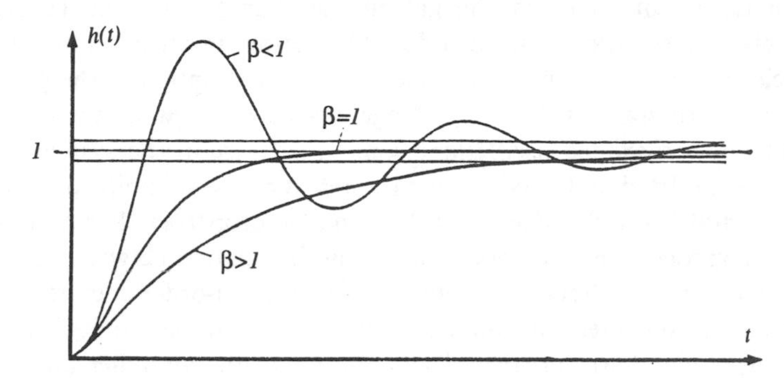

FIG. 2.4. The transient response of an instrument described by a second-order differential equation; β is the damping ratio.

Here $\tau = \omega_0 t$ and the steady-state value of the output signal is taken to be equal to unity, i.e., $h(t) = y(t)/K$. Thanks to this, Eqs. (2.8) and the graphs corresponding to them, presented in Fig. 2.4, are universal in the sense that they do not depend on the specific values of ω_0 and K.

The impulse characteristic $g(t)$ is found from the transient response in accordance with its definition:

$$g(t) = \frac{dh(t)}{dt}.$$

It should be noted that some types of measuring instruments do not have dynamic characteristics at all: measures of length, weights, vernier calipers, etc. Some measuring instruments, such as measuring capacitors (measures of capacitance), do not have an independent dynamic characteristic. But when they are connected into an electric circuit, which always has some resistance and sometimes an inductance also, the circuit always acquires, together with a capacitance, definite dynamic properties.

Measuring instruments are very diverse. Occasionally, in order to describe adequately their dynamic properties, it is necessary to resort to linear equations of a higher order, nonlinear equations, or equations with distributed parameters. However, complicated equations are used very rarely. This is not an accident. After all, measuring instruments are created specially to perform measurements, and their dynamic properties are made so as to guarantee convenience of use. For example, in designing an automatic-plotting instrument, the transient response is made to be quite short, approaching the established level monotonically or oscillating insignificantly. In addition, the instrument scale is made to be linear. But when these requirements are met, the dynamic properties of the instrument can be described by one of the characteristics corresponding to a linear differential equation of order no higher than second.

A differential equation of high order is most often obtained when synthesizing the dynamic characteristics of an instrument based on the dynamic characteristics of its subunits. Thus, for example, calculating the dynamic characteristic of a galvanometric amplifier with a photoelectric converter that converts the angle of rotation of the moving part of the galvanometer into a voltage (current), we formally obtain an equation of third order: The galvanometer gives two orders and the photoelectric converter gives one order. Such a description of the properties of the amplifier is necessary at the design stage, because otherwise it is impossible to understand why self-excited oscillations sometimes arise in the system. But when the design is completed and reasonable parameters of the subunits are chosen, it is desirable to simplify the description of the dynamic properties. For this the amplifier must be regarded as a black box. Analyzing the relation between the input and output, we find that it is described quite well by a second-order equation. The same result can also be obtained informally, as done, for example, in Ref. 49. Decomposing the dynamic characteristics of all subunits of the amplifier into first-order characteristics and comparing them, we can see that one of them can be neglected.

Standardization of the dynamic characteristics of measuring instruments is performed for a specific type of instrument. The problem is solved in two stages. First an appropriate dynamic characteristic must be chosen, after which the nominal dynamic characteristic and the permissible deviations from it must be established. Thus for recording instruments and universal measuring transducers one of the complete dynamic characteristics must be standardized: without having the complete dynamic characteristic a user cannot effectively use these instruments.

For indicating instruments, it is sufficient to normalize the response time. In contrast to the complete characteristics, this characteristic is a partial dynamic characteristic. The dynamic error is another form of a partial dynamic characteristic. Standardization of the limits of permissible dynamic error is very convenient for the measuring instruments employed, but it is justified only when the form of the input signals does not change much.

For measuring instruments described by linear first- and second-order differential equations, the coefficients of all terms in the equations can be standardized. In the simplest cases, the time constant is standardized in the case of a first-order differential equation, and the natural frequency and the damping ratio of the oscillations are standardized in the case of a second-order differential equation.

When imposing requirements on the properties of measuring instruments it is always necessary to keep in mind how compliance will be checked. For dynamic characteristics the basic difficulties are connected with creating test signals of predetermined (with sufficient accuracy) form or with recording the input signal with a dynamically more accurate measuring instrument than the measuring instrument whose dynamic properties are being checked.

If test signals with adequate accuracy can be created and the dynamic characteristic is found with the help of the corresponding signal, i.e., a transient response as a response of a unit step function signal, and frequency response as a response of a sinusoidal test signal, then in principle the obtained experimental data can be processed without any difficulties.

But sometimes the problem must be solved using a test signal that does not correspond to the signal intended for determining the complete dynamic characteristic. For example, one would think that the problem can be solved given the tracing of signals at the input and output of a measuring instrument. In this case, however, special difficulties arise because of the fact that small errors in recording the test signal and reading the values of the input and output signals often lead to the fact that the dynamic characteristic obtained based on them does not correspond to the dynamic properties of the measuring instrument and is physically meaningless. Such an unexpected effect is explained by the fact that the problem at hand is a so-called improperly posed problem. A great deal of attention is currently being devoted to such problems in mathematics, automatics, geophysics, and other disciplines. Improperly posed problems are solved by the methods of regularization, which essentially consist of the fact that the necessary degree of filtering (smoothing) of the obtained solution is determined based on *a priori* information about the true solution.

Improperly posed problems in dynamics in application to measurement engineering are reviewed in Ref. 32.

A separate problem, which is important for some fields of measurement, is the determination of the dynamic properties of measuring instruments directly when the instruments are being used. An especially important question here is the question of the effect of random noise on the accuracy with which the dynamic characteristics are determined.

This then is a brief review of the basic aspects of the problem of standardizing and determining the dynamic properties of measuring instruments.

2.6. Statistical analysis of the errors of measuring instruments based on data provided by calibration laboratories

A general characteristic of the errors of the entire population of measuring instruments of a specific type could be their distribution function. I made an attempt to find such functions for several types of measuring instruments. The results of these investigations, which were performed together with T. L. Yakovleva, were published in Refs. 50 and 61.

The errors of measuring instruments are determined by calibration, and a decision was made to use the data provided by calibration laboratories. Because it is impossible to obtain the errors of all instruments of a given type that are in use, the use of a sampling method is unavoidable.

In order to establish a property of an entire group (general collection) based on a sample, the samples must be representative. Sample homogeneity is a necessary indicator of representativeness. In the case of two samples, in order to be sure that the samples are homogeneous it is necessary to check the hypothesis H_0: $F_1 = F_2$, where F_1 and F_2 are distribution functions corresponding to the first and, correspondingly, the second sample.

The results of the check, as is well known, depend not only on the error of the measuring instrument being calibrated but also on the error of the standard. For this reason, measuring instruments that are checked with not less than a fivefold margin of accuracy were selected for analysis.

In addition, in order to ensure that the samples are independent they were formed either based on data provided by calibration laboratories in different regions of the former USSR or, if data from a single laboratory were used, the data were separated by a significant time interval. The sample size was maintained approximately constant.

We shall discuss first Ref. 50. Table 2.2 gives the basic statistical characteristics

TABLE 2.2. The example of main statistical characteristics of errors for six types of measuring instruments.

Type of measuring instrument	Year of calibration	Point of check	Sample size	Moment: First initial	Moment: Second central	Coefficient: Asymmetry skewness-3	Coefficient: Excess
Э59 ammeter	1974	80 Divi-	160	0.163	0.074	−0.40	3.56
	1976	sions	160	0.180	0.042	−1.33	7.27
Э59 voltmeter	1974	150 Divi-	120	0.050	0.063	−0.47	2.71
	1976	sions	108	0.055	0.065	−0.18	3.15
Д566 wattmeter	1974	150 Divi-	86	0.088	0.024	−0.50	2.46
	1976	sions	83	0.062	0.021	0.05	3.81
TH-7 thermo-	1975	100 °C	92	−0.658	0.198	0.14	2.86
meter	1976		140	−0.454	0.128	0.45	4.57
Standard spring	1973	9.81 kPa	250	0.158	0.012	0.55	3.54
manometer	1976		250	0.128	0.012	0.59	2.87
P331 resistance	1970	100 Ω	400	0.33×10^{-3}	1.6×10^{-2}	0.82	4.08
measure	1975		400	0.1×10^{-3}	1.2×10^{-3}	0.44	5.02

of the samples for six types of different instruments. Two samples, obtained at different times, are presented for each of them. For brevity, the data referring to only one numerical scale marker are presented. The arithmetic mean of the values obtained by continuously approaching the marker checked from both sides was taken as the value of the error. The first initial and second central moments are given in the same units in which the value of the point of checking is presented, i.e., in fractions of a scale graduation, in degrees Celsius, etc. (in the corresponding power). Errors exceeding twice the limit of permissible error were eliminated from the analysis.

The test was made with the help of the Wilcoxon–Mann–Whitney and Siegel–Tukey criteria with a significance level $q=0.05$. The technique of applying these criteria is described below in Chap. 4.

The results of the analysis are presented in Table 2.3. Rejection of the hypothesis is indicated by a minus sign and acceptance is indicated by a plus sign. The symbol 0 means that a test based on the given criterion was not made.

The Wilcoxon–Mann–Whitney and Siegel–Tukey criteria are substantially different: The former is based on comparing averages and the latter is based on comparing variances. For this reason it is not surprising that cases when the hypothesis H_0 is rejected according to one criterion but accepted according to the other are encountered. It is more important that the hypothesis of sample homogeneity must be rejected irrespective of the criterion. Both samples were found to be homogeneous only for the Д566 wattmeters and standard manometers. For other measuring instruments, the compared samples were often found to be nonhomogeneous. It is interesting that on one scale marker they can be homogeneous while on

TABLE 2.3. The results of testing the hypothesis of homogeneity for samples of six types of measuring instruments.

Type of measuring instrument	Year of calibration	Point of check	Result of testing the hypothesis based on the criterion of	
			Wilcoxon—Mann—Whitney	Siegel–Tukey
Э59 ammeter	1974	30 Divisions	+	–
	1976	60	0	–
		80	0	–
		100	+	+
Э59 voltmeter	1974	70 Divisions	–	0
	1976	150	+	+
Д566 wattmeter	1974	70 Divisions	+	+
	1976	150	+	+
TH-7 thermometer	1975	100 °C	0	–
	1976	150 °C	–	+
		200 °C	+	+
Standard spring manometer	1973 1976	9.81 kPa	+	+
P331 resistance measure	1970	10 kΩ	0	–
	1975	100 Ω	0	–
		10 Ω	0	–

another they are inhomogeneous (э59 voltmeters and ammeters). TH-7 thermometers had homogeneous samples in one range of measurement and nonhomogeneous in a different range. The calculations were repeated for significance levels of 0.01 and 0.1, but on the whole the results were the same in both cases.

The results obtained show that samples of measuring instruments are frequently nonhomogeneous with respect to errors. For this reason, they cannot be used to determine the distribution function of the errors of the corresponding instruments. The facts that we excluded from analysis very large errors and in the case of instruments in which the indications varied we averaged the errors could only make it easier to accept the zero hypothesis regarding homogeneity of samples and for this reason could not affect the result obtained.

The experiment described was formulated in order to check the temporal stability of the distribution functions of the errors, but since in the samples compared the numbers of the instruments themselves were not recorded and they, of course, changed, the result obtained has a different but no less important meaning: It indicates that the also are nonhomogeneous.

This is indicated also by the results of Ref. 61, in which samples obtained based on data provided for э59 ammeters by four calibration laboratories in different regions of the former USSR were compared. The number of samples was equal to 150–160 everywhere. The errors were recorded at the numerical markers 30, 60, 80, and 100 graduations. The samples were assigned the numbers 1, 2, 3, and 4 and the hypotheses $H_0:F_1 = F_2, F_2 = F_3, F_3 = F_4$, and $F_4 = F_2$ were checked. The combinations of samples were arbitrary. The hypothesis testing was based on the Wilcoxon–Mann–Whitney criterion with $q = 0.05$. The analysis showed that we can accept the hyphothesis $H_0:F_1=F_2$ only and only at the marker 100. In all other cases the hypothesis had to be rejected.

Thus the sample method does not permit finding the distribution function of the errors of measuring instruments. There are evidently two reasons for this. The first reason is that the stock of instruments of each type is not constant. On the one hand, new instruments that have just been manufactured are added to it. On the other hand, in the verification some instruments are rejected, some instruments are replaced, while others are discarded. The ratio of the numbers of old and new instruments is constantly changing. The second reason is that the errors of the instruments unavoidably change with time. Moreover, many instruments are used under very different conditions, and the conditions of use affect differently the rate at which the instrumental errors change.

The temporal instability of measuring instruments raises the question of whether or not the errors of measuring instruments are in general sufficiently stable so that a collection of measuring instruments can be described by some distribution function. At a fixed moment in time this is in principle always possible. The other problem is how to find this distribution function. The simple sampling method, as we saw above, is not suitable. But even if the distribution function can be found by some complicated method, after some time it would have to be redetermined, since the errors themselves and the composition of the stock of measuring instruments change. All this is too complicated. It must be concluded that the distribution of errors of measuring instruments cannot be found based on the experimental data.

The results presented above were obtained in the former USSR and instruments manufactured in the former USSR were studied. However, there are no grounds for expecting that instruments manufactured in other countries will have different statistical properties.

Chapter 3

Prerequisites for the analysis of the inaccuracy of measurements and for synthesis of their components

3.1. Preliminary remarks

As has already been mentioned, a measurement error cannot be found directly from its definition, i.e., using the definition as an algorithm, since the true value of the measured quantity is unknown. The problem must be solved by performing calculations based on estimates of all components of the measurement inaccuracy. This is why the problem of analysis—the identification of the sources and the reasons for the appearance of the components of the measurement error and estimation of these components—is so important.

We shall call the smallest components of the measurement error, based on whose estimates the total measurement error or uncertainty of measurement is calculated, the elementary errors.

If in the analysis it is possible to find for some elementary errors concrete specific values, i.e., in the language of mathematical statistics, to find point estimates of these errors, then these components are immediately eliminated by introducing the corresponding corrections. Of course, this is possible only in the case of elementary systematic errors. However, no corrections can make the measurement result absolutely accurate; there always remains an uncertainty. In particular, the corrections themselves cannot be absolutely accurate, and after they are introduced there remain residuals of the corresponding errors that have not been eliminated and that later play the role of elementary errors.

Let us turn back once more to the terms *error* and *uncertainty*. For the last two decades or perhaps a bit longer, both were used in the U.S. with the same meaning. Thus they were synonyms. But synonyms are not allowed in a proper system of terms, and this situation had to be improved.

One way to solve this problem was proposed in Ref. 18. The main idea of that paper is that the term "measurement error" appears to be used in two different senses. In one sense, in the opinion of the authors of Ref. 18, it expresses the fact that the measurement result is different from the true value of the measured quantity, while in the other sense it reflects the uncertainty of the measurement result. For example, in the first case one would use the expression "the error $+1\%$," while in the second case one would say "the error $\pm 1\%$." In order to distinguish the meaning of the word "error" in these cases, it is proposed that in the second case the word "uncertainty" be used instead of the word "error."

In order to understand the essential significance of this proposition, it is first necessary to check the correctness of the examples presented. The first example is obvious. The second example requires some analysis. To be precise, the expression "the error $\pm 1\%$" means that the measurement error is simultaneously both $+1\%$ and -1%. But this cannot be, since there can only be one result of a measurement, a fixed numerical value, i.e., this expression is incorrect. In this case one should say "the error falls within the range $\pm 1\%$" or "the limits of error are $\pm 1\%$." If the correct expression were used, then the contradiction mentioned in Ref. 18 would not occur.

Thus the problem lies not in the fact that the term *measurement error* has two meanings but rather that this term is not used correctly. Nevertheless, it may appear that this proposal eliminates the synonymy. But as a matter of fact it replaces the term *error* with the term *uncertainty* because the first case mentioned above is very rare.

A different solution is given in Ref. 1. The term *uncertainty* is used there for inaccuracy of measurement results only, while the term *error* is kept for all other cases. The shortcomings of this solution were studied in paper.*

A more grounded solution is suggested in Refs. 2 and 13. An inaccuracy of measurement results is expressed there with the term *uncertainty* but every numerical value of uncertainty is accompanied with a corresponding confidence probability. The latter is very important, and it is a good reason to have a special term in this case. Of course, a new term could be constructed by adding special adjectives to the root word *error*. For example, there is a term *confidence limits of error* in Ref. 3. But shorter terms are preferable. Therefore the term *uncertainty* is used in the present book for this case.

Thus the imperfection of measurement results can be quantitatively described using two terms: *limits of error* and *uncertainty*. Yet another term is needed to refer to the qualitative side of the phenomenon of imperfection of measurements. In this book, the term *inaccuracy* is used for this purpose. This term is short and expressive. Moreover, it can be used not only for measurements but for measuring instruments too. It appears that this term has the potential to become widely accepted.

3.2. Classification of elementary errors

The classification of measurement errors presented in Chap. 1 also applies, of course, to elementary errors. Continuing the analysis, this classification must be further developed.

Most elementary errors are estimated by analysis and definite limits are found for them. We shall divide elementary errors that have definite limits into absolutely constant and conditionally constant errors.

By *absolutely constant elementary errors*, we mean errors that, although they

---

*S. G. Rabinovich and G. M. Deich, "Uncertainty" versus "error"—a problem of terminology, in Proceedings of the 33rd International Instrumentation Symposium, Las Vegas, Instrument Society of America, Research Triangle Park, North Carolina, 1987.

have definite limits, remain the same in repeated measurements performed under the same conditions as well as for all measuring instruments of the same type. An example of such an error is the error owing to the inaccuracy in the formula used to determine the quantity being measured, if the limits of this error have been established. Another example is the error of digital thermometers, for which the temperature dependence of the emf of the thermocouple is linearized with the help of a polynomial of a fixed degree (see Sec. 9.5). Thus absolutely constant elementary errors are, based on their properties, purely systematic errors.

By *conditionally constant errors*, we mean errors that have certain limits but can vary within these limits both owing to the nonrepeatability and nonreproducibility of the results. A typical example of such an error is the measurement error due to the intrinsic error of the measuring instrument.

The intrinsic error, by its nature, can be a purely systematic error, but it can also have a random component. For example, for weights the intrinsic error does not have a random component, but its actual magnitude varies from one weight to another. The intrinsic error of an electric measuring instrument with an indicator needle has both systematic and random components, but on the whole the intrinsic error has definite limits that are the same for any instrument of a given type.

A conditionally constant error can even be purely random. Examples are the roundoff error in reading the indications of analog instruments and the error owing to the limited resolution of digital instruments.

Thus a fundamental property of conditionally constant elementary errors is that although they have certain limits, they can vary within these limits. An elementary error that does not have estimatable limits is the common random error.

A random error, as is well known, is estimated after a measurement is performed. The estimate is based on data obtained in the course of the measurements. If the random error is significant, then the measurement is performed many times. The primary characteristic of a random error is usually the standard deviation, which is calculated from the experimental data, and the entire error, and not its separate components, is estimated directly. For this reason, there is no need to add to the term *random component of the measurement error* or briefly *random measurement error* the additional word *elementary*.

When performing an analysis, however, it is important to distinguish purely random and quasirandom errors. Purely random errors can arise from different reasons. For example, they can arise owing to noise or small (regarded as permissible) variations in the influence quantities or the random components of the errors of the measuring equipment.

Quasirandom errors appear in measurements of quantities that are by definition averages when the quantities appearing in the group being averaged are of different size. The difference is not random but rather is regarded as random and is characterized, just as in the case of a purely random error, by an estimate of the standard deviation.

The error classification studied above is reminiscent of the classification contained in the recommendations of the International Bureau of Weights and Measures (BIPM).[30] BIPM distinguishes errors that are estimated by statistical methods (type A errors) and errors that are estimated by nonstatistical methods (type B errors). These names are purely arbitrary, and this is one of their deficiencies. It is more important that the classification indicator here refers not to the object of classification itself and to its properties but rather to the method employed to estimate them, and in general it is a secondary indicator that follows from something that has not been identified.

The proposed classification does not have this deficiency, but it also does not fit within the framework of the division of errors solely into systematic and random categories.

3.3. Mathematical models of elementary errors

A measurement error is calculated based on data of its components, i.e., this is a problem of synthesis, performed mathematically. Correspondingly, elementary errors must be represented by mathematical models. We shall examine all four types of elementary errors from this viewpoint.

Absolutely constant errors. Each such error has a constant value that is the same in any measurement, although it is unknown. Only the limits of these errors are known. A mathematical model of such errors could be a determinate quantity whose magnitude has an interval estimate, i.e., it lies within an interval of known limits. We shall use this model for absolutely constant elementary errors.

We can foresee an objection to this model. Some people think that if the value of the error is unknown, then it can be regarded as a random quantity. However, this is not correct. A model of an object can be constructed only based on what we know about it and not based on what we do not know.

There is another objection. If a determinate model is adopted, then when several absolutely constant errors are summed, their limits must be added arithmetically. This is equivalent to the assumption that all terms have limiting values and the same sign, which is unlikely. This objection also is invalid. First of all, the argument "unlikely" is not correct here, since we are not using a probabilistic model. Second, the fact that we do not like the result—the answer seems exaggerated—is also not an argument. In mathematics, precisely the same situation arises in methods of approximate calculations and the limits of errors are added arithmetically.

Fortunately, in a measurement there are rarely more than one or two absolutely constant errors and they are, as a rule, insignificant.

Conditionally constant errors. The values of these errors characteristically vary from one measurement to another and from one measuring apparatus to another, and they are different under different conditions. In all cases, however, in each such error the limits of the interval containing any possible realization of such an error remain unchanged.

As a mathematical model of conditionally constant errors one would like to use a random quantity. For this, however, it is necessary to know the probability distribution function corresponding to this random quantity. Best of all, one would like to be able to find this function based on the experimental data. Such an attempt was made for the intrinsic error of measuring instruments. The results of such an investigation were presented in Chap. 2. They showed that the distribution function of the intrinsic error and, especially, the distribution function of the additional errors cannot be found from selective data.

Thus in order to adopt a probability model, the form of the distribution function, in this case, must be prescribed. It is well known that among distributions with fixed limits, the uniform distribution has the highest uncertainty (in the sense of information theory). The roundoff error also has known limits, and in mathematics this error has for a long time been regarded as a random quantity with a uniform probability distribution. For this reason we shall also assume that the model of conditionally constant errors will be a random quantity with a uniform probability distribution within prescribed limits.

This suggestion was made comparatively a long time ago.[48] At the present time this model is widely employed in the theory of measurement errors.[4,5,30]

Purely random errors. Such errors appear in multiple measurements. They are characterized by the standard deviation that is computed from the experimental data. Such errors are assumed to be normally distributed random quantities.

The form of the distribution function can, in principle, be found based on the data from each multiple measurement. In practice, however, the number of measurements performed in each experiment is insufficient for this. Every time measurements are performed, it is often assumed that the hypothesis of a normal distribution was checked in the preceding experiment. For example, when measures of mass are compared on the standard balances at the D. I. Mendeleev All-Union Scientific-Research Institute of Metrology (USSR) it is assumed that the distribution is normal, but this is not directly checked. True, the results obtained are not inconsistent with the practice so that this assumption is evidently justified.

In general, I have never encountered a case when the normality of the distribution of a random error was checked mathematically and when this has led to misunderstandings.

Quasirandom errors. As noted above, these errors occur when measuring quantities that are averages by definition and the value of each separate quantity in the group of quantities being averaged remains constant. These quantities are essentially not random, but in aggregate they can be regarded as a general collection of quantities. This is possible in accordance with the goal of the measurement, based on agreement of experts. The parameters of the distribution that characterize this distribution should be determined by agreement. Most often the standard deviation is chosen as this parameter.

We shall now discuss the question of interdependence and correlation of elementary errors. Mathematically, it is preferable to regard these errors as correlated quantities, since this approach has great generality. However, such an approach complicates the problem, and most of the time it is not justified. Under reference conditions all elementary errors are independent and they are uncorrelated. Exceptions can be encountered in measurements performed under normal operating conditions, especially in the case of indirect measurements and measurements performed with the help of measuring systems, when one and the same influence quantity gives rise to appreciable additional errors in several instruments or components in the measuring channel of the system. An example is a measurement in which a measuring transducer, amplifier, and automatic-plotting instrument is employed. A change in the temperature of the medium can cause all of these devices to acquire an additional temperature-induced error. Obviously, all these additional errors will be interrelated.

3.4. Methods for describing random quantities

Random quantities are studied in the theory of probability, a well-developed field of mathematics. The properties of a random quantity are completely described by the distribution function $F(x)$, which determines the probability that a random quantity X will assume a value less than x:

$$F(x)=P\{X<x\}.$$

The distribution function is a nondecreasing function, defined so that $F(-\infty)=0$ and $F(+\infty)=1$.

Together with the distribution function $F(x)$, which is said to be cumulative or integral, the differential function, usually called the probability density $f(x)$, is also widely employed:

$$f(x)=\frac{dF(x)}{dx}.$$

We call attention to the fact that the probability density is a dimensional function:

$$\dim f(x)=\dim \frac{1}{X}.$$

In the practice of precise measurements one most often deals with normal and uniform distributions. Figure 3.1(a) shows integral functions of these distributions and Fig. 3.1(b) shows the probability densities of the same distributions.

For the normal distribution we have

$$f(x)=\frac{1}{\sigma\sqrt{2\pi}}e^{-(x-A)^2/2\sigma^2},$$

$$F(x)=\frac{1}{\sigma\sqrt{2\pi}}\int_{-\infty}^{x} e^{-(x-A)^2/2\sigma^2}dx. \tag{3.1}$$

The parameter σ^2 is the variance and A is the mathematical expectation of the random quantity.

Calculation of $F(x)$ for some fixed x_f gives the probability $P\{X < x_f\} = P_f$. When the graph of $f(x)$ is used to calculate this probability, it is necessary to find the area under the curve to the left of the point x_f in Fig. 3.1(b).

The normal distribution function obtained by transforming to the random quantity $z=(X-A)/\sigma$ is widely employed in calculations:

$$f(z)=\frac{1}{\sqrt{2\pi}}e^{-z^2/2}, \quad F(z)=\frac{1}{\sqrt{2\pi}}\int_{-\infty}^{z} e^{-y^2/2}dy. \tag{3.2}$$

Tables of values of the function $\Phi(z)$ defined by the expression

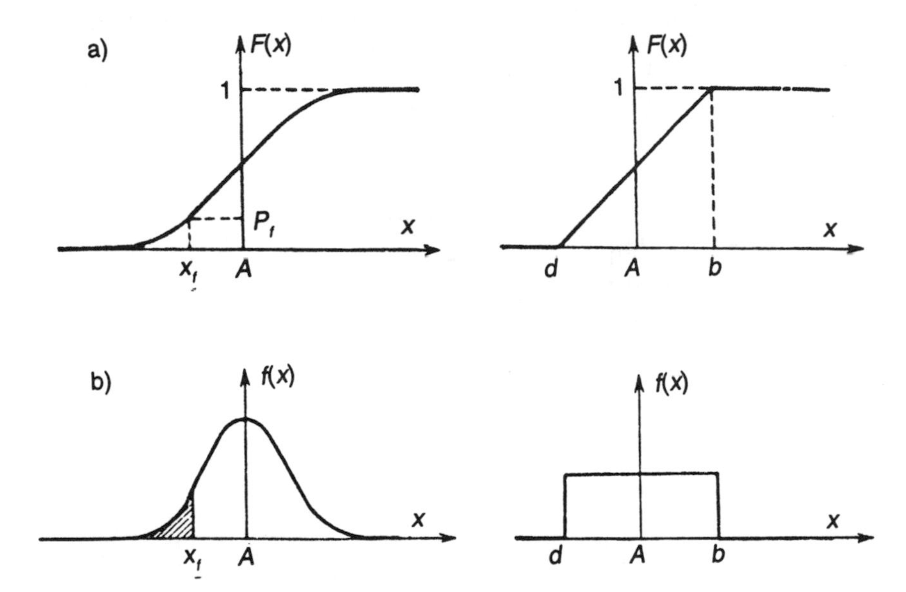

FIG. 3.1. (a) The probability distribution and (b) the probability density for a normal distribution (on the left) and uniform distribution (on the right) of continuous random quantities.

$$\Phi(z)=\frac{1}{\sqrt{2\pi}}\int_0^z e^{-y^2/2}dy \tag{3.3}$$

and called the normalized Laplace function are often given.

It is obvious that for $z\geqslant 0$

$$F(z)=0.5+\Phi(z).$$

The branch for $z<0$ is found based on symmetry considerations:

$$F(z)=0.5-\Phi(z).$$

Tables of the functions $f(z)$ and $\Phi(z)$ are given in the Appendix (Tables A.1 and A.2).

The normal distribution is remarkable in that according to the central limit theorem, a sum of an infinite number of infinitesimal random quantities with an arbitrary distribution has a normal distribution. In practice, the distribution of the sum of a comparatively small number of random quantities already is found to be close to a normal distribution.

The uniform distribution is defined as

$$f(x)=\begin{cases} 0; & x<d, \\ \dfrac{1}{b-d}; & d\leqslant x\leqslant b, \\ 0; & x>b, \end{cases}$$

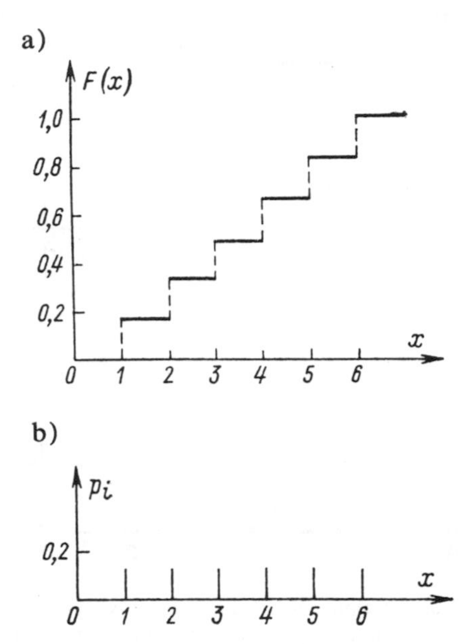

FIG. 3.2. (a) The probability distribution and (b) the distribution of probabilities of a discrete random quantity.

$$F(x)=\begin{cases} 0; & x<d, \\ \dfrac{x-d}{b-d}; & d\leqslant x\leqslant b, \\ 1; & x>b. \end{cases} \tag{3.4}$$

We shall also use the uniform distribution often.

In addition to continuous random variables, discrete random variables are also encountered in metrology. An example of an integral distribution function and the probability distribution of a discrete random variable are given in Fig. 3.2.

Distribution functions are complete characteristics of random quantities, but they are not always convenient to use in practice. For this reason, random quantities are also described by their numerical characteristics. For this the moments of random quantities are employed.

The initial moments m_k (moments about zero) and central moments μ_k (moments about the mean value) of order k are defined by the formulas

$$m_k=M[X^k]=\int_{-\infty}^{\infty} x^k f(x)dx,$$

$$m_k=M[X^k]=\sum_{i=1}^{n} x_i^k p_i. \tag{3.5}$$

$$\mu_k=M[X-M[X]]^k=\int_{-\infty}^{\infty} (x-M[X])^k f(x)dx,$$

$$\mu_k = M[X - M[X]]^k = \sum_{i=1}^{n} (x_i - M[X])^k p_i. \qquad (3.6)$$

In the relations (3.6)–(3.8) the first formulas refer to continuous and the second to discrete random quantities.

Of the initial moments, the first moment ($k = 1$) is most often employed. It gives the mathematical expectation of the random quantity

$$m_1 = M[X] = \int_{-\infty}^{\infty} x f(x) dx,$$

$$m_1 = M[X] = \sum_{i=1}^{n} x_i p_i. \qquad (3.7)$$

It is assumed that $\Sigma_{i=1}^{n} p_i = 1$, i.e., the complete group of events is studied.

Of the central moments, the second moment ($k = 2$) plays an especially important role. It is the variance of the random quantity

$$\mu_2 = D[X] = M[(X - m_1)^2] = \int_{-\infty}^{\infty} (x - m_1)^2 f(x) dx,$$

$$\mu_2 = D[X] = M[(X - m_1)^2] = \sum_{i=1}^{n} (x_i - m_1)^2 p_i. \qquad (3.8)$$

The positive square root of the variance is called the standard deviation of the random quantity

$$\sigma = + \sqrt{D[X]}. \qquad (3.9)$$

Correspondingly, $D[X] = \sigma^2$.

The third and fourth central moments are also used in applications. They are used to characterize the symmetry and sharpness of distributions. The symmetry is characterized by the skewness $a = \mu^3/\sigma^3$, and the sharpness is characterized by the excess $e = \mu^4/\sigma^4$.

The normal distribution is completely characterized by two parameters: $m_1 = A$ and σ. For it, characteristically, $a = 0$ and $e = 3$. The uniform distribution is also determined by two parameters: $m_1 = A$ and $l = d - b$. It is well known that

$$m_1 = \frac{d+b}{2}, \quad D[X] = \frac{(d-b)^2}{12} = \frac{l^2}{12}. \qquad (3.10)$$

Instead of l, the quantity $h = l/2$ is often used. Then $D[X] = h^2/3$ and $\sigma(X) = h/\sqrt{3}$.

3.5. Construction of the composition of uniform distributions

So, we have adopted the uniform distribution as a mathematical model of conditionally constant elementary errors. In solving the problem of synthesis of these errors one must know how to construct the composition of uniform distributions.

The theoretical solution of this problem is well known and is presented, for example, in Ref. 58. For our purposes it is interesting to clarify the possibility of constructing a simplified solution for the applied problem at hand.

Consider n random quantities x_i $(i = 1,...,n)$, each of which has a uniform distribution centered at zero in the interval $[-\frac{1}{2}; +\frac{1}{2}]$. We introduce the notation $\vartheta = \Sigma_{i=1}^{n} x_i$. The probability densities of the sum of these random quantities has the form

$$f_n(\vartheta)=\frac{1}{(n-1)!}\left[\left(\vartheta+\frac{n}{2}\right)^{n-1}-C_n^1\left(\vartheta+\frac{n}{2}-1\right)^{n-1}+C_n^2\left(\vartheta+\frac{n}{2}-2\right)^{n-1}+\cdots\right],$$

where the sum must include only those terms in which the base of power $(\vartheta + n/2)$, $(\vartheta + n/2 - 1)$, etc. is nonnegative for a given value of n; for example, if $n = 2$, then

$$f_2(\vartheta)=(\vartheta+1)-2\vartheta=\begin{cases}\vartheta+1, & -1<\vartheta\leqslant 0\\ 1-\vartheta, & 0\leqslant\vartheta<1.\end{cases}$$

The probability density of the sum of two terms has the form of a triangle. For $n = 3$ the graph of $f_3(\vartheta)$ consists of three segments of a quadratic parabola and looks very much like the curve of a normal distribution. For $n = 4$ this distribution function is almost indistinguishable from the normal distribution.

Given the equation for the probability density, it is not difficult to find the probability distribution function

$$F_n(\vartheta)=\frac{1}{n!}\left[\left(\vartheta+\frac{n}{2}\right)^{n}-C_n^1\left(\vartheta+\frac{n}{2}-1\right)^{n}+C_n^2\left(\vartheta+\frac{n}{2}-2\right)^{n}+\cdots\right]. \quad (3.11)$$

In practice, however, it is desirable to have a simpler and more convenient solution. Such a solution can be found by taking into account the fact that in accordance with the principle of error estimation from above we are interested in limits $\pm\theta$ for the sum of the components such that the probability $P\{|\vartheta|\leqslant\theta\}>0.9$.

Bearing the last remark in mind, we shall examine the distribution function $F_n(\vartheta)$ in the extreme intervals $[-n/2, -n/2+1]$ and $[n/2-1, n/2]$.

For one section Eq. (3.11) assumes the form

$$F_n(\vartheta)=\frac{1}{n!}\left(\vartheta+\frac{n}{2}\right)^{n}.$$

The composition of the distributions is symmetric relative to the ordinate axis.

We shall discuss how to calculate, given the probability distribution, the limits of the confidence interval corresponding to a fixed value α of the confidence probability. The limits of the confidence interval corresponding to α are $\pm\theta$.

By definition, the probability that a random quantity lies within the confidence interval $[-\theta, +\theta]$ is α. Therefore the probability that the random quantity does not lie in the confidence interval is $(1-\alpha)$. If the distribution is symmetric relative to 0 (and we are studying a symmetric distribution), then the probability that the random quantity will take on a value less than $-\theta$ will be equal to the probability that it will take on a value greater than $+\theta$. These probabilities are obviously equal to $(1-\alpha)/2$.

Consider first the left-hand branch of the distribution function. The probability corresponding to the point $-\theta$ [the arguments (points) of the distribution function are also called quantiles of the distribution] is equal to $P\{\vartheta \leqslant -\theta\} = (1-\alpha)/2$. We shall now consider the right-hand branch. The probability that $\vartheta \leqslant +\theta$ will obviously be equal to $1-[(1-\alpha)/2] = (1+\alpha)/2$.

We shall now return to our problem. Given $F_n(\vartheta)$ and α, we are required to find the quantiles $-\theta$ and $+\theta$. Their moduli are equal. For this reason we shall only calculate $-\theta$. For this we have the condition

$$P\{\vartheta \leqslant -\theta\} = F_n(-\theta) = \frac{1}{n!}\left(-\theta + \frac{n}{2}\right)^n = \frac{1-\alpha}{2}, \tag{3.12}$$

from which θ can be calculated. We shall represent in the following form the values of θ found from the formula (3.12):

$$\theta = k\sqrt{\sum_{i=1}^{n} \theta_i^2}, \tag{3.13}$$

where θ_i is the limit of the range of values of x_i ($-\theta_i \leqslant x_i \leqslant +\theta_i$), where k is a correction factor.

In the case at hand $\theta_i = 1/2$ for all $i = 1,\ldots,n$, i.e.,

$$\theta = k\sqrt{n}/2, \quad k = 2\theta/\sqrt{n}. \tag{3.14}$$

Formula (3.13) is convenient for calculations, and for this reason we shall investigate the dependence of the coefficient k on α and n. The calculations are performed as follows. Given α and n, we find θ from Eq. (3.12). Next, the correction factor k is found for the given values of α and n from formula (3.13) or (3.14).

For example, let $\alpha = 0.99$ and $n = 4$. Then $(1-\alpha)/2 = 0.005$. Let us check whether or not the value of θ corresponding to this probability falls within the left extreme interval $[-2, -1]$. For this, we shall find the probability corresponding to the highest value of ϑ in this interval, i.e., $F_4(-1)$:

$$F_4(-1) = \frac{1}{4!}(-1+2)^4 = \frac{1}{1\times2\times3\times4} = 0.041.$$

Since $0.005 < 0.041$, the value of θ of interest to us lies in this interval.

Substituting the initial data into Eq. (3.12) we find θ:

$$\frac{1}{4!}(-\theta+2)^4 = 0.005, \quad -\theta+2 = \sqrt[4]{24\times0.005}, \quad \theta = 1.41.$$

Having found θ, we obtain from formula (3.14):

$$k = \frac{2\times1.41}{\sqrt{4}} = 1.41.$$

The values of k for other values of α and n were calculated analogously and are presented in Table 3.1. The value of k for $n \to \infty$ was found using the fact that by virtue of the central limit theorem, the resulting distribution may be regarded as being normal.

TABLE 3.1. Values of the coefficient k as a function of the number of terms and confidence level.

	Values of the coefficient k for confidence level α			
Number of terms n	0.90	0.95	0.99	0.9973
2	0.97	1.10	1.27	1.34
3	0.96	1.12	1.37	1.50
4	*	1.12	1.41	1.58
5	*	*	*	1.64
...	...	...	...	...
∞	0.95	1.13	1.49	1.73

*Cases for which the coefficient k is not calculated, since θ for the given value of n falls outside the limits of the extreme interval.

We can write the next:

$$\vartheta = \sum_{i=1}^{n} x_i,$$

$$D[\vartheta] = D\left[\sum_{i=1}^{n} x_i\right] = \sum_{i=1}^{n} D[x_i], \quad M[x_i] = 0.$$

But, as is well known, $D[x_i] = \theta_i^2/3$. Therefore

$$D[\vartheta] = \frac{\sum_{i=1}^{n} \theta_i^2}{3}, \quad \sigma[\vartheta] = \sqrt{\frac{1}{3}\sum_{i=1}^{n} \theta_i^2}. \tag{3.15}$$

Thus if $n \to \infty$ we have a random quantity with a normal distribution $N(0,\sigma)$. We shall calculate the modulus of the limits of the confidence interval from its upper limit $\theta = z_{1+\alpha/2}\sigma$, where $z_{1+\alpha/2}$ is the quantile of the normal distribution corresponding to the probability $p = (1+\alpha)/2$ (see above). Thus we obtain

$$\theta = \frac{z_{1+\alpha/2}}{\sqrt{3}}\sqrt{\sum_{i=1}^{n} \theta_i^2}. \tag{3.16}$$

Comparing Eq. (3.16) with Eq. (3.13) we find

$$\underset{n\to\infty}{k} = \frac{z_{1+\alpha/2}}{\sqrt{3}}.$$

For $\alpha = 0.9973$ we obtain $z_{1+\alpha/2} = 3$ and $k = 1.73$.

Looking at the table obtained in this manner (Table 3.1), it should be noted that the correction factor k has the interesting property that for $\alpha \leqslant 0.99$ it is virtually independent of the number of terms. We can make use of this property, and take for k the average values:

α	0.90	0.95	0.98	0.99
k	0.95	1.1	1.3	1.4

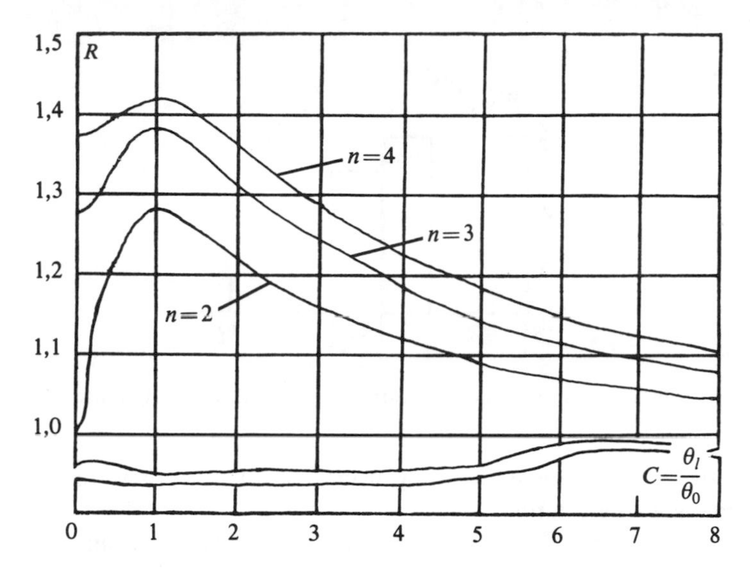

FIG. 3.3. The coefficient k as a function of the change in one of the terms relative to the other terms ($n = 2,3,4$).

The error caused by using the average values of k, as one can see by comparing them with the exact values given in Table 3.1, does not exceed 10%.

The small effect of the number of terms indicates indirectly that it is not necessary to assume, as done above, that all θ_i are equal. Thus if one of the terms θ_l is reduced, then in the limit instead of n we obtain $n-1$ terms. The value of the factor k, however, in the process remains practically unchanged. If, on the other hand, θ_l is gradually increased, then the factor k will decrease.

The dependence of k on the ratio $c = \theta_l/\theta_0$ for $\alpha = 0.99$ is given in Fig. 3.3; θ_0 is the modulus of the limiting values of the remaining terms, which are assumed to be equal.

The factor k for $\alpha = 0.99$ and $n = 4$ can also be calculated using the formula approximating the right-hand branch of the curve presented in Fig. 3.3:

$$k = 1.45 - 0.05\,\frac{\theta_l}{\theta_0}. \tag{3.17}$$

Formula (3.16) can be used instead of Eq. (3.13) to calculate θ when the number of terms is large. However, as follows from the above-presented estimate of the error of calculations based on formula (3.13), the accuracy cannot be increased by more than 10% (for $\alpha = 0.99$). At the same time formula (3.13) is also useful for summing a small number of terms. For this reason, for practical calculations relation (3.13) is preferable.

3.6. Universal method for constructing the composition of distributions

In the general case, in order to combine random quantities it is necessary to construct the composition of the distributions of the terms. If the distribution functions are given analytically, then their composition is found either by direct integration of

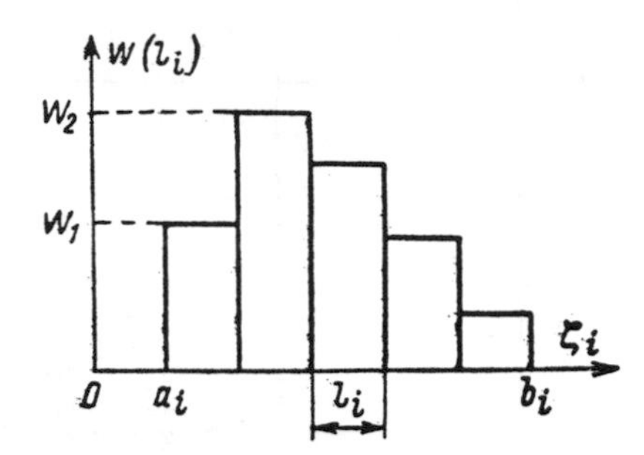

FIG. 3.4. Histogram of the distribution of some random quantity.

the derivatives of the functions or by using the characteristic functions, which usually simplifies the solution.

In practice, however, the analytical form of the distribution functions is usually unknown. Based on the experimental data, it is possible only to construct a histogram, and an error is unavoidably made by passing from the histogram to the distribution function. For this reason, we shall study the summation of random quantities whose distribution is given by histograms and not by distribution functions.[33]

Suppose that we are required to find the distribution function of the random quantity $\zeta = \zeta_1 + \cdots + \zeta_n$, where ζ_i is a random quantity given by a histogram with m_i intervals in the region of possible values of ζ_i with the limits a_i and b_i.

Thus the interval

$$[a_i, b_i] = l_{i1} + l_{i2} + \cdots + l_{im_i}, \quad i = 1, \ldots, n.$$

We shall assume that the probability that the random quantity falls within each interval of the histogram is equal to the area of the part of the histogram that corresponds to this interval (the area of the corresponding column of the histogram):

$$P\{\zeta_i \in l_{ik}\} = p_{ik},$$

where $k = 1, \ldots, m_i$ is the number of the interval of the histogram of the distribution of the random quantity ζ_i.

Figure 3.4 shows as an example a histogram with five intervals of equal length $l_i = l$, so that $b_i - a_i = 5l$. For this histogram

$$p_{i1} = W_1 l, \quad p_{i2} = W_2 l, \quad \ldots, \quad p_{i5} = W_5 l,$$

where $W_1, \ldots, W_5$ are the heights of the columns of the histogram; by construction the area of the entire histogram is equal to unity, i.e., $\Sigma_{k=1}^{5} P_{ik} = 1$.

We recall that in constructing histograms (which are constructed based on empirical data) the height of the column of each interval is found by dividing the frequency with which the values fall within the corresponding interval by the length of this interval. The frequency is an empirically obtained estimate of the probability of the corresponding event.

Next, we shall represent continuous random quantities by discrete random quantities corresponding to them. For this, we denote by a_{ik} the center of each interval

l_{ik} and we introduce a new random quantity η_i, which corresponds to the random quantity ζ_i so that η_i assumes the value a_{ik} with probability p_{ik}. This is possible, since from what we have said above it is obvious that

$$\sum_{k=1}^{m_i} p_{ik}=1, \quad \text{for all } i=1,\dots,n.$$

It is desirable to represent the obtained data for each random quantity η_i by a table of the following form:

$$\eta_i \quad \left| \frac{a_{i1} \quad a_{i2} \quad \dots \quad a_{im_i}}{p_{i1} \quad p_{i2} \quad \dots \quad p_{im_i}} \right. .$$

We shall now study the random variable $\eta = \eta_1 + \eta_2 + \cdots + \eta_n$. We obtain all its possible values by sorting through all combinations of the obtained realizations of a_{ik} of the components η_i.

For the calculations it is convenient to write out the possible values of the random quantities in a single table of the form

$$\begin{array}{ll} \eta_1 & a_{11}\dots a_{1m_1} \\ \eta_2 & a_{21}\dots a_{2m_2} \, . \\ \eta_n & a_{n1}\dots a_{nm_n} \end{array}$$

Next we calculate the values of the random quantity η that correspond to each possible combination of realizations of the random quantities η_i,

$$\eta_i=a_{1k_1}+a_{2k_2}+\cdots+a_{nk_n}$$

and the corresponding probabilities, which we find from the formula

$$p_t=P\{\eta_1=a_{1k_1},\ \eta_2=a_{2k_2},\dots\}=\prod_{i=1}^{n} p_{ik_i}. \tag{3.18}$$

Adding the probabilities that correspond to one and the same realization $\eta_t = a_t$, we obtain the probability that the random quantity η assumes each possible value from series of $a_1,\dots,a_N$.

The number of combinations of terms will be $\Pi_{i=1}^{n} m_i$, but since among them there are terms whose values are the same,

$$N \leqslant \prod_{i=1}^{n} m_i. \tag{3.19}$$

The obtained data make it possible to construct a step function of the distribution $F_1(x)$ of the random quantity η:

$$F_1(x)=\sum_t P\{\eta=a_t\}, \quad a_t \leqslant x. \tag{3.20}$$

The curve $F_1(x)$ is the first approximation to the distribution function $F(x)$ sought. The obtained step function can be smoothed by the method of linear interpolation as follows.

We find the center of the intervals $[a_t, a_{t+1}]$ with $t = 1, ..., N-1$:

$$\beta_t = \frac{a_{t+1} + a_i}{2}. \tag{3.21}$$

From the points β_t we raise perpendiculars up to the broken line $F_1(x)$. We obtain points with the coordinates $(\beta_t, F_1(x))$ for $t = 1, ..., N-1$. To the points obtained we associate points at which the distribution function assumes the values $F_1(\beta_0) = 0$ and $F_1(\beta_n) = 1$:

$$\beta_0 = \sum_{i=1}^{n} a_i, \quad \beta_n = \sum_{i=1}^{n} b_i. \tag{3.22}$$

Joining the $N + 1$ points so obtained with straight lines, we obtain the function $F_2(x)$, which is the approximation sought.

The method presented above gives a solution of the problem using all available information and does not introduce any distortions. In the general case, however, $D[\zeta_i] \neq D[\eta_i]$ and the variance of the random quantity with the distribution $F_1(x)$ or $F_2(x)$ can differ from the variance of the random quantity ζ. For this reason, if the terms are independent, the variance of their sum must be calculated in the standard manner using the formula

$$D[\zeta] = D\left[\sum_{i=1}^{n} \zeta_i\right] = \sum_{i=1}^{n} D(\zeta_i).$$

We note that for $n > 5$ the distribution of the sum of terms can be regarded as a normal distribution, which is completely determined by the variance and the mathematical expectation. Both parameters can be easily calculated from the parameters of the terms

$$M[\zeta] = \sum_{i=1}^{n} M[\zeta_i], \quad D[\zeta] = \sum_{i=1}^{n} D[\zeta_i],$$

and for this reason the calculations presented above are useful only if the number of terms $n < 5$.

It should also be noted that the method presented above for constructing a composition of distributions is also useful in the case when the distributions of the random quantities are given in analytic form. The smooth curve expressing the density of the distribution of the random quantity ζ_i is replaced by a step curve with m_i steps, in a manner so that the area it bounds, as also the area under the smooth curve, is equal to unity. If the branches of the smooth curve of the distribution function approach the abscissa axis asymptotically, this distribution is replaced by a truncated distribution.

It is also obvious that this method is useful both for the case of discrete quantities ζ_i and for the mixed case.

In general, the method examined above is essentially an algorithm for constructing numerically the composition of distributions and can be easily implemented with the help of computers.

We shall illustrate the method with an example. Let $\zeta = \zeta_1 + \zeta_2$, where ζ_1 has a normal distribution with the density

$$f_1(x)=\frac{1}{\sqrt{2\pi}}e^{-(x-2)^2/2},$$

and ζ_2 has a distribution with a uniform density $f_2(x) = 1/6$.

For a normal distribution with the parameters $A=2$ and $\sigma=1$ we shall take the domain of ζ_1 to be $[A-3\sigma, A+3\sigma]=[-1,5]$. We divide this interval into five intervals ($m_1 = 5$), symmetrically arranged relative to the point 2—the mathematical expectation:

$$[-1,5]=[-1,0.5]+[0.5,1.5]+[1.5,2.5]+[2.5,3.5]+[3.5,5].$$

For the random quantity ζ_2, whose domain is the interval $[-3,3]$ and which has a distribution with a uniform density, we assume $m_2 = 3$:

$$[-3,3]=[-3,-1]+[-1,1]+[1,3].$$

Next we calculate the probability that the random quantities fall into the corresponding intervals. For the normal distribution we have

$$p_{11}=\int_{-1}^{0.5}\frac{1}{\sqrt{2\pi}}e^{-(x-2)/2}dx=0.067,$$

$$p_{12}=\int_{0.5}^{1.5}\frac{1}{\sqrt{2\pi}}e^{-(x-2)^2/2}dx=0.242,$$

$$p_{13}=\int_{1.5}^{2.5}\frac{1}{\sqrt{2\pi}}e^{-(x-2)^2/2}dx=0.382.$$

In view of the symmetry of the normal distribution

$$p_{14}=p_{12}=0.242, \quad p_{15}=p_{11}=0.067.$$

For the uniform distribution

$$p_{21}=\int_{-3}^{-1}\frac{1}{6}dx=\frac{1}{3}, \quad p_{22}=\int_{-1}^{1}\frac{1}{6}dx=\frac{1}{3}, \quad p_{23}=\int_{1}^{3}\frac{1}{3}dx=\frac{1}{3}.$$

Next we find the centers of the constructed intervals:

$$a_{11}=\frac{-1+0.5}{2}=-0.25, \quad a_{12}=\frac{0.5+1.5}{2}=1,$$

$$a_{13}=\frac{1.5+2.5}{2}=2, \quad a_{14}=\frac{2.5+3.5}{2}=3, \quad a_{15}=\frac{3.5+5}{2}=4.25,$$

$$a_{21}=\frac{-3-1}{2}=-2, \quad a_{22}=\frac{-1+1}{2}=0, \quad a_{23}=\frac{1+3}{2}=2.$$

This determines η_1, which assumes values a_{1k} with probabilities p_{1k}, where $k = 1,...,5$, and η_2, which assumes values a_{2k} with probabilities p_{2k}, where $k=1$, 2, and 3. As a result of the calculations we have obtained

TABLE 3.2. Data for sorting through variants of sums of the random quantities η_1 and η_2 and the corresponding probabilities.

η	p
$-0.25-2=-2.25$ $-0.25+0=-0.25$ $-0.25+2=1.75$	$0.067\times0.333=0.022$
$1-2=-1$ $1+0=1$ $1+2=3$	$0.242\times0.333=0.081$
$2-2=0$ $2+0=2$ $2+2=4$	$0.382\times0.333=0.127$
$3-2=1$ $3+0=3$ $3+2=5$	$0.242\times0.333=0.081$
$4.25-2=2.25$ $4.25+0=4.25$ $4.25+2=6.25$	$0.067\times0.333=0.022$

$$\eta_1\begin{cases} a_{1k} & -0.25 \quad 1 \quad 2 \quad 3 \quad 4.25 \\ p_{1k} & 0.067 \quad 0.242 \quad 0.382 \quad 0.242 \quad 0.067, \end{cases}$$

$$\eta_2\begin{cases} a_{2k} & -2 \quad 0 \quad 2 \\ p_{2k} & 0.333 \quad 0.333 \quad 0.333. \end{cases}$$

Next we transfer to the random quantity $\eta=\eta_1+\eta_2$. We estimate the number of different terms η from formula (3.19). In our case $m_1=5$, $m_2=3$, and $N\leqslant 15$.

We shall represent the values obtained for η_1 and η_2 in the form of a table:

$$\begin{array}{cccccc} \eta_1 & -0.25 & 1 & 2 & 3 & 4.25\,, \\ \eta_2 & -2 & 0 & 2 & — & —\,. \end{array}$$

We find the values of $\eta=\eta_1+\eta_2$ with the help of this table. The order of the calculations is explained in Table 3.2.

Next we arrange the values of η_t in increasing order. To each value of η_t there corresponds a unique probability p_t. If one and the same value of η_t is encountered several times, then the probability of this value is taken to be the sum of these probabilities. We obtain

$$\begin{array}{cccccccc} \eta_t & -2.25 & -1 & -0.25 & 0 & 1 & 1.75 & 2 \\ p_t & 0.022 & 0.081 & 0.022 & 0.127 & 0.162 & 0.022 & 0.127\,, \end{array}$$

$$\begin{array}{ccccccc} \eta_t & 2.25 & 3 & 4 & 4.25 & 5 & 6.25 \\ p_t & 0.22 & 0.162 & 0.127 & 0.022 & 0.081 & 0.022\,. \end{array}$$

Since $\eta=1$ and $\eta=2$ were encountered twice, $N=13$.

TABLE 3.3. Data for the stepped approximation to the distribution function of a sum of two random quantities studied.

x		$F_1(x)$
$-\infty$	-2.25	0.0
-2.25	-1.00	0.022
-1.00	-0.25	0.103
-0.25	0.00	0.125
0.0	1.00	0.252
1.00	1.75	0.414
1.75	2.00	0.436
2.00	2.25	0.563
2.25	3.00	0.585
3.00	4.00	0.747
4.00	4.25	0.874
4.25	5.00	0.896
5.00	6.25	0.978
6.25	∞	1.00

Based on the data obtained, using Eq. (3.20) it is not difficult to construct $F_1(x)$. The values of this function in the intervals found are presented in Table 3.3 and the corresponding graph is given in Fig. 3.5 in the form of a stepped line.

We find β_t for $t = 1,...,12$ from Eq. (3.21) and we determine β_0 and β_{13} from Eq. (3.22). Using these calculations as well as the data of Table 3.3 we construct the distribution function $F_2(x)$. The function $F_2(x)$ is plotted in Fig. 3.5 as a broken line connecting the points $(\beta_t, F_1(\beta_t))$ for $t = 0,...,13$. The numerical values of $F_2(x)$ for $x = \beta_t$, where $t = 0,...,13$ are presented in Table 3.4.

As we have already mentioned, the approximation of the limiting distribution function $F(x)$ by $F_2(x)$ can be improved by reducing the subdivisions of the domains of the starting random quantities.

It is interesting to note that the solution given above makes it possible to find the edges of the distribution function $F_2(x)$ without constructing the entire function. In many cases this is the main problem.

3.7. Natural limits of measurements

For metrology as the science of measurements, it is of fundamental interest to estimate the limiting possibilities of measurements. First, extremely small and extremely large measurable quantities must be estimated. Next, in the case when instantaneous values are measured, there arises the question of the maximum rate of change of the quantity. For functionals, such as the effective current, it is important to establish both the maximum and minimum frequency of initial process. It is obviously possible to add to this list.

Among all limiting parameters, the lower limits of measurements are of greatest interest, since they are determined by the physical, i.e., natural, limitations. Com-

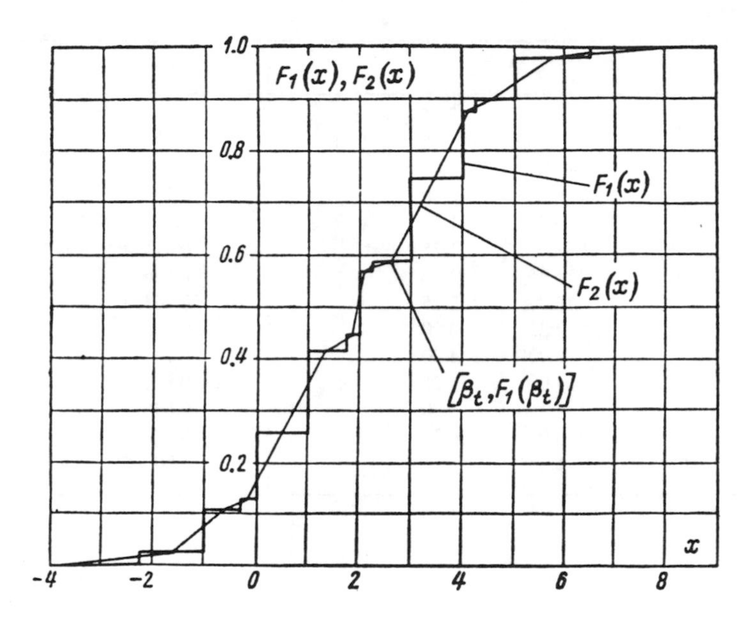

FIG. 3.5. Step and linear approximations of the distribution function.

paring these limits with the limits that the real measuring instruments permit makes it possible to judge the level of development of measuring instruments and stimulates improvements in their construction. For this reason, we shall confine our attention to the natural limits of measurements.

TABLE 3.4. Data for the linear apporximation to the distribution function of the sum of two random quantities studied.

t	β_t	$F_2(x)$
0	– 4.00	0.000
1	– 1.62	0.022
2	– 0.62	0.103
3	– 0.14	0.125
4	0.50	0.252
5	0.14	0.414
6	1.87	0.436
7	2.12	0.563
8	2.62	0.585
9	3.50	0.747
10	4.12	0.874
11	4.62	0.896
12	5.62	0.978
13	8.00	1.000

3.7.1 Limitations imposed by thermal noise

Measurements are always accompanied by an interaction of the object of study and the measuring instruments. For this reason, the limiting possibilities of measurements must be estimated for the measuring instruments together with the object of study—the carrier of the physical quantity.

We shall first study instruments that have an inertial moving system and elastic elements that keep it in a position of equilibrium. Examples are galvanometers, some types of balances, etc. Such systems are modeled mathematically by Eqs. (2.3) and (2.4). The moving part of these instruments is continuously bombarded by air molecules (and molecules of liquid, if the oscillations are damped by a liquid). On average, the number of impacts and their effect are the same on all sides of the system. But at any given moment in time the effect of impacts from one side can be greater than from another side, while at the next instant the situation is reversed. As a result, careful observations reveal continuous oscillations of the movable part of the instrument around the position of equilibrium. These fluctuations limit the possibilities of instruments.[59]

According to a well-known theorem of statistical physics, in the state of thermal equilibrium with a medium at temperature T, to each degree of freedom of the body there is associated an average energy of fluctuations equal to $\bar{\varepsilon} = kT/2$, where k is Boltzmann's constant. The movable part of an instrument has one degree of freedom. For this reason, the average energy of the fluctuations of the movable part is equal to $\bar{\varepsilon}$. But this energy is equal to the average strain energy of the elastic elements $\bar{P} = W\overline{\alpha^2}/2$, where W is the stiffness of the elastic elements and α is the strain, i.e., the displacement of the moving part. From the equality $\bar{\varepsilon} = \bar{P}$ it follows that

$$\overline{\alpha^2} = \frac{kT}{W}. \tag{3.23}$$

We shall transform this formula, introducing into it the measured quantities. It is obvious that such a transformation cannot be universal: it depends on the type of measured quantity and on the principle of operation of the instrument. Consider, for example, a moving-coil galvanometer. From the relation between the current strength I in the steady-state regime we have $\alpha W = \Phi I$, where Φ is the magnetic constant of the galvanometer. In addition, the operating parameters β and ω_0 are related to the structural parameters J, P, and W, where J is the moment of inertia of the moving part and P is the damping constant [see Eq. (2.3) and for example Ref. 49], $\omega_0^2 = W/J$, and $2\beta/\omega_0 = P/W$.

With the help of these relations Eq. (3.23) can be transformed as follows:

$$\overline{I^2} = \frac{kTW}{\Phi^2} = \frac{kT}{\Sigma R}\frac{W}{P} = \frac{kT}{\Sigma R}\frac{\omega_0}{2\beta}.$$

Here ΣR is the sum of the resistances of the moving coil of the galvanometer and the external circuit. The damping constant P is related to ΣR by the equation $P = \Phi^2/\Sigma R$.

The mean-square fluctuations are usually written not in terms of the angular frequency ω_0 but rather in terms of the period of free oscillations of the system $T_0 = 2\pi/\omega_0$. Then we obtain

$$\overline{I^2} = \frac{\pi kT}{\Sigma R} \frac{1}{\beta T_0}. \tag{3.24}$$

For the measured voltage E, we obtain analogously

$$\overline{E^2} = \pi kT \frac{\Sigma R}{\beta T_0}. \tag{3.25}$$

It is well known that for $\beta > 0.8$, $\beta T_0 \approx t_r$[49] where t_r is the response time of the moving part of the instrument. For this reason, instead of Eqs. (3.24) and (3.25) we can write

$$\overline{I^2} = \frac{\pi kT}{\Sigma R} \frac{1}{t_r}, \quad \overline{E^2} = \pi kT (\Sigma R) \frac{1}{t_r}.$$

Analogous arguments for torsion balances give a relation that is similar to Eqs. (3.24) and (3.25): $\overline{m_x^2} = (kTP/g^2)(\omega_0/2\beta)$, where g is the acceleration of gravity. In general, we can write

$$\overline{x^2} = kTC_x \frac{1}{t_r}, \tag{3.26}$$

where x is the measured quantity and C_x is a constant, determined by the principle of operation and the construction of the instrument. For balances $C_x = \pi P/g^2$, while for a galvanometer

$$C_I = \frac{\pi}{\Sigma R}, \quad C_E = \pi \Sigma R.$$

The expressions (3.24)–(3.26) show that the mean-square flucuations of the indications of instruments with an inertial moving part are inversely proportional to their response time. But the response time is also the minimum measurement time (approximately). Moreover, it is obvious that the minimum value of the measured quantity is related with $\sqrt{\overline{x^2}}$ by the fixed accuracy of the measurement. For this reason, it follows from expressions (3.24)–(3.26) that

$$x_{\min} = K \frac{1}{\sqrt{t_r}}, \tag{3.27}$$

where $x_{\min}$ is the minimum value of the measured quantity and K is a constant.

The moving part of balances can interact with the medium in only one way: by means of collisions with molecules of the medium (for example, air). For a galvanometer the situation is different. Apart from mechanical contact with the medium, the moving part of these instruments also interacts with the medium through the electric circuit, in which electron velocity fluctuations occur. Relations (3.24) and (3.25) were derived based on the fact that the galvanometer interacts with the medium only by means of collisions of air molecules with the movable part of the galvanometer. But these relations can be derived under the assumption that the entire interaction occurs as a result of fluctuations of the electrons' velocity in the input circuit. In general, in statistical physics it has been established that for fluctuations of a system, it makes no difference how the system interacts with the medium; only the temperature of the system is important.

Thermal fluctuations in electric circuits are usually calculated with the help of Nyquist's formula:

$$\overline{E^2}=4kTR\Delta f, \tag{3.28}$$

where $\overline{E^2}$ is the mean-square noise of emf that is brought in a circuit with an active resistance R and Δf is the frequency band in which the noise emf is calculated. Nyquist's formula is very often presented as an expression for the spectral density of the square of the noise emf:

$$F_E(f)=4kTR. \tag{3.29}$$

The spectral density may be assumed to be constant up to very high frequencies, corresponding to the collision frequency of the charge carriers.

Consider one of the best galvanometers—a galvanometer of the type Z_c manufactured by Kipp Company. For $T_0 = 7$ s, $\beta = 1$, $\Sigma R = 50\ \Omega$, and $S_u = 1.7 \times 10^7$ mm m/V. Therefore at room temperature ($T = 293$ K) we shall have ($k = 1.38 \times 10^{-23}$ J/K) $\overline{E^2} = \pi \times 1.38 \times 10^{-23} \times 293 \times 50/1 \times 7 = 9.06 \times 10^{-20}\ \text{V}^2$, i.e., $\overline{E} = 3 \times 10^{-10}$ V.

The galvanometer constant $C_u = 1/S_u = 6 \times 10^{-8}$ V/mm m. If the beam length is taken to be 2 m rather than 1 m, then we find that to a displacement of 1 mm there corresponds a voltage of about $E_m = 3 \times 10^{-8}$ V. Further increasing of the beam length has no effect, because the effect of the shaking of the ground and the base to which the galvanometer is fastened usually increases correspondingly. The difference between $E_m = 3 \times 10^{-8}$ V and $\overline{E} = 3 \times 10^{-10}$ V is very large, and it is obvious that the galvanometer does not exhibit thermal noise. In precisely the same way, thermal noise is also usually not observed when using balances. The situation is different in the case of instruments with amplifiers. For example, consider an electronic-measuring instrument having an input circuit with resistance R and a wideband amplifier with a quite large gain. The thermal noise can now be appreciable. An analogous situation exists with electromechanical devices.

The sensitivity of modern galvanometers to shaking has been radically reduced by using taut bands and liquid dampers for damping the transverse oscillations of the moving part. This makes it possible to increase the beam length. But, instead of this, the rotation of the moving part is indicated photoelectro-optically, which is more efficient. The noise associated with the electronic circuit is suppressed with the help of negative feedback. Such devices can be made to be so sensitive that it is possible to observe thermal noise in their input circuits. To calculate this noise it is necessary to take into account the fact that the photoelectric amplifier and the instrument at the output have certain inertial properties. We shall examine, as an example, photogalvanometric self-balancing amplifiers.[49]

The structural arrangement of the self-balancing amplifier is shown in Fig. 3.6. Block 1 is a galvanometer together with its input circuit; block 2 is a photoelectro-optic transducer–amplifier; block 3 is the feedback block (balancing resistor); block 4 is the output device.

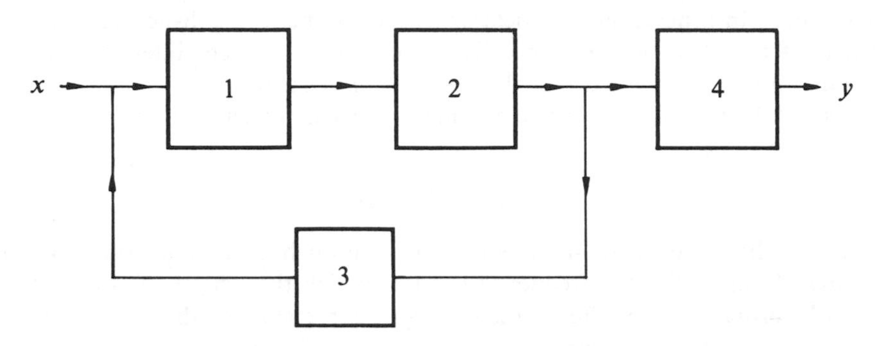

FIG. 3.6. Structural layout of a photogalvanometric self-balancing amplifier.

We are interested in the noise at the output of the self-balancing amplifier. As is well known, the spectral density $F(\omega)$ of the noise at the output is related to the spectral density of the noise at the input $F_{E,I}(\omega)$ by the relation

$$F(\omega)=F_{E,I}(\omega)\,|\,W(j\omega)\,|^2, \tag{3.30}$$

where $|\,W(j\omega)\,|$ is the modulus of the amplitude–frequency response of the system and ω is the angular frequency. For the angular frequency $\omega=2\pi f$, formula (3.29), determining the spectral density of the thermal noise at the input of the system, assumes the form

$$F_E(\omega)=\frac{2kTR}{\pi}. \tag{3.31}$$

The self-balancing amplifier's transfer function, which relates the indications of the output instrument and the measured emf, is expressed by the following formula, if the output instrument is calibrated based on the input:

$$W(p)=\frac{1}{[\varkappa p^3+(1+2\beta\varkappa)p^2+2\beta p+1](q^2p^2+2\beta_{\rm out}q+1)}. \tag{3.32}$$

Here $p = s/\omega_0$, $\varkappa = \tau\omega_0$, $q = \omega_0/\omega_{\rm out}$, ω_0 is the angular frequency of the characteristic oscillations of the self-balancing amplifier, $\omega_{\rm out}$ is the angular frequency of the characteristic oscillations of the output instrument, s is the Laplace operator, and τ is the time constant of the photoelectro-optic transducer amplifier. In addition, $\beta_{\rm out}$ is the damping ratio of the output instrument. The transfer function (3.32) does not take into account the steady-state residual error of the self-balancing amplifier, since it is always small. It is also assumed that the damping of the output device does not depend on resistance of its circuit.

The operator p is dimensionless and corresponds to the relative frequency $\eta = \omega/\omega_0$. Substituting $p=j\eta$ into Eq. (3.32), we obtain the amplitude–frequency response in complex form. Its modulus can be represented in the form

$$|\,W(j\eta)\,|=\frac{1}{|\,C_5(j\eta)^5+C_4(j\eta)^4+C_3(j\eta)^3+C_2(j\eta)^2+C_1(j\eta)+C_0|}.$$

In our case $C_0 = 1$, $C_1 = 2\beta + 2\beta_{out}q$, $C_2 = 2\beta\varkappa + 4\beta\beta_{out}q + q^2 + 1$, $C_3 = \varkappa + 4\beta\beta_{out}\varkappa q + 2\beta_{out}q + 2\beta q^2$, $C_4 = 2\beta_{out}\varkappa q + 2\beta\varkappa q^2 + q^2$, $C_5 = \varkappa q^2$.

The spectral density of the noise at the input must also be expressed in terms of the relative frequency. Since $F_E(\omega)\Delta\omega = F_E(\eta)\Delta\eta$,

$$F_E(\eta)=F_E(\omega)\omega_0=\frac{2}{\pi}kTR\omega_0$$

or, since $\omega_0 = 2\pi/T_0$, $F_E(\eta) = 4kTR/T_0$.

So, the general expression for the spectral density of the noise in the indications of the photogalvanometric self-balancing amplifier has the form

$$F(\eta)=\frac{4kTR}{T_0}\frac{1}{|C_5(j\eta)^5+C_4(j\eta)^4+C_3(j\eta)^3+C_2(j\eta)^2+C_1(j\eta)+C_0|^2}. \tag{3.33}$$

Now we can calculate the mean-square fluctuations of the indications of the self-balancing amplifier: $\overline{E^2} = \int_0^\infty F(\eta)d\eta$. Integrals of expressions of the type (3.33) for stable systems are well known; they can be found in books on automatic control,* and for the specific formula (3.23) the integral is given in Ref. 49:

$$\overline{E^2}=\frac{2\pi kT\Sigma R}{T_0}$$
$$\times\frac{C_2^2C_5+C_1C_4^2-C_2C_3C_4-C_0C_4C_5}{C_0C_3^2C_4-C_1C_2C_3C_4-2C_0C_1C_4C_5+C_0^2C_5^2+C_0C_2C_3C_5+C_1C_2^2C_5+C_1^2C_4^2}. \tag{3.34}$$

Formula (3.34) is complicated, but for a specific instrument all coefficients are simply numbers, so that this formula is easy to use. As shown in Ref. 49, Eq. (3.34) yields formulas for all particular cases: an individual galvanometer, a galvanometric amplifier with noninertial transducer–amplifier ($\varkappa = 0$), and a galvanometric amplifier with current output. If measurement of the current strength in the circuit with the resistance R (including the resistance of the galvanometer) and not the emf is studied, then bearing in mind that $I = E/R$, the transfer from Eq. (3.34) to the mean-square current does not present any difficulties.

Consider a specific instrument, a HΦK-1 photogalvanometric self-balancing nanovoltmeter.[49] The modifications HΦK-2 and HΦK-3 of this instrument have the same parameters and differ from HΦK-1 only by external finishing. One graduation of the instruments is equal to 4×10^{-10} V. The parameters of the instrument are as follows: $T_0 = 0.37$ s, $\beta = 12$, $\beta_{out} = 1$, $\tau = 0.35$ s, $\Sigma R = 11\ \Omega$, and $T_{out} = 0.8$. Therefore, $\omega_0 = 2\pi/T_0 = 17$ rad/s, $q = 0.8/0.37 = 2.2$, $\varkappa = 0.35\times 17 = 6$, $C_0=1$, $C_1=24+2\times 2.2=28.4$; $C_2=2\times 12\times 6+4\times 12\times 1\times 2.2 +2.2^2+1=255$; $C_3 = 6 + 4 \times 12 \times 6 \times 2.2 + 2 \times 2.2 + 2 \times 12 \times 2.2^2 = 755$, $C_4 = 2 \times 6 \times 2.2 + 2 \times 12 \times 6 \times 2.2^2 + 2.2^2 = 693$, $C_5 = 6 \times 2.2^2 = 27.6$.

*These integrals were first presented in the book *Theory of Servomechanisms* by H. M. James, N. B. Nichols, and R. S. Phillips (McGraw-Hill, New York, 1947).

The cofactor in formula (3.24), consisting of the coefficients C_0–C_5, is equal to 4.2×10^{-2}. At $T = 293$ K we obtain

$$\overline{E^2} = \frac{2\pi \times 1.38 \times 10^{-23} \times 293 \times 11}{0.37} \times 4.2 \times 10^{-2} = 3.2 \times 10^{-20}\ \text{V}^2,$$

$$\sqrt{\overline{E^2}} = 1.8 \times 10^{-10}\ \text{V}.$$

The experimentally estimated mean-square voltage of the fluctuations of the nanovoltmeter is equal to 2.8×10^{-10} V, which exceeds the theoretical value by only a factor of less than 2. Therefore it can be considered that the sensitivity of the НФК instruments has practically reached its theoretical limit.

3.7.2. Restrictions imposed by shot noise

An electric current, as is well known, consists of a flow of electrons. When the current becomes very weak, i.e., the number of electrons passing per second through the transverse cross section of the conductor becomes small, it is observed that their number fluctuates randomly. The random oscillation of the current strength arising in this manner is caused by the randomness of the moments at which electrons appear in the electric circuit. The noise arising is called the shot noise or Schottky noise (sometimes this noise is also called generation-recombination noise). Shot noise is characterized by the fact that it is observed only when current flows along the circuit. Thermal noise, however, does not depend on whether or not a current flows in the circuit; it also occurs in the absence of current.

The mean-square fluctuation produced in the current strength owing to shot noise is given by the formula[59]

$$\overline{I_{sn}^2} = 2eI\Delta f_{sn}, \tag{3.35}$$

where I is the current flowing in the circuit, e is the electron charge, and Δf_{sn} is the equivalent bandwidth of the shot noise. Formula (3.35) was derived by replacing the real spectrum of the squared current, owing to the shot noise, which has a maximum at zero frequency, by the equivalent spectrum that is uniform in the band Δf_{sn}; in addition, $\Delta f_{sn} = 1/2t_0$, where t_0 is the average transit time of an electron or, in the case of the generation-recombination noise, the average lifetime of an electron.

The fluctuations of the indications of the measuring instrument connected in a circuit with current I can be calculated based on Eq. (3.35) in precisely the same manner as was done for thermal noise. If the transmission band of the device is Δf and $\Delta f \leqslant \Delta f_{sn}$, then the mean-square fluctuations produced by shot noise in the indications of the instrument can be estimated from the formula $\overline{I_{sn}^2} = 2eI\Delta f$.

It is sometimes convenient to estimate the shot noise based on the variance of the number of electrons forming the current I: $\overline{\Delta n^2} = n$. Thus a current of 1×10^{-14} A is formed by the passage of approximately 6×10^4 electrons per second through the transverse cross section of the conductor. The mean-square fluctuations of this number of particles is equal to 6×10^4, and the relative value of the mean-square deviation will be

$$\frac{\sqrt{\overline{\Delta n^2}}}{n}=\frac{1}{\sqrt{n}}\approx\frac{1}{2.4\times 10^2}$$

or 0.4%.

A current of 1×10^{-16} A is equal to approximately only 600 electrons per second, and the mean square of its fluctuations is already equal to 4%.

Shot and thermal noise are independent of one another. For this reason, when both types of noise occur, the variance of the fluctuations of the measuring instrument is calculated as the sum of the variances of the shot and thermal components.

3.7.3. Estimate of the minimum measurable value

Shot and thermal noise in the input circuit of a measuring instrument are essentially indicators of the fact that the model of the object does not correspond to the object itself. In accordance with what was said in Sec. 1.3, measurement with a prescribed accuracy is possible only if the error due to this discrepancy (threshold discrepancy) will be less than the permissible prescribed measurement error. But how does one compare these errors?

The measurement error is usually established in the form of a limit of permissible error (absolute or relative). The error due to natural noise is usually calculated as the mean square of the fluctuations of the indications of the instrument or the mean-square deviation. These are entirely different indicators of the error. The permissible limit of the measurement error must be compared not with the mean-square deviation but rather the width of the confidence interval for fluctuations of the indications. The latter width is not so easy to calculate, since the random process at the output of the instrument can hardly be regarded as being white noise. It is clear, however, that for a confidence level of the order of 0.95, the width of the confidence interval will be more than 1 mean-square deviation on each side of the central line. The limit of permissible error, even for measurement of the smallest values, cannot exceed 50%. The width of the confidence interval can be expressed as $k\sqrt{\overline{x^2}}$ (more accurately, this is the half-width of this band), and the limit of permissible error can be expressed as $0.5x_{\min}$. If it is assumed that these quantities are equal, then we obtain

$$x_{\min}=2k\sqrt{\overline{x^2}}.$$

If $k=2$, then $x_{\min}=4\sqrt{\overline{x^2}}$.

The foregoing arguments, however, are unsatisfactory in certain respects, because not all frequency components of the noise are equivalent to the observer: high-frequency noise can be easily averaged when reading the indications of the instrument, while noise of very low frequency can in many cases be neglected. In addition, it was necessary to make an assumption and use the relation between the measured quantity and the limit of permissible measurement error. Is it possible to approach this question in some other manner?

The mean-square fluctuations of the indications is an integral characteristic, which is not related either to the time or the frequency. Meanwhile, measurements require some time, so that the temporal characteristics of these fluctuations are of interest. In the theory of random processes, the average frequency of excesses above a prescribed level is used as a temporal characteristic of the process. The physical meaning of this characteristic is very clear and simple. For example, let the response time of the instrument and therefore the measurement time be equal to 5 s, and let one overshoot per graduation of the instrument occur on the average once per 60 s. It is clear that an instrument with such noise can clearly measure a quantity corresponding to one graduation of the instrument scale. But if one overshoot occurred on the average over a time close to the response time of the instrument, then such a small quantity could no longer be measured.

The overshoots above a prescribed level provide a very convenient characteristic, one that is easy to estimate experimentally. Mean-square fluctuations are difficult to estimate experimentally. But it is easier to calculate the mean-square fluctuations than the average frequency of overshoots.

We shall now calculate the average frequency of overshoots above a prescribed level for a normal stationary process. Thermal noise corresponds to all these conditions. Let σ be the mean-square deviation of the random process and c the threshold level for determining overshoot values. As is shown, for example, in Ref. 41, the average frequency of overshoots $\overline{N}$ can be calculated from the formula

$$\overline{N}=e^{-\gamma^2/2}\sqrt{\frac{1}{\sigma^2}\int_0^\infty f^2F(f)df},$$

where $\gamma=c/\sigma$ and $F(f)$ is the spectral density of the process.

Consider once again a photogalvanometric self-balancing amplifier. The variance of the indications of the instrument is given by formula (3.34) and the spectral density is given by formula (3.33). The latter formula, however, pertains to the relative frequency η. For this reason, we shall transform the integrand in the expression so that f is replaced by η: $f^2F(f)df = (1/4\pi^2)\times\omega^2F(\omega)d\omega = (\omega_0^2/4\pi^2)\times\eta^2F(\eta)d\eta$.

Referring once again to the tables of integrals mentioned above, we find[49]

$$\overline{N}=\frac{1}{T_0}e^{-\gamma^2/2}\Phi,\quad \Phi=\sqrt{\frac{C_0(C_4C_3-C_2C_5)}{C_0C_4C_5+C_2C_3C_4-C_2^2C_5-C_1C_4^2}}.$$

For the HΦK-1 instrument, $\sqrt{\overline{E^2}} = 1.8\times10^{-10}$ V and $c = 4\times10^{-10}$ V. Hence, $\gamma = 4\times10^{-10}/1.8\times10^{-10} = 2.2$. The coefficients C_0–C_5 were presented above, and knowing them we calculate $\Phi = 0.065$. Substituting the numerical values into the working formula we obtain

$$\overline{N}=\frac{1}{0.37}\times0.09\times0.065=1.6\times10^{-2}\ \text{overshoots/s}.$$

This means that one overshoot per graduation occurs on the average once per minute. The response time of the instrument is 5 s. It is clear that noise does not prevent indicating, with the help of the HΦK-1 instrument, a quantity corresponding to one graduation.

Chapter 4

Statistical methods for analysis of multiple measurements in the absence of systematic errors

4.1. Formulation of the problem

Multiple measurements are a classical object of mathematical statistics and the theory of measurement errors. Under certain restrictions on the starting data, mathematical statistics gives a number of elegant methods for analyzing observations and estimating measurement errors.[52,60]

Unfortunately, the restrictions required by the mathematics are not often satisfied in practice. Then these methods cannot be used, and practical methods for solving the problems must be developed. But even in this case the methods of mathematical statistics provide a point of reference and a theoretical foundation.

In this chapter the principal mathematical methods for solving the problem mentioned above are presented. The situation corresponding to direct multiple measurements, free of systematic errors, i.e., having only random errors, is studied. Under this restriction mathematical methods can be fully employed.

A separate result, i.e., a separate value of the random error, cannot be predicted. But a large collection of random errors of some measurement satifies definite laws. These laws are statistical (probabalistic). They are established and proved in metrology based on the methods of mathematical statistics and the theory of probability.

The problem can be solved best if the distribution function of the observations is available. In practice, however, distribution functions are, as a rule, unavailable.

If the random character of the observational results is caused by the measurement errors, it is usually assumed that the observations have a normal distribution. The computational results based on this assumption do not, as a rule, lead to contradictions. This probably happens for two reasons. First, the measurement errors consist of many components. According to the central limit theorem this leads, in the limit, to the normal distribution. In addition, the measurements for which accuracy is important are performed under controlled conditions, as a result of which the distributions of their errors turn out to be bounded. For this reason, their approximation by a normal distribution, for which the random quantity can take on arbitrary values, results in some spare room, for example, it leads to wider confidence intervals than the intervals that would be obtained if the true distribution were known.

There are examples, however, when the observational results do not correspond to the normal distribution. In addition, when the measured quantity is an average value, the distribution of the observations can have any form. For this reason, the hypothesis that the distribution of the observations is normal must, in principle, be checked.

The methods of statistical calculations for observations that are distributed normally have been well developed and the required tables have been constructed. If, however, the hypothesis that the distribution is normal must be rejected, then the statistical analysis of the observations becomes much more complicated. Mathematicians have been working to find, if not better than, at least satisfactory estimates for parameters of distributions whose form has not been precisely established.

Random and quasirandom errors of multiple measurements are always estimated based on the experimental data obtained in the course of the measurements, i.e., *a posteriori.*

As was shown in Chap. 1, in spite of the existence of random errors, a measured quantity can only be a quantity that is defined in a model as nonrandom and constant. The problem is to find from the experimentally obtained data the best estimate of the measured quantity. The estimates obtained from statistical data must be consistent, unbiased, and effective.

An estimate $\tilde{A}$ is said to be consistent if, as the number of observations increases, it approaches the true value of the estimated quantity A (it converges in probability to A):

$$\tilde{A}(x_1,\ldots,x_n) \underset{n\to\infty}{\to} A.$$

The estimate of A is said to be unbiased, if its mathematical expectation is equal to the true value of the estimated quantity:

$$M[\tilde{A}]=A.$$

In the case when several unbiased estimates can be found, the estimate that has the smallest variance is, naturally, regarded as the best estimate. The smaller the variance of an estimate, the more effective the estimate is.

Methods for finding estimates of a measured quantity and indicators of the quality of the estimates depend on the form of the distribution function of the observations.

For a normal distribution of the observations, the arithmetic mean of the observations, as well as their median, can be taken as an estimate of the true value of the measured quantity. The ratio of the variances of these estimates is well known[21]:

$$\sigma_{\bar{x}}^2/\sigma_m^2=0.6,$$

where $\sigma_{\bar{x}}^2$ is the variance of the arithmetic mean and σ_m^2 is the variance of the median.

Therefore the arithmetic mean is a more effective estimate of A than the median.

In the case of a uniform distribution the arithmetic mean of the observations or the half-sum of the minimum and maximum values can be taken as an estimate of A:

$$\tilde{A}_1=\frac{1}{n}\sum_{i=1}^{n} x_i,\quad \tilde{A}_2=\frac{x_{\min}+x_{\max}}{2}.$$

The ratio of the variances of these estimates is also well known[21]:

$$\frac{D[\tilde{A}_1]}{D[\tilde{A}_2]}=\frac{(n+1)(n+2)}{6n}.$$

For $n=2$ this ratio is equal to unity, and it increases for $n>2$. Thus for $n=10$ it is already equal to 2.2. Therefore the half-sum of the minimum and maximum values is, in this case, already a more effective estimate than the arithmetic mean.

4.2. Estimation of the parameters of the normal distribution

If the available data are consistent with the hypothesis that the distribution of the observations is normal, then in order to describe fully the distribution, the expectation $M[X]=A$ and the variance σ must be estimated.

When the probability density of a random quantity is known, its parameters can be estimated by the method of maximum likelihood. We shall use this method to solve our problem.

The elementary probability of obtaining some result of an observation x_i in the interval $x_i \pm \Delta x_i/2$ is equal to $f_i(x_i,A,\sigma)\Delta x_i$. All observational results are independent. For this reason, the probability of encountering all experimentally obtained observations with $\Delta x_i = \cdots = \Delta x_n$ is equal to

$$P_l=\prod_{i=1}^{n} f_i(x_i,A,\sigma)\Delta x_1\cdots\Delta x_n.$$

The idea of the method is to take for the estimate of the parameters of the distribution (in our case these are the parameters A and σ) the values that maximize the probability P_l. The problem is solved, as usual, by equating to 0 the partial derivatives of P_l with respect to the parameters being estimated. The constant cofactors do not affect the solution, and for this reason only the product of the functions f_i is studied; this product is called the likelihood function:

$$L(x_1,\ldots,x_n;A,\sigma)=\prod_{i=1}^{n} f_i(x_1,\ldots,x_n;A,\sigma).$$

We now return to our problem. For the available group of observations $x_1,\ldots,x_n$ the values of the probability density will be

$$f_i(x_i,A,\sigma)=\frac{1}{\sigma\sqrt{2\pi}}e^{-(x_i-A)^2/2\sigma^2}.$$

Therefore

$$L=\left(\frac{1}{\sigma\sqrt{2\pi}}\right)^n \exp\left(-\frac{1}{2\sigma^2}\sum_{i=1}^{n}(x_i-A)^2\right).$$

To find the maximum of L it is convenient to investigate $\ln L$:

$$\ln L=-\frac{n}{2}\ln 2\pi-\frac{n}{2}\ln\sigma^2-\frac{1}{2\sigma^2}\sum_{i=1}^{n}(x_i-A)^2.$$

The maximum of L will occur when $\partial L/\partial A=0$ and $\partial L/\partial\sigma^2=0$:

$$\frac{\partial L}{L\partial A}=\frac{1}{\sigma^2}\sum_{i=1}^{n}(x_i-A)=0,$$

$$\frac{\partial L}{L\partial(\sigma^2)}=-\frac{n}{2\sigma^2}+\frac{1}{2\sigma^4}\sum_{i=1}^{n}(x_i-A)^2=0.$$

From the first equation we find an estimate for A:

$$\tilde{A}=\frac{1}{n}\sum_{i=1}^{n}x_i. \tag{4.1}$$

The second equation gives the estimate $\tilde{\sigma}^2=(1/n)\Sigma_{i=1}^{n}(x_i-A)^2$. But A is unknown; taking instead of A its estimate $\bar{x}$, we obtain

$$\tilde{\sigma}_*^2=\frac{1}{n}\sum_{i=1}^{n}(x_i-\bar{x})^2.$$

We shall check to see whether the obtained estimates are consistent and unbiased. The mathematical expectation $M(x_i)=A$, since all x_i refer to one and the same distribution. For this reason

$$M[\tilde{A}]=\frac{1}{n}\sum_{i=1}^{n}M(x_i)=A.$$

Therefore $\tilde{A}$ is an unbiased estimate of A. It is also a consistent estimate, since as $n\to\infty$ $\tilde{A}\to A$ according to the law of large numbers.

We shall now investigate $\tilde{\sigma}_*^2$. In the formula derived above the random quantities are x_i and $\bar{x}$. For this reason, we shall rewrite it as follows:

$$\begin{aligned}\tilde{\sigma}_*^2&=\frac{1}{n}\sum_{i=1}^{n}(x_i-A+A-\bar{x})^2\\&=\frac{1}{n}\sum_{i=1}^{n}[(x_i-A)^2-2(x_i-A)(\bar{x}-A)+(\bar{x}-A)^2]\\&=\frac{1}{n}\sum_{i=1}^{n}(x_i-A)^2-\frac{2}{n}\sum_{i=1}^{n}(x_i-A)(\bar{x}-A)+\frac{1}{n}\sum_{i=1}^{n}(\bar{x}-A)^2\\&=\frac{1}{n}\sum_{i=1}^{n}(x_i-A)^2-(\bar{x}-A)^2,\end{aligned}$$

since

$$\frac{1}{n}\sum_{i=1}^{n}(\bar{x}-A)^2=(\bar{x}-A)^2$$

and

$$\frac{2}{n}\sum_{i=1}^{n}(x_i-A)(\bar{x}-A)=\frac{2}{n}(\bar{x}-A)\sum_{i=1}^{n}(x_i-A)=2(\bar{x}-A)^2.$$

We shall find $M[\tilde{\sigma}_*^2]$. For this the following relations must be used. By definition, according to Eq. (3.8) we have $M(x_i - A)^2 = \sigma^2$. Therefore

$$M\left[\frac{1}{n}\sum_{i=1}^{n}(x_i-A)^2\right]=\frac{1}{n}M\left[\sum_{i=1}^{n}(x_i-A)^2\right]=\sigma^2.$$

For the random quantity $\bar{x}$ we can write analogously $M(\bar{x} - A)^2 = D[\bar{x}]$. We shall express $D[\bar{x}]$ in terms of $\sigma^2 = D[X]$:

$$D[\bar{x}]=D\left[\frac{1}{n}\sum_{i=1}^{n}x_i\right]=\frac{1}{n^2}\sum_{i=1}^{n}D(x_i)=\frac{1}{n}D[X]=\frac{\sigma^2}{n}.$$

Thus

$$M[\tilde{\sigma}_*^2]=\sigma^2-\frac{\sigma^2}{n}=\frac{n-1}{n}\sigma^2.$$

Therefore the obtained estimate $\tilde{\sigma}_*^2$ is biased. But as $n\to\infty$ $M[\tilde{\sigma}_*^2] \to \sigma^2$, and therefore this estimate is consistent.

To correct the estimate, i.e., to make it unbiased, $\tilde{\sigma}_*^2$ must be multiplied by the correction factor $n/(n-1)$. Then we obtain

$$\tilde{\sigma}^2=\frac{1}{n-1}\sum_{i=1}^{n}(x_i-\bar{x})^2. \tag{4.2}$$

This estimate is also consistent, but, as one can easily check, it is now unbiased. Some deviation from the maximum of the likelihood function is less important for us than the biasness of the estimate.

The standard deviation of the random quantity X is $\sigma = \sqrt{D[X]}$ and it is not the random quantity. Instead of σ^2 we must use the estimate of the variance from formula (4.2)—a random quantity. The extraction of a square root is a nonlinear procedure; it introduces bias into the estimate $\tilde{\sigma}$. To correct this estimate a factor k_n, depending on n as follows, is introduced:

n	3	4	5	6	7	10
k_n	1.13	1.08	1.06	1.05	1.04	1.03 .

So,

$$\tilde{\sigma}=k_n\sqrt{\frac{1}{n-1}\sum_{i=1}^{n}(x_i-\bar{x})^2}. \tag{4.3}$$

We have obtained estimates of the parameters of the normal distribution, but they are also random quantities: When the measurement is repeated, we obtain a different group of observations with different values of $\bar{x}$ and $\tilde{\sigma}$. The spread in these estimates can be characterized by their standard deviations $\sigma(\bar{x})$ and $\sigma(\tilde{\sigma})$. We already obtained above that $D[\bar{x}] = \sigma^2/n$. Therefore

$$\sigma(\bar{x}) = \sqrt{D[\bar{x}]} = \frac{\sigma}{\sqrt{n}}. \tag{4.4}$$

Since instead of $\sigma(\bar{x})$ we shall take $\tilde{\sigma}(\bar{x})$, we obtain the estimate $\tilde{\sigma}(\bar{x})$. Neglecting the value of k_n we arrive at the well-known formula

$$\tilde{\sigma}(\bar{x}) = \sqrt{\frac{\sum_{i=1}^{n} (x_i - \bar{x})^2}{n(n-1)}}. \tag{4.5}$$

Often $\tilde{\sigma}(\bar{x})$ is denoted by the symbol $S_{\bar{x}}$.

Uncertainty of estimation given in (4.5) depends on the number of measurements n and of the confidance probability α. E.g. for $n = 25$ and $\alpha = 0.80$ the uncertainty of this estimation is about 20%, for $n = 15$ and $\alpha = 0.80$ it is about 30%. The method of this computation is described below in item 4.4 (page 96).

Since the number of observations is rarely very large, the error in the determination of the standard deviation can be very significant. In any case, this error is significantly larger than the error owing to the biasness introduced into the estimate by extraction of the square root and eliminated by the correction factor k_n. For this reason, in practice this biasness can usually be neglected and the formula

$$\tilde{\sigma} = \sqrt{\frac{\sum_{i=1}^{n} (x_i - \bar{x})^2}{n-1}} \tag{4.6}$$

can be used instead of formula (4.3).

4.3. Outlying results

If in the group of measurement results, one or two differ sharply from the rest, and no slips of the pen, reading errors, and similar blunders have not been found, then it is necessary to check whether they are extreme events that should be excluded. The problem is solved by statistical methods based on the fact that the distribution to which the group of observations under study refers can be regarded as a normal distribution. The methodology for solving the problem and the computed tables are presented in the standard Ref. 10.

The solution scheme is as follows. An ordered series $x_1 < x_2 < \cdots < x_n$ is constructed from the obtained results. From all x_i we calculate $\bar{x}$ and S, and then t given by

$$t = \frac{\max|x_i - \bar{x}|}{S}, \tag{4.7}$$

where S is calculated using (4.6). Obviously, this will either be

$$t_1=\frac{\bar{x}-x_1}{S}, \tag{4.8a}$$

or

$$t_n=\frac{x_n-\bar{x}}{S}. \tag{4.8b}$$

Table A.3 of the Appendix gives the 0.5, 1, and 5 percentage points t_q of the corresponding unilateral check of the series $x_1,...,x_n$.

In performing the measurements we cannot foresee whether x_1 or x_n will be checked. Therefore we are interested in a bilateral check. In this case the critical value of t_q must be taken from the column of Table A.3 in which the significance level is two times lower than the value we adopted for checking our data.

If the value of $t_{1,n}$ that we calculated from Eq. (4.8) is greater than t_q, then the corresponding value of x_1 or x_n must be discarded: the probability of an observation giving $t > t_q$ is small; it is less than or equal to the adopted significance level.

Example 1. Current measurements gave the following data (the current strength in mA): 10.07, 10.08, 10.10, 10.12, 10.13, 10.15, 10.16, 10.17, 10.20, and 10.40. The value 10.40 differs sharply from the other values. We shall check to see whether or not it can be discarded. We shall use the criterion presented, though we do not have the data that would allow us to assume that these observations satisfy the normal distribution:

$$\bar{x}=10.16 \text{ mA},$$

$$S=0.094 \text{ mA},$$

$$t=\frac{10.40-10.16}{0.094}=2.55.$$

Let $q = 1\%$. In the column of Table A.3 with the significance level 0.5% for $n = 10$ we find $t_q = 2.48$. Since $2.55 > 2.48$, the observation 10.40 mA can be discarded.

4.4. Construction of confidence intervals

Having obtained the estimate $\tilde{A}$, it is of interest to determine by how much it can change in repeated measurements performed under the same conditions. This question is clarified by constructing the confidence interval for the true value of the measured quantity.

The *confidence interval* is the interval that includes, with a prescribed probability called the *confidence probability,* the true value of the measured quantity.

In our case the confidence interval can be constructed based on the Chebyshev inequalities[21]:

$$P\{|X-A|\geqslant t\sigma\}\leqslant\frac{1}{t^2}.$$

For the random quantity $\bar{x}$ we have

$$P\left\{|\bar{x}-A|\geqslant\frac{t\sigma}{\sqrt{n}}\right\}\leqslant\frac{1}{t^2}. \tag{4.9}$$

It is not necessary to know the form of the distribution of the observations, but it is necessary to know $\sigma[X]$. However, the intervals obtained with the help of the Chebyshev inequalities are found to be too large for practice, and they are not used.

If the distribution of the observations can be regarded as normal with a known standard deviation, then the confidence interval is constructed based on the expression

$$P\left\{|\bar{x}-A|\leqslant z_p\frac{\sigma}{\sqrt{n}}\right\}=\alpha,$$

where z_p is the quantile of the normalized normal distribution, corresponding to the selected confidence probability.

We shall show how to find the value of z_p, using the normalized Laplace function (Table A.2 of the Appendix).

Let $\alpha = 0.95$. With this probability the interval

$$\left[\bar{x}-z_p\frac{\sigma}{\sqrt{n}},\bar{x}+z_p\frac{\sigma}{\sqrt{n}}\right]$$

should include the true value of A. The probability that A falls outside this interval is equal to $1-\alpha = 0.05$. Since the normal distribution is symmetric, the probabilities that A exceeds the upper and lower limits of the interval are the same. Each of these probabilities is equal to $(1-\alpha)/2 = 0.025$. Therefore the quantile z_p is the quantile corresponding to the probability $(1-0.025) = 0.975$. It is obvious that this probability can be calculated based on the formula

$$P=1-\frac{1-\alpha}{2}=\frac{1+\alpha}{2}.$$

The Laplace function $\Phi(z)$ is related to the distribution function $F(z)$ by the relation $F(z) = 0.5 + \Phi(z)$.

Therefore in our example $\Phi(z) = F(z) - 0.5 = 0.975 - 0.5 = 0.475$. In Table A.2 we find the quantile $z_p = 1.96$ corresponding to the argument 0.475.

Often, on the other hand, the value of the quantile z_p is given and the corresponding probability p is found. For example, for $z_p = 1$ $\Phi(z) = 0.3413$ and $F(z) = \Phi(z) + 0.5 = 0.841$. Then $(1+\alpha)/2 = 1 - F(z) = 0.159$ and $\alpha = 0.682$. Analogously, for $z_p = 3$ we find $\Phi(z) = 0.49865$, $F(z) = 0.99865$, $(1+\alpha)/2 = 0.00135$, and $\alpha = 0.9973$.

In practice, however, the standard deviation is rarely known. Usually we know only its estimate S and, correspondingly, $S_{\bar{x}} = S/n$. Then the confidence intervals are constructed based on Student's distribution, which is the distribution of the random quantity

$$t=\frac{\bar{x}-A}{S_{\bar{x}}}, \tag{4.10}$$

where $S_{\bar{x}}$ is the estimate of the standard deviation of the arithmetic-mean value $\bar{x}$, calculated from formula (4.5).

The confidence interval $[\bar{x} - t_q S_{\bar{x}}, \bar{x} + t_q S_{\bar{x}}]$ corresponds to the probability

$$P\{|\bar{x} - A| \leqslant t_q S_{\bar{x}}\} = \alpha,$$

where t_q is the q percent point of Student's distribution; the value of t_q is found from Table A.4 based on the degree of freedom $\nu = n - 1$ and the significance level $q = 1 - \alpha$.

The confidence probability should not be too low, but even for a value that is very high there are usually not enough reliable starting data available. In measurement practice, the confidence probability is increasingly often set equal to 0.95.

Existing methods make it possible to check the admissibility of the hypothesis that the observations are described by a normal distribution and therefore the hypothesis that the Student's distribution is admissible (see Sec. 4.5). The significance level q, used for constructing the confidence probability, has to be consistent with the significance level adopted when checking the normality of the distribution, but this problem does not yet have a definite solution.

In practice, confidence intervals are constructed based on Student's distribution, often without checking its admissibility. The fact that, in the process, as a rule, no misunderstandings arise indirectly confirms the opinion stated above that real distributions are truncated distributions that are narrower than normal distributions.

Sometimes confidence intervals are constructed for the standard deviation. For this the χ^2 distribution, presented in Table A.5, is employed. The confidence interval with the limits $(\sqrt{n-1}/\chi_L)\tilde{\sigma}$ and $(\sqrt{n-1}/\chi_U)\tilde{\sigma}$ for the probability

$$P\left\{\frac{\sqrt{n-1}}{\chi_L}\tilde{\sigma} < \sigma < \frac{\sqrt{n-1}}{\chi_U}\tilde{\sigma}\right\} = \alpha$$

is found as follows. Table A.5 gives the probabilities $P\{\chi^2 > \chi_q^2\}$. The value of χ_U^2 is found from the table for $p_U = (1 + \alpha)/2$ and the value of χ_L^2 is found for $p_L = (1 - \alpha)/2$.

For example, let $\tilde{\sigma} = 1.2 \times 10^{-5}$ and $n = 10$. Take $\alpha = 0.90$. Then $p_U = (1 + 0.9)/2 = 0.95$ and $p_L = (1 - 0.9)/2 = 0.05$. The degree of freedom $\nu = 10 - 1 = 9$. From Table A.5 we find $\chi_U^2 = 3.325$ and $\chi_L^2 = 16.92$. The confidence interval will be

$$\left[\frac{\sqrt{10-1}}{\sqrt{16.92}} \times 1.2 \times 20^{-5}, \frac{\sqrt{10-1}}{\sqrt{3.325}} \times 1.2 \times 10^{-5}\right]$$

i.e., $[0.88 \times 10^{-5}, 2.0 \times 10^{-5}]$. The confidence probability in this case can be taken to be less than the confidence probability when constructing the confidence interval for the true value of the measured quantity. Often $\alpha = 0.70$ is sufficient.

We return to Chebyshev's inequality, which is very attractive because it is not related to the form of the distribution function of the observations. The measured quantity can practically always be estimated by the arithmetic mean (though in the case when the distribution differs from a normal distribution, the estimate will not be the most effective estimate), and if instead of the standard deviation its estimate is employed, then the limits of the error of this result can be estimated with the help of Chebyshev's inequality.

We shall transform the inequality (4.9) so that it would determine the probability that a deviation of the random quantity from its true value is less than $t\sigma$. The random quantity here is the arithmetic mean $\bar{x}$. After simple transformations we obtain

$$P\left\{|\bar{x}-A| \leqslant t\frac{\sigma}{\sqrt{n}}\right\} \geqslant 1-\frac{1}{t^2}.$$

The variance of the results of measurements can be estimated using formula (4.6).

The coefficient t can be calculated based on a prescribed confidence probability α from the relation $\alpha = 1 - 1/t^2$, which gives

$$t=\frac{1}{\sqrt{1-\alpha}}.$$

If the distribution of the random errors can be assumed to be symmetric relative to A, then the confidence interval can be narrowed somewhat,[21] using the inequality

$$P\left\{|\bar{x}-A| \leqslant t\frac{\sigma}{\sqrt{n}}\right\} \geqslant 1-\frac{4}{9}\frac{1}{t^2}.$$

Now

$$t=\frac{2}{3\sqrt{1-\alpha}}.$$

Unfortunately, the confidence intervals constructed in this manner are still only approximate, since the effect of replacing the standard deviation by its estimate is not taken into account. Moreover, as we have already mentioned, the intervals obtained in the process are too wide, i.e., the uncertainty is exaggerated.

Confidence intervals should not be confused with statistical tolerance intervals.

The interval that with prescribed probability α contains not less than a prescribed fraction p_0 of the entire collection of values of the random quantity (population) is said to be the *statistical tolerance interval.* Thus the statistical tolerance interval is the interval for a random quantity, and this distinguishes it in principle from the confidence interval that is constructed in order to cover the values of a nonrandom quantity.

If, for example, the sensitivity of a group of strain gauges is measured, then the obtained data can be used to find the interval with limits l_1 and l_2 in which with prescribed probability α the sensitivity of not less than the fraction p_0 of the entire batch (or the entire collection) of strain gauges of the given type will fail. This is the statistical tolerance interval. Methods for constructing this tolerance interval can be found in books on the theory of probability and mathematical statistics, for example, Refs. 34 and 60.

One must also guard against confusing the limits of statistical tolerance and confidence intervals with the tolerance range for the size of some parameter. The tolerance or the limits of the tolerance range are, as a rule, determined prior to the fabrication of a manufactured object, so that the objects for which the value of the parameter of interest falls outside the tolerance range are unacceptable and are discarded. In other words, the limits of the tolerance range are strict limits that are not associated with any probabilistic relations.

The statistical tolerance interval, however, is determined by objects that have already been manufactured, and its limits are calculated so that with a prescribed probability the parameters of a prescribed fraction of all possible manufactured objects fall within this interval. Thus the limits of the statistical tolerance interval, as also the limits of the confidence interval, are random quantities, and this is what distinguishes them from the tolerance limits or tolerance which are nonrandom quantities.

4.5. Methods for testing hypotheses about the form of the distribution function of a random quantity

The problem is usually posed as follows: for a group of measurement results it is hypothesized that these results can be regarded as realizations of a random quantity with a distribution function having a chosen form. Then this hypothesis is checked by the methods of mathematical statistics and is either accepted or rejected.

For a large number of observations ($n > 50$) Pearson's test (χ^2 test) for grouped observations and the Kolmogorov–Smirnov test for nongrouped observations are regarded as the best tests. These methods are described in many books devoted to the theory of probabilities and statistics. For example, see Refs. 21, 34, 52, and 60.

We shall discuss the χ^2 test, and for definiteness we shall check the data on the corresponding normal distribution.

The idea of this method is to monitor the deviations of the histogram of the experimental data from the histogram with the same number of intervals that is constructed based on the normal distribution. The sum of the squares of the differences of the frequencies over the intervals must not exceed the values of χ^2 for which tables were constructed as a function of the significance level of the test q and the degree of freedom $\nu = L - 3$, where L is the number of intervals.

The calculations are performed as follows.

(1) The arithmetic mean of the observations and an estimate of the standard deviations are calculated.

(2) Measurements are grouped according to intervals. For about 100 measurements, five to nine intervals are normally taken. For each interval the number of measurements $\tilde{\varphi}_i$ falling within the interval is calculated.

(3) The number of measurements for each interval, which theoretically corresponds to the normal distribution, is calculated. For this the range of data is first centered and normalized.

Let $x_{\min} = a_0$ and $x_{\max} = b_0$, and divide the range $[a_0, b_0]$ into L intervals. The length of each interval is $h_0 = (b_0 - a_0)/L$.

Centering and normalization are done according to the formula

$$x_{ic} = \frac{x_{i0} - \bar{x}}{\tilde{\sigma}}.$$

For example, the transformed limits of the range of the data for us will be as follows:

$$a_c = \frac{a_0 - \bar{x}}{\tilde{\sigma}}, \quad b_c = \frac{b_0 - \bar{x}}{\tilde{\sigma}}.$$

The length of the transformed interval $h_c = (b_c - a_c)/L$. Then we mark the limits $\{z_i\}$, $i = 0,1,...,L$, of all intervals of the transformed range $[a_c, b_c]$:

$$z_0 = a_c, \ z_1 = a_c + h_c, \ z_2 = a_c + 2h_c, \ ..., \ z_L = a_c + Lh_c = b_c.$$

Now we calculate the probability that a normally distributed random quantity falls within each interval:

$$p_i = \frac{1}{2\pi} \int_{z_i}^{z_{i+1}} e^{-x^2/2} dx.$$

After this we calculate the number of measurements that would fall within each interval if the population of measurements is normally distributed:

$$\varphi_i = p_i n.$$

(4) If less than five measurements fall within some interval, then this interval in both histograms is combined with the neighboring interval. Then the degree of freedom $\nu = L - 3$, where L is the total number of intervals (if the intervals are enlarged, then L is the number of intervals after the enlargement), is determined.

(5) The indicator χ^2 of the difference of frequencies is calculated:

$$\chi^2 = \sum_{i=1}^{L} \chi_i^2, \quad \chi_i^2 = \frac{(\tilde{\varphi}_i - \varphi_i)^2}{\varphi_i}.$$

(6) The significance level of the test q is chosen. The significance level must be sufficiently small so that the probability of rejecting the correct hypothesis (committing an error of the first kind) would be small. On the other hand, too small a value of q increases the probability adopted for the incorrect hypothesis, i.e., for committing an error of the second kind.

From the significance level q and a degree of freedom ν in Table A.5, we find the limit of the critical region χ_q^2, so that $P\{\chi^2 > \chi_q^2\} = q$.

The probability that the value obtained for χ^2 exceeds χ_q^2 is equal to q and is small. For this reason, if it turns out that $\chi^2 > \chi_q^2$, then the hypothesis that the distribution is normal is rejected. If $\chi^2 < \chi_q^2$, then the hypothesis that the distribution is normal is accepted.

The smaller the value of q, the larger the value of χ_q^2 for the same value of ν is and the more easily the condition $\chi^2 < \chi_q^2$ is satisfied and the hypothesis being tested is accepted. But, in this case, the probability of committing an error of the second kind increases. For this reason, q should not be taken to be less than 0.01. For too large a value of q, as pointed out above, the probability of an error of the first kind increases and, in addition, the sensitivity of the test decreases. For example, for $q = 0.5$ the value of χ^2 can be both greater and less than χ_q^2 with equal probability, and therefore it is impossible to accept or reject the hypothesis.

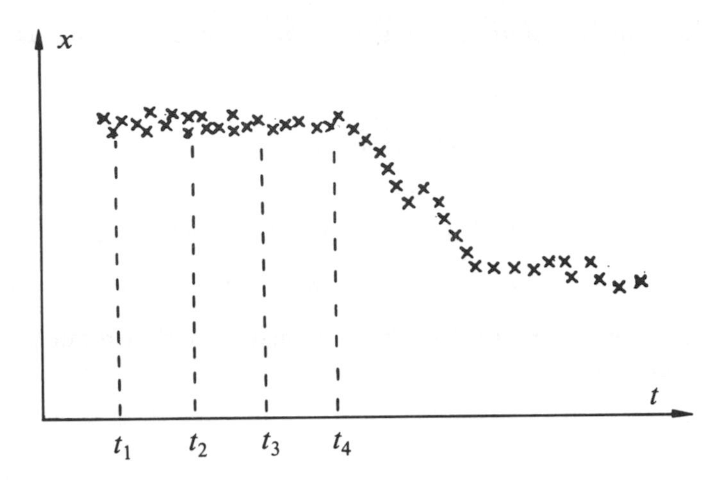

FIG. 4.1. Example of a sequence of single measurement results obtained in an unstable measurement.

In order to achieve a uniform solution of the problem at hand, it is desirable to unify the significant levels adopted in metrology. To this end, we can try to limit the choice of significant level to the interval $0.01 \leqslant q \leqslant 0.1$.

It should be noted that the test examined above makes it possible to check the correspondence between the empirical data and any theoretical distribution, not only a normal distribution. This test, however, as also with, by the way, other goodness-of-fit tests, does not make it possible to establish the form of the distribution of the observations; it only makes it possible to check whether or not the observations conform to a normal or some other previously selected distribution.

4.6. Methods for testing sample homogeneity

Measurements with large random errors require careful attention. One must make sure that the obtained results are statistically under control, stable, i.e., that the measurement results cluster around the same central value and have the same variance. If the measurement method and the object of investigation have been little studied, then the measurements must be repeated until one is sure that the results are stable.[28] This determines the duration of the investigation and the required number of measurements.

The stability of measurements is often estimated intuitively based on prolonged observations. However, there exist mathematical methods that are useful for solving this problem, so-called methods for testing homogeneity.

A necessary condition is that indications of homogeneity must be present, but this is not sufficient for homogeneity in reality, since groups of measurements can be incorrectly or unsuccessfully chosen.

Figure 4.1 shows the results of measurements of some quantities, presented in the sequence in which they were obtained. Consider three groups of measurements performed in the time intervals $t_2 - t_1$, $t_3 - t_2$, and $t_4 - t_3$. They apparently will be homogeneous. Meanwhile, subsequent measurements would differ significantly from the first measurements, and on the whole the results obtained from the first group of measurements will give a picture of a stable, statistically under control, measurement, which is actually not the case.

The choice of groups for monitoring homogeneity remains a problem for the specialist–experimenter, just as does the problem of separating one group from another. In general, it is best to have of the order of 10 measurements in a group (according to Ref. 23, from 5 to 10 measurements) and it is better to have several such groups than two groups with a large number of measurements.

Once the groups have been reliably determined to be homogeneous, they can be combined and later regarded as one group of data.

We shall consider first the most common parametric methods for testing homogeneity. These methods are based on the normal distribution of a population. For this reason, each group of data must first be checked for normality.

The admissibility of differences between estimates of the variances is checked with the help of R. Fisher's test in the case of two groups of observations and M. Bartlett's test if there are more than two groups. We shall present both methods.

Let the unbiased estimates of the variances of these groups be S_1^2 and S_2^2, where $S_1^2 > S_2^2$. The number of observations in the groups is n_1 and n_2, so that the degrees of freedom are, respectively, $\nu_1 = n_1 - 1$ and $\nu_2 = n_2 - 1$. We form the ratio

$$F = \frac{S_1^2}{S_2^2}.$$

Next from Tables A.6 and A.7, where the probabilities

$$P\{F > F_q\} = q$$

for different degrees of freedom ν_1 and ν_2 are presented, we choose the value F_q.

The hypothesis is accepted, i.e., estimates of the variances can be regarded as corresponding to one and the same variance, if $F < F_q$. The significance level is equal to $2q$.

Now assume that there are L groups and for them unbiased estimates of the variances of groups of observations are known, $S_1^2, \ldots, S_L^2$ $(L > 2)$, and each group has $\nu_i = n_i - 1$ degrees of freedom; in addition, all $\nu_i > 3$. The test of the hypothesis that the variances of the groups are equal is based on the statistics

$$M = N \ln\left(\frac{1}{N} \sum_{i=1}^{L} \nu_i S_i^2\right) - \sum_{i=1}^{L} \nu_i \ln S_i^2,$$

where $N = \sum_{i=1}^{L} \nu_i$.

If the hypothesis that the variances are equal is correct, then the ratio

$$\chi_1^2 = \frac{M}{1 + \dfrac{1}{3(L-1)}\left(\sum_{i=1}^{L} \dfrac{1}{\nu_i} - \dfrac{1}{N}\right)}$$

is distributed approximately as χ^2 with $\nu = L - 1$ degrees of freedom.[34]

Given the significance level q, from Table A.5 we find χ_q^2, such that $P\{\chi^2 > \chi_q^2\} = q$. If the inequality $\chi_1^2 < \chi_q^2$ is satisfied, then differences between the estimates of the variances are admissible.

The admissibility of differences between the arithmetic means is also checked differently in the case of two or more groups of observations. We shall first examine the comparison of the arithmetic means for two groups of observations, when there are many observations, so that each estimate of the variances can be assumed to be equal to its own variance.

We denote by $\bar{x}_1$, σ_1^2, and n_1 the data belonging to one group and by $\bar{x}_2$, σ_2^2, and n_2 the data belonging to the other group. We form the difference $\bar{x}_1 - \bar{x}_2$ and estimate its variance:

$$\sigma^2(\bar{x}_1-\bar{x}_2)=\frac{\sigma_1^2}{n_1}+\frac{\sigma_2^2}{n_2}.$$

Next, having chosen a definite significance level q we find $\alpha = 1 - q$ and from the Table A.2 we find the argument $z_{\alpha/2}$ of the Laplace function corresponding to the probability $\alpha/2$. A difference between the arithmetic means is regarded as acceptable, if

$$|\bar{x}_1-\bar{x}_2| \leqslant z_{\alpha/2}\sigma(\bar{x}_1-\bar{x}_2).$$

If the variances of the groups are not known, then the problem can be solved only if both groups have the same variances (the estimates of this variance $\tilde{\sigma}_1^2$ and $\tilde{\sigma}_2^2$ can, naturally, be different). In this case t is calculated as

$$t=\frac{|\bar{x}_1-\bar{x}_2|}{\sqrt{(n_1-1)\tilde{\sigma}_1^2+(n_2-1)\tilde{\sigma}_2^2}}\sqrt{\frac{n_1n_2(n_1+n_2-2)}{n_1+n_2}}.$$

Next, given the significance level q, from Table A.4 for Student's distribution with $\nu = n_1 + n_2 - 2$ degrees of freedom, we find t_q. A difference between the arithmetic means is regarded as admissible if $t < t_q$.

If the number of groups is large, the admissibility of differences between the arithmetic means is checked with the help of R. Fisher's test. It is first necessary to check that all groups have the same variance.

Fisher's method consists of comparing estimates of the intergroup variance S_L^2 and the average variance of the groups $\overline{S^2}$:

$$S_L^2=\frac{1}{L-1}\sum_{i=1}^{L} n_i(\bar{x}_i-\bar{x})^2,$$

where

$$\bar{x}=\frac{\sum_{i=1}^{L} n_i\bar{x}_i}{N}, \quad N=\sum_{i=1}^{L} n_i$$

(the estimate S_L^2 has $\nu_1 = L - 1$ degrees of freedom);

$$\overline{S}^2=\frac{1}{N-L}\sum_{i=1}^{L}\sum_{j=1}^{n_i}(x_{ji}-\bar{x}_i)^2$$

(the degree of freedom $\nu_2 = N - L$).

Both estimates of the variances have a χ^2 distribution with ν_1 and ν_2 degrees of freedom, respectively. Their ratio has Fisher's distribution with the same degrees of freedom.

A spread of the arithmetic means is acceptable if $F = S_L^2/\bar{S}^2$ for the selected probability α lies within the interval from F_L to F_U:

$$P\{F_L \leqslant F \leqslant F_U\} = \alpha.$$

The upper limits of Fisher's distribution F_U are presented in Tables A.6 and A.7; the lower limits are found from the relation $F_L = 1/F_U$. If the significance levels in finding F_U and F_L are taken to be the same $q_1 = q_2 = q$, then the common significance level of the test will be $2q$ and

$$\alpha = 1 - 2q.$$

A method for checking the admissibility of the spread in the arithmetic means of the groups when the variances of the groups are different has also been developed, but it is more complicated.

It should be noted that a significant difference between the arithmetic means could indicate that there exists a constant systematic error in the observational results of one or another group as well as the fact that the interesting parameter of the model used to describe the object of investigation is variable. The latter means that the postulate β is not satisfied and, therefore, measurements cannot be performed with the required accuracy.

We shall now discuss nonparametric methods for testing homogeneity. These methods do not require any assumptions about the distribution function of the population and are widely used in mathematical statistics.[35,52,53,60]

Wilcoxon and Mann–Whitney tests. Assume that we have two samples: $\{x_i\}$, $i = 1,...,n$, and $\{y_j\}$, $j = 1,...,n_2$, and let $n_1 \leqslant n_2$. We check the hypothesis H_0: $F_1 = F_2$, where F_1 and F_2 are the distribution functions of the random quantities X and Y, respectively.

The sequence of steps in checking H_0 is as follows. Both samples are combined and an ordered series is constructed from $N = n_1 + n_2$ elements, i.e., all x_i and y_j are arranged in increasing order, irrespective of the sample to which one or another value belongs. Next, all terms of the ordered series are enumerated. The numbers of terms are called their ranks.

When the numerical values of several elements are the same, to each of them a rank equal to the arithmetic mean of the ranks of the corresponding values is assigned.

Next the sum of the ranks of all elements of the smaller sample is calculated. The sum T obtained is then compared with the critical value T_q.

For small values of n_1 and n_2 tables $T_q(n_1,n_2)$ are given in most modern books on statistics. For $n_1,n_2 > 25$ the critical value T_q can be calculated using the normal distribution $N(m_1,\sigma^2)$: $z = (T - m_1)/\sigma$, where $m_1 = [n_1(N + 1) - 1]/2$ and $\sigma^2 = [n_1n_2(N + 1)]/12$. Thus

$$T_q = \frac{n_1(N+1)-1}{2} + z_{1-q}\sqrt{\frac{n_1n_2(N+1)}{12}},$$

where z_{1-q} is the quantile of the level $(1-q)$ of the standard normal distribution $N(0,1)$.

The hypothesis H_0 is rejected with significance level q if $T > T_q$. This is Wilcoxon's test.

A variant of Wilcoxon's test is the Mann–Whitney test. This test is based on calculations of the so-called inversions U, which are more difficult to calculate than the ranks T. But U and T are uniquely related, and the values of U can be found from the values of T:

$$U_1 = T_1 - \frac{n_1(n_1+1)}{2}, \quad U_2 = n_1 n_2 + \frac{n_2(n_2+1)}{2} - T_2.$$

The parameter U is compared with the critical value U_q, for which tables or a calculation similar to the one performed above are employed.

Given q, we find from the standard normal distribution $N(0,1)$ the quantile z_{1-q} and calculate U_q:

$$U_q = \frac{n_1 n_2 - 1}{2} + z_{1-q}\sqrt{\frac{n_1 n_2 (N+1)}{12}}.$$

The hypothesis H_0 is rejected if the smaller of U_1 and U_2 is greater than U_q.

For the Siegel–Tukey test, as in the case of Wilcoxon's test, two samples x_i and y_j are studied and likewise the hypothesis H_0: $F_1 = F_2$ is checked. All $N = n_1 + n_2$ values of x_i and y_j are likewise arranged in increasing order, with an indication that each term belongs to the sequence X or Y. The ranks are assigned as follows: rank 1 to the first term, rank 2 to the last (Nth) term, rank 3 to the $(N-1)$st term, rank 4 to the second term, rank 5 to the third term, rank 6 to the $(N-2)$nd term, etc. Values that are equal to one another are assigned the average rank. Next, the sums of the ranks R_1 and R_2, referring to the samples $\{x_i\}$ and $\{y_j\}$ and the normalized variable z defined as

$$z = \frac{\left| R_1 - \frac{n_1(N+1)}{2} \right| - 0.5}{\sqrt{\frac{n_1 n_2 (N+1)}{12}}}$$

are calculated. For significance level q, the hypothesis H_0 is rejected if $z > z_{1-q}$, where z_{1-q} is a quantile of order $(1-q)$ of the standard normal distribution $N(0,1)$.

The calculations of R_1 and R_2 can be checked with the help of the relation

$$R_1 + R_2 = \frac{N(N+1)}{2}.$$

The Wilcoxon test is based on comparing the average values of two samples, while the Siegel–Tukey test is based on estimates of the variances. For this reason, these two tests supplement one another.

As an example of the application of these tests, Table 4.1 gives data from calculations on a check of the homogeneity of two batches of 160 ammeters for a

TABLE 4.1. The example of rank determination.

	Number of instruments with a fixed error in the sample			Wilcoxon's test		Siegel–Tukey test	
Value of the error	x	y	$x+y$	Average rank of a given value of the error	Sum of ranks for a given value of the error in the sample x	Average rank of a given value of the error	Sum of ranks for a given value of the error in the sample x
−0.5	1	1	2	1.5	1.5	2.5	2.5
−0.4	3	0	3	4	12	7.3	22
−0.3	3	0	3	7	21	13.7	41
−0.25	1	0	1	9	9	17	17
−0.2	13	5	18	18.5	240.5	36.5	474.5
−0.15	2	2	4	29.5	59	58.5	117
−0.1	10	8	18	40.5	405	80.5	805
−0.05	3	2	5	52	156	103.6	310.8
0	15	28	43	76	1140	151.5	2272.5
0.05	5	5	10	102.5	512.5	204.5	1022.5
0.1	26	35	61	138	3588	573.5	7108.4
0.15	7	4	11	174	1218	293.5	2054.5
0.2	34	41	75	217	7378	207.5	7055
0.25	1	3	4	256.5	256.5	128.5	128.5
0.3	17	11	28	272.5	4632.5	96.5	1640.5
0.4	13	11	24	298.5	3880.5	44.5	578.5
0.45	1	1	2	311.5	311.5	18.5	18.5
0.5	4	2	6	315.5	1262	10.5	42
0.6	0	1	1	319	0	3	0
0.8	1	0	1	320	320	2	2

moving-iron instrument э59 with respect to the error at marker 30 of the graduated scale.[50] The experiment is described in Sec. 2.6.

Following the recommendations made above, we obtain $T = 25\,403$. Let $\alpha = 0.05$. Then $z_{0.95} = 1.96$ and

$$T_q = \frac{160 \times 321}{2} - 0.5 + 1.96\sqrt{\frac{160 \times 160 \times 321}{12}} = 27\,620.$$

Since $25\,403 < 27\,620$, the hypothesis that the samples are homogeneous is accepted based on Wilcoxon's test.

Consider now the Siegel–Tukey test. According to the data in the table $R_1 = 23\,713$. Analogous calculations give $R_2 = 27\,647$. Taking R_1 ($R_1 < R_2$) we obtain

$$z=\frac{\left|23\,713-\dfrac{160\times 321}{2}\right|-0.5}{\sqrt{\dfrac{160\times 160\times 321}{12}}}=2.3.$$

We chose $q=0.05$ and therefore $z_{0.95}=1.96$. Since $z>z_{0.95}$, the hypothesis that the samples are homogeneous is rejected based on the Siegel–Tukey test.

4.7. Estimation of the parameters of non-normal distributions

In studying the analysis of non-normally distributed observational results, it should be noted that, working with the effectiveness, the arithmetic mean of the observational results can practically always be used as an estimate of the measured quantity. The error of the result obtained in this manner can in principle be estimated with the help of the Chebyshev inequality (see above).

If the distribution function of the observations is known, then the problem of estimating the true value of the measured quantity can be solved based on the principle of maximum likelihood. In the general case, when the distribution function is known, the best estimate of the measured quantity is given by Pitman's formula:

$$\tilde{A}=\frac{\displaystyle\int_{-\infty}^{+\infty} uf(x_1-u)\cdots f(x_n-u)du}{\displaystyle\int_{-\infty}^{+\infty} f(x_1-u)\cdots f(x_n-u)du}.$$

The variance of Pitman's estimate does not exceed the variance of the estimate obtained based on the maximum likelihood principle.

Most often, however, if the distribution is not a normal distribution, then the distribution function cannot be determined accurately. For this reason, in the last few years methods for solving this problem that do not require precise knowledge of the distribution function have been under development in mathematics. These methods are reviewed in Ref. 37. Based on this work as well as Ref. 45 the following scheme for solving the problem can be suggested.

An ordered series, i.e., a series in which the observational results are arranged in increasing order

$$x_1\leqslant\cdots\leqslant x_i\leqslant\cdots\leqslant x_n,$$

is constructed from the observational results. The distribution of the experimental results is characterized by the number $\varkappa$ calculated according to the formula

$$\varkappa=\frac{\sum_{i=1}^{n}(x_i-\bar{x})^4}{nS^4},$$

where S^2 is the sample variance.

For $\varkappa\geqslant 4$ it is best to take as an estimate of the measured quantity an estimate of the median, obtained based on the ordered series constructed:

$$\tilde{A}=\begin{cases}\frac{1}{2}(x_{n/2}+x_{n/2+1}), & \text{if } n \text{ is odd number,}\\ x_{(n+1)/2}, & \text{if } n \text{ is even number.}\end{cases}$$

If $2.5<\varkappa<4$, then the measured quantity is best estimated by the arithmetic mean of the observational results.

For distributions approaching the uniform distribution $(1.8\leqslant\varkappa\leqslant2.5)$, the most useful estimate of the measured quantity is the half-sum of the maximum and minimum values:

$$\tilde{A}=\frac{x_1+x_n}{2}.$$

The critical values of $\varkappa$ were chosen somewhat arbitrarily.

The accuracy of the result can be characterized by the length of the nonparametric confidence interval.

If x_k and x_{n-k+1} are, respectively, the kth and $(n-k+1)$st ordinal statistics, formed based on a sample belonging to a population with a continuous symmetric distribution, then the interval $[x_k, x_{n-k+1}]$ is the confidence interval for the center of the distribution of the random quantity, and the confidence probability is equal to

$$P\{x_k<A<x_{n=k+1}\}=\frac{1}{2^n}\sum_{i=k}^{n-k}C_n^i.$$

In particular, for

$$k=2,\quad P\{x_2<A<x_{n-1}\}=1-\frac{n-1}{2^{n-1}},$$

$$k=3,\quad P\{x_3<A<x_{n-2}\}=1-\frac{n^2+n+2}{2^n}.$$

For $k>3$ the working formulas become much more complicated. But for $k=4$ and 5, approximate relations, presented in the paper Ref. 45 mentioned above, can be used:

$$k=4,\quad P\{x_4<A<x_{n-3}\}\approx1-\frac{0.17n^3}{2^{n-1}},$$

$$k=5,\quad P\{x_5<A<x_{n-4}\}\approx1-\frac{0.037n^4}{2^{n-1}}.$$

The only assumption used in these formulas is the assumption that the distribution functions of the observations are symmetric. This is very often a natural assumption.

TABLE 4.2. For calculation of the statistical characteristics based on the results of comparison of measures of mass.

x_i (g)	$x_{i0}\times 10^6$	$(x_{i0}-\bar{x}_{i0})\times 10^6$	$(x_{i0}-\bar{x}_{i0})^2\times 10^{12}$
999.998 738	738	+ 17	289
999.998 699	699	− 22	484
999.998 700	700	− 21	441
999.998 743	743	+ 22	484
999.998 724	724	+ 3	9
999.998 737	737	+ 16	256
999.998 715	715	− 6	36
999.998 738	738	+ 17	289
999.998 703	703	− 18	324
999.998 713	713	− 8	64
Sum	7210	0	2676

4.8. Example: Analysis of measurement results in comparisons of measures of mass

As a result of comparing the measure of a 1 kg mass with the reference standard measure of mass with the same rating, the group of measurement results presented in column 1 of Table 4.2 was obtained.[51] Column 2 gives the values of $x_{i0} = (x_i - 999.998\,000) \times 10^3$, and columns 3 and 4 give the results of auxiliary calculations.

The measurement was performed by the methods of precise weighing, which eliminated the error caused by the fact that the arms of the balances were not equal. Thus it can be assumed that there are no systematic errors. The random errors are assumed to correspond to the normal probability distribution.

We assume the measured mass is equal to the arithmetic mean determined according to the formula

$$\bar{x}=999.998\,000+\bar{x}_{i0}=999.998\,721 \text{ g}.$$

Next we calculate using formula (4.6) and the data in column 4 an estimate of the standard deviation of the observations

$$S=\sqrt{\frac{2676}{9}\times 10^{-12}}\approx 17\times 10^{-6} \text{ g}.$$

From formula (4.5) we calculate an estimate of the standard deviation of the arithmetic mean

$$S_{\bar{x}}=\frac{17\times 10^{-6}}{\sqrt{10}}=5\times 10^{-6} \text{ g}.$$

We shall now find the uncertainty of the result. Let $\alpha = 0.95$, then, using Student's distribution (Table A.4), we find the coefficient t_q. The degree of freedom

$\nu = 10 - 1 = 9$, and $q = 1 - \alpha = 0.05$. Therefore $t_{0.05} = 2.26$. In accordance with the formula (4.10) we obtain the uncertainty of measurement result:

$$U_{0.95} = 2.26 \times 5 \times 10^{-6} = 11 \times 10^{-6} \text{ g}.$$

Thus the mass m of the measure studied lies in the interval

$$999.998\,710 \text{ g} \leqslant m \leqslant 999.998\,732 \text{ g}.$$

The result obtained can be written more compactly as

$$m_{0.95} = 999.998\,721 \pm 11 \times 10^{-6} \text{ g}.$$

Chapter 5

Direct measurements

5.1. Relation between single and multiple measurements

The classical theory of measurement errors is constructed based on the well-developed statistical methods and pertains to multiple measurements. In practice, however, the overwhelming majority of measurements are single measurements, and however strange it may seem, for this class of measurements there is no accepted method for estimating errors.

In searching for a well-founded method for estimating errors in single measurements it is first necessary to establish the relation between single and multiple measurements.[24,25]

At first glance it seems natural to regard single measurements as a particular case of multiple measurements, when the number of measurements is equal to 1. Formally this is correct, but it does not give anything, since statistical methods do not work when $n = 1$. In addition, the question of when one measurement is sufficient remains open. In the approach examined, to answer this question—and this is the fundamental question—it is first necessary to perform a multiple measurement, and then, analyzing the results, to decide whether or not a single measurement was possible. But such an answer is in general meaningless: a multiple measurement has already been performed and nothing is gained by knowing, for example, that in a given case one measurement would have sufficed. Admittedly, it can be objected that such an analysis will make it possible not to make multiple measurements when future such measurements are performed. Indeed, that it is what is done, but only when preliminary measurements are performed, i.e., in scientific investigations when some new object is studied. This is not done in practical measurements.

When it is necessary to measure, for example, the voltage of some source with a given accuracy, a voltmeter with suitable accuracy is chosen and the measurement is performed. If, however, the numbers on the voltmeter indicator dance about, then it is impossible to perform a measurement with the prescribed accuracy, and the measurement problem must be reexamined rather than performing a multiple measurement.

For practical applications, we can state the opinion that single measurements are well founded by experience, concentrated in the construction of the corresponding measuring instruments, and that measuring instruments are manufactured so that single measurements could be performed.

From the foregoing assertion there follows a completely different point of view

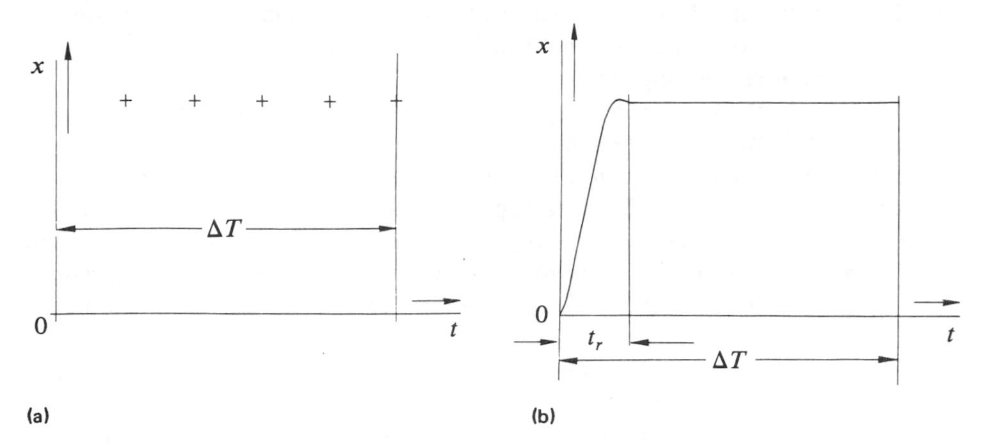

FIG. 5.1. Results of measurements in the case of (a) a multiple measurement and in (b) a single measurement (a motion picture film of the indications of the measuring instrument).

regarding the relation between single and multiple measurements, namely, that single measurements are the primary, basic form of measurement while multiple measurements are derived from single measurements. Multiple measurements are performed when necessary, based on the formulation of the measuring problem. It is interesting that these problems are known beforehand; they can even be enumerated. Namely, multiple measurements are performed as follows:

(a) When investigating a new phenomenon or a new object and relationships between the quantities characterizing the object as well as their connection with other physical quantities are being determined, or briefly, when preliminary measurements, according to the classification given in Chap. 1, are performed.

(b) When measuring the average value of some parameter, corresponding to the goal of the measuring problem itself.

(c) When the effect of random errors of measuring instruments must be reduced.

(d) In measurements for which measuring instruments have not yet been developed.

Of the four cases presented above, the first is typical for investigations in science and the third is typical for calibration practice.

There is another point of view, namely, that any measurement must be a multiple measurement, since otherwise it is impossible to judge the measurement process and its stability and to estimate its inaccuracy.

We cannot agree with this opinion. First of all, it contradicts practice. Second, it also does not withstand fundamental analysis. Imagine that one and the same constant quantity is measured simultaneously with the help of multiple and single methods of measurement. In both cases the measurements are performed using the same instrument whose response time is t_r.

In Fig. 5.1(a) the dots represent the results of single measurements comprising a multiple measurement and Fig. 5.1(b) shows a photorecording of the indications of the instrument in a single measurement.

A single measurement makes it possible to obtain the value of the measurand immediately after the response time t_r of the measuring instrument. If it is desirable to check the stability of the measurement, then the observation must be continued. The process of measurement is stable if the readings of the instrument over a chosen time ΔT do not change appreciably.

The reading of the instrument gives the estimation $\tilde{A}$ of a measurand A. Of course, $\tilde{A} \neq A$. Methods for calculating errors and uncertainty of the results of single measurements are given later in this chapter. Thus, in this case a single measurement is sufficient to obtain the measurement result and to estimate its inaccuracy. As to the stability of the measurement process, a single measurement allows one to make a better judgment than a multiple measurement because the latter represents only separate moments of the process, while the former gives the whole continuous picture.

The foregoing example does not say that a single measurement is better than a multiple measurement. It says only that a multiple measurement should not be performed when a single measurement can be performed. But when a multiple measurement is necessary, one single measurement is useless, and in this case and in this sense a multiple measurement is better than a single measurement.

So, single measurements must be regarded as independent and, in addition, the basic form of measurement. Correspondingly, the problem of developing methods for estimating the errors of single measurements must be regarded as an independent and important problem of the theory of measurement errors.

This is a good point at which to discuss another aspect of the question at hand. In many fields of measurements, modern measuring devices can operate so fast that over the time allotted for measurements, say, 1 s, hundreds of measurements can be performed. Carrying out all these measurements and averaging their results, we employ all of the time allotted for measurement and, thanks to this, we reduce correspondingly the effect of interference and noise.

Consider now an analog or a digital instrument having the same accuracy as a fast measuring device, but with a response time equal to the time allotted to the measurement, i.e., in our case, 1 s. Owing to the time constant of the instrument, the effect of interference and noise will be suppressed to the same degree as for discrete averaging in the first case, i.e., we shall obtain the same result.

In other words, the measurement time is of fundamental importance, and there is no significance in how the interference and noise are filtered—in the discrete or analog form—over this time. In practice, discrete averaging is often more convenient, since in this case the averaging time can be easily changed.

5.2. Identification and elimination of systematic errors

Taking into account and eliminating systematic errors are a very important problem in every accurate measurement. In the theory of errors, however, little attention has been devoted to systematic errors.

In most books on methods of data processing, the question of systematic errors either is neglected or it is stipulated that these errors are assumed to have been eliminated.

In reality, however, systematic errors cannot be completely eliminated; some unexcluded residuals always remain. These residuals must be taken into account in order to estimate the limits of the unexcluded systematic error of the result, which determine its systematic error.

In addition, many measurements are performed without special measures taken to eliminate systematic errors, since either it is known *a priori* that they are quite small or the measurement conditions make it impossible to eliminate them. For example, suppose the mass of a body is being measured and corrections are not made for the balances employed either because the corrections are small or because the actual values of the masses are not known (only the limits of their errors are known).

Sometimes the unexcluded residuals of the systematic errors are assumed to be random errors based on the fact that their values are unknown. We cannot agree with this point of view. When classifying errors as systematic or random, attention should be focused on their properties rather than on whether or not their values are known.

For example, suppose that the resistance of a resistor is being measured and a correction is made for the influence of the temperature. The systematic error would be eliminated if we knew exactly the temperature coefficient of the resistor and the temperature. We know both quantities with limited accuracy, and for this reason we cannot completely eliminate this error. An unexcluded residual of the error will remain. It can be small or large; this we can and should estimate, but its real value remains unknown. Nonetheless this residual error has a definite value, which remains the same when the measurement is repeated under the same conditions, and for this reason it is a systematic error.

Errors that have been eliminated are no longer errors. For this reason, as we have already mentioned, the systematic error in a measurement also should be understood to be the unexcluded residual of the systematic error, if it cannot be neglected.

The error in a measurement can be both systematic and random, but after the measurement has already been performed, the measurement error becomes a systematic error. Indeed, the result of a measurement has a definite numerical value, and its difference from the true value of the measured quantity is also constant. Even if the entire error in a measurement were random, for a measurement result it becomes systematic, i.e., it seemingly freezes.

We shall now discuss the classification of systematic errors. We shall base our discussion on the work of M. F. Malikov, and following this work we shall distinguish systematic errors according to their sources and properties.[43]

The sources of systematic errors can be three components of the measurement: the method of measurement, the measuring instrument, and the experimenter. Correspondingly, methodological, instrumental, and personal systematic errors are customarily distinguished.

Methodological errors arise owing to imperfections of the method of measurement and owing to the limited accuracy of the formulas used to describe the phenomena on which the measurement is based. We shall classify as methodological errors the errors arising as a result of the influence of the measuring instrument on the object whose property is being measured.

For example, the moving-coil voltmeter draws current from the measurement circuit. Because of the voltage drop on the internal resistance of the source of the voltage being measured, the voltage on the terminals of the voltmeter will be less than the measured value. The indications of the voltmeter, however, are proportional to the voltage on its terminals. The error that arises—a methodological error—should be insignificant if the measurement is performed properly.

A methodological error can also arise in connection with the use of the measuring instrument. The gain of a voltage amplifier is determined by measuring the voltages at the input and output. If these voltages are measured successively using the same voltmeter, as is often done in practice, then, aside from the voltmeter error, the measurement error will include the error owing to the uncontrollable change in voltage at the amplifier input over the time during which the voltmeter is switched and the voltage at the output is measured (or vice versa). This error does not arise when two voltmeters are employed. When the measurement is performed using one voltmeter, however, the effect of the voltmeter error decreases.

We note that the error owing to the threshold discrepancy between the model and the object is also a methodological error.

Instrumental systematic errors are errors caused by imperfections of the measuring instrument. Classical examples of such errors are errors of a measuring instrument that are caused by imprecise calibration of the instrument scale and the error of a resistive voltage divider owing to the inaccurate adjustment of the resistances of its resistors.

Another group of such errors are additional and dynamic errors. These errors also depend on the imperfections of the measuring instruments, but they are caused by influence quantities and noninformative parameters of the input signal as well as by the change in the input signal in time. Most often the additional and dynamic errors are systematic errors. When the influence quantities and the forms of the input signal are unstable, however, they can become random errors.

Setup errors, i.e., errors arising owing to the arrangement of the measuring instruments and their effect on one another, are also instrumental errors.

Personal systematic errors are systematic errors connected with the individual characteristics of the observer.

We shall discuss the errors in the reading of the indications of indicating instruments. Such errors were investigated by H. Bäkström.[15] Although real reading devices were simulated in this work by blanks with lines, depicting the edge of a scale graduation and the indicator of the instrument, drawn on them, the results obtained are quite plausible.

The results of the investigation consist of the following.

The systematic errors made by every observer when estimating tenths of the graduation of an instrument scale can reach 0.1 graduations and are much larger than the random errors. These systematic errors are manifested by the fact that for different tenths of a graduation, different observers characteristically make estimates with different frequencies, and in addition the distribution characteristic of every observer remains constant for a long period of time. Thus one observer refers, more often than one would think, indications to the lines forming the edges of graduation, and to the value 0.5 of a graduation. Another observer refers indications to the values 0.4 and 0.6 of a graduation. A third observer prefers 0.2 and 0.8 graduations, etc. Tenths of a graduation, which are arranged symmetrically in the space between scale markers, are estimated with the same frequency.

The error in estimation of tenths of graduations depends on the thickness of the markers—the lines forming the scale. The optimal thickness of these markers is 0.1 of the length of a graduation. The length of a graduation significantly affects the error in reading tenths of a graduation. Instrument scales for which tenths of a graduation can be read are usually made so that the length of a graduation is equal to about 1 mm (not less than 0.7 mm and not more than 1.2 mm).

On the whole, for a random observer, the distribution of systematic errors in the readings of tenths of a graduation can be assumed to be uniform with limits of ± 0.1 graduations.

It is interesting that the components of the random error are usually not singled out. This is because the random error in a measurement is, as a rule, estimated from the experimental data and the entire error is measured at once, while the systematic error is measured by components.

Constant systematic errors are distinguished from regularly varying systematic errors. The latter errors, in turn, are subdivided into progressing and periodic errors and errors that vary according to a complicated law.

A constant systematic error is an error that remains constant and for this reason is repeated in each observation or measurement. For example, such an error will be present in measurements performed using the same material measures that have a systematic error: balances, resistors, etc. The personal errors made by experienced experimenters can also be classified as constant (for inexperienced experimenters they are usually of a random character).

Progressing errors are errors that increase or decrease throughout the measurement time. Such errors are caused, for example, by the change in the working current of a potentiometer owing to the voltage drop of the storage battery powering it.

Periodic errors are errors that vary with a definite period.

In the general case a systematic error can vary according to a complicated aperiodic law.

The discovery of systematic errors is a complicated problem. It is especially difficult to discover a constant systematic error. To solve the problem, in this case several measurements (at least two) should be performed by fundamentally different methods. This method is ultimately decisive. It is often realized by comparing the results of measurements of one and the same quantity that were obtained by different experimenters in different laboratories.

It is easier to discover variable systematic errors. This can be done with the help of statistical methods, correlation, and regression analysis. But nonmathematical possibilities also should not be avoided. Thus, in the process of performing a measurement it is helpful to employ a graph on which the results of the measurements are plotted in the sequence in which they were obtained. The overall arrangement of the points obtained makes it possible to discover the presence of a systematic change in the results of observations without mathematical analysis. The human capability of perceiving such regularities is widely employed in metrology, although this capability has apparently still not been thoroughly studied.

If a regular change in observational results has been found and it is known that the measured quantity did not change in the process, then this indicates the presence of a regularly varying systematic error.

It is also helpful to measure the same quantity using two different instruments (methods) or to measure periodically a known quantity instead of the unknown quantity.

If the presence of a systematic error has been discovered, then it can usually be estimated and eliminated. In precise measurements, however, this often presents great difficulties and is not always possible.

In most fields of measurements the most important sources of systematic errors are known and measurement methods have been developed that eliminate the appearance of such errors or prevent them from affecting the result of a measurement. In other words, systematic errors are eliminated not by mathematical analysis of experimental data but rather the use of appropriate measurement methods. The analysis of measurement methods and the systematization and generalization of measurement methods are very important problems, but they fall outside the scope of this book, which is devoted to the problem of analysis of experimental data. For this reason, we shall confine our attention to a brief review of the most widely disseminated general methods for studying such problems.

Elimination of constant systematic errors:

Method of replacement.

This method gives the most complete solution of the problem. It is a version of the method of comparison, when the comparison is made by replacing the measured quantity by a known quantity and in a manner so that in the process no changes occur in the state and operation of all measuring instruments employed.

Consider, for example, weighing performed by Borda's method. The method is designed so as to eliminate the error owing to the inequality of the arms of the balances. Let x be the measured mass, P the mass of the balancing weights, and l_1 and l_2 the lengths of the arms of the balances. The measurement is performed as follows. First the weighed body is placed in one pan of the balances and the balances are balanced with the help of a weight with mass T. Then

$$x=\frac{l_2}{l_1}T.$$

Next, the mass x is removed and a mass P that once again balances the pans is placed in the empty pan:

$$P=\frac{l_2}{l_1}T.$$

Since the right-hand sides of both equations are the same, the left sides are also equal to one another, i.e., $x = P$, and the fact that $l_1 \neq l_2$ has no effect on the result.

The resistance of a resistor can be measured in an analogous manner with the help of a sensitive but inaccurate bridge and an accurate magazine of resistances. A number of other quantities can be measured analogously.

Method contraposition. This measurement method is a version of the comparison method. The measurement is performed with two observations, performed so that the reason for the constant error would affect the results of observations differently but in a known, regular fashion.

An example of this method is Gauss's method of weighing. First the weight being weighed is balanced by balance weights P_1. Using the notation of the preceding example, we have

$$x = \frac{l_2}{l_1} P_1.$$

Next the unknown weight is placed into the pan that held before the balancing weights, and the two loads are once again balanced. Now we have

$$x = \frac{l_1}{l_2} P_2.$$

We now eliminate the ratio l_2/l_1 from these two equalities and find

$$x = \sqrt{P_1 P_2}.$$

The sign method of error compensation. This method involves two measurements performed so that the constant systematic error would appear with different signs in each measurement.

For example, consider the measurement of an emf x with the help of a dc potentiometer that has a parasitic thermo-emf. One measurement gives E_1. Next the polarity of the measured emf is reversed, the direction of the working current in the potentiometer is changed, and once again the measured emf is balanced. This gives E_2. If the thermo-emf gives the error ϑ and $E_1 = x + \vartheta$, then $E_2 = x - \vartheta$. From here

$$x = \frac{E_1 + E_2}{2}.$$

Elimination of progressing systematic errors. The simplest, but frequent case of a progressing error is an error that varies linearly, for example, in proportion to the time.

An example of such an error is the error in the measurement of voltage with the help of a potentiometer, if the voltage of the storage battery, generating the working current, drops appreciably. Formally, if it is known that the working current of the potentiometer varies linearly in time, then to eliminate the error arising it is sufficient to perform two observations at times after the working current along the normal element is regulated. Let

$$E_1 = x + Kt_1, \quad E_2 = x + Kt_2,$$

where t_1 and t_2 are the time intervals between regulation of the working current and the observations, K is the coefficient of proportionality between the measurement error and the time, and E_1 and E_2 are the results of the observations. From here

$$x = \frac{E_1 t_2 - E_2 t_1}{t_2 - t_1}.$$

For accurate measurements, however, it is best to use a somewhat more complicated method of symmetric observations. In this method, several observations are performed equally separated in time and then the arithmetic means of the symmetric observations are calculated. Theoretically all these averages must be equal, and this makes it possible to control the course of the experiment, and also to eliminate these errors.

When the errors vary according to more complicated laws, the methods for eliminating the errors become more complicated, but the problem can always be solved if these laws are known. If, however, the law is so complicated that it is pointless or impossible to find it, then the systematic errors can be reduced to random or quasirandom errors. This requires a series of observations, arranged in a manner so that the observational errors would be as diverse as possible and look like random errors. However, this technique is not as effective as finding the error and eliminating it directly.

The methods listed above do not exhaust all possibilities for eliminating systematic errors. Thus, to eliminate the systematic error of a measuring instrument from the result of a measurement, the measurement can be performed not by one but rather by several instruments simultaneously (if the errors of the instruments are uncorrelated). Taking for the result of the measurement a definite combination of indications of all instruments, we can make their systematic errors, which are different for the different instruments, compensate one another somehow, and the error of the result obtained will be less than for an individual instrument. In this case, the systematic errors of the instruments can be regarded as a realization of a random quantity.

In those cases when for the measured quantity several exact relations between it and other quantities are known, these relations can be used to reduce the measurement error. For example, if the angles of a plane triangle are measured, then the fact that their sum is equal to 180° must be taken into account.

5.3. Estimation of elementary errors

It is difficult to describe in a generalized form a method for estimating elementary errors, since these errors are by their very nature extremely diverse. The general rules for solving this problem can nonetheless be formulated.

To estimate measurement errors it is first necessary to determine all their possible sources. If it is known that some corrections will be introduced (or corrections have been introduced), then the errors in determining the corrections must be included among the elementary errors.

All elementary measurement errors must be estimated in the same manner, i.e., in the form of either absolute or relative errors. Relative errors are usually more convenient for *a posteriori* error estimation, and absolute errors are more convenient for *a priori* error estimation. However, the tradition of each field of measurement should be kept in mind. Thus for lineal–angular measurements absolute errors are preferred, while for measurements of electromagnetic quantities relative errors are preferred.

An unavoidable elementary error in any measurement is the intrinsic error of the measuring instrument. If the limits of this error are given in the form of absolute or relative errors, then conversions are not required and these limits are the limits of the given elementary error. But often the limits of intrinsic error of a measuring instrument are given in the form of a fiducial error, i.e., as a percentage of the normalizing fiducial value. The conversion into relative error is made using the formula

$$\delta_{\text{in}} = \gamma \frac{x_N}{x},$$

where δ_{in} is the limit of the intrinsic error in relative form, γ is the limit of the fiducial error, x_N is the fiducial value, and x is the value of the measurand. Conversion into the form of absolute errors is done according to the formula

$$\Delta_{\text{in}} = \delta_{\text{in}} x = \gamma x_N.$$

Very often the environmental conditions, characterized by the temperature, pressure, humidity, vibrations, etc., affect the result of a measurement. Each influence quantity, in principle, engenders its own elementary error. In order to estimate it, it is first necessary to estimate the possible value of the corresponding influence quantity and then compare it with the limits of the range of values of this quantity concerning the reference condition. If the influence quantity falls outside the limits of reference values, then it engenders a corresponding additional error; this error is also an elementary error.

Consider an error of the temperature. Let the temperature of the medium exceed its reference values by ΔT. If $T_1 \leqslant \Delta T \leqslant T_2$ and the limit of the additional error for the interval $[T_1, T_2]$ has the same modulus, then this limit is the limit of the given additional error. If, however, for this interval the limiting value of the temperature coefficient is given, then the limits of temperature error are calculated according to the formula

$$\delta_T = \pm W_T \Delta T,$$

where δ_T is the limit of additional temperature error and W_T is the limiting value of the modulus of the temperature coefficient of the instrument.

In the general case, the dependence of the limit of additional error δ_i or Δ_i on the deviations of the influence quantity outside the limits of its reference values can be given in the form of a graph or expressed analytically (see Chap. 2). If the influence function of some influence quantity is indicated in the specifications provided by the manufacturer of the instrument, then a deviation of this quantity outside the limits of its reference values can be taken into account by the corresponding correction. In the process, the elementary error decreases significantly, even if the influence function is given with a large margin of error.

Suppose, for example, instead of the limiting value of the temperature coefficient $W_T = \pm b/T$ the influence function $W'_T = (1 \pm \varepsilon) b/T$ is given. To calculate the correction one must know the specific value of the temperature during the measurement and therefore its deviation ΔT from the reference value. Then the additional temperature error will be

$$\zeta_T = (1 \pm \varepsilon)\, b \frac{\Delta T}{T}.$$

We eliminate the constant part of this error with the help of the correction

$$c = -\frac{b\Delta T}{T}.$$

There then remains the temperature error

$$\delta'_T = \pm\varepsilon\frac{b\Delta T}{T}.$$

If the influence function is given comparatively inaccurately, for example, $\varepsilon = 0.2$ (20%), then the temperature error even in this case decreases by a factor of 4–6:

$$\frac{\delta_T}{\delta'_T} = \frac{1 \pm 0.2}{0.2} = 4 \text{ or } 6.$$

It should also be kept in mind that if the influence quantity is estimated with an appreciable error, then this error must also be taken into account when calculating the corresponding additional error.

In many cases the input signal in a measurement is a function of time. It can be represented by some model, characterized by a series of parameters. One of these parameters is the informative, measured parameter while the other parameters are noninformative. Measuring instruments are constructed so as to make them insensitive to all noninformative parameters of the input signal. This, however, cannot be achieved completely, and in the general case the effect of the noninformative parameters is only decreased. Furthermore, for all noninformative parameters it is possible to determine limits such that when the noninformative parameters vary within these limits, the total error of the measuring instrument will change insignificantly. This makes it possible to establish the reference ranges of the values of the noninformative parameters.

If some noninformative parameter falls outside the normal limits, then the error arising is regarded as an *additional error*. The effect of each noninformative parameter is normalized separately, as for influence quantities.

Normalization of the effect of the noninformative parameters and estimation of the errors arising owing to them are performed based on the same assumptions as those used for taking into account the additional errors caused by the external influence quantities.

The errors introduced by changes in the noninformative parameters of the input signals are occasionally called *dynamic errors*. In the case of several parameters, however, this provides little information. It is more informative to give each error a characteristic name, as usually done in electric and radio measurements. For example, the change produced in the indications of an ac voltmeter by changes in the frequency of the input signal is called the frequency error. In the case of a voltmeter, for measurements of the peak variable voltages, apart from the frequency errors, the errors caused by changes in the widths of the pulse edges, the decay of the flat part of the pulse, etc., are taken into account.

There are a number of peculiarities in estimating dynamic errors. The most important of these peculiarities must be discussed.

First of all, it should be noted that although for a long time now the dynamic errors have been taken into account in particular situations, the general theory of estimation of dynamic errors of measurements, as the theory of dynamic measurements in general, is still in the formative stage. In Ref. 55 an attempt is made to formulate the basic concepts of the theory of dynamic measurements. In studying methods for estimating dynamic errors below, we shall adhere to the concepts presented in Ref. 55.

Next, we have to mention that measuring instruments do not have dynamic errors but they may have additional errors in the dynamic regime. These errors are a special type of elementary errors of measurements.

A typical example of a measurement for which the dynamic error is significant is a measurement in which a time-varying signal is recorded. In this case, in accordance with the general definition of absolute error, the dynamic error can be written as

$$\zeta_d(t)=\frac{y(t)}{K}-x(t), \tag{5.1}$$

where $\zeta_d(t)$ is the dynamic error, $x(t)$ and $y(t)$ are the signals at the input and output, respectively, of the measuring instrument, and K is the transduction constant.

The relation between the signals at the input and output of the measuring instrument can be represented by an operator equation

$$y=Bx, \tag{5.2}$$

where B is the operator of the instrument.

The operator expresses in general form the entire aggregate of dynamic properties of a measuring instrument. These properties depend on the particular action with respect to which they are studied. Thus the dynamic properties with respect to a variable influence quantity or interference that does not act at the input of the measuring instrument can be different from the dynamic properties with respect to the input signal. In Eq. (5.2) the operator B pertains to the input signal.

When measuring instruments are constructed, the transduction constant is usually made to be independent of the strength of the input. Then the measuring instruments can be described by a linear model, and as a rule linear models can have lumped parameters.

Substituting Eq. (5.2), Eq. (5.1) can be represented in the operator form

$$\zeta_d=\left(\frac{B}{K}-I\right)x,$$

where I is the identity operator, $Ix\equiv x$.

The input and output signals vary in time, and therefore the dynamic error is a function of time.

One would think that given the output signal $y(t)$ and the operator of the measuring instrument it is possible using (5.2) to find the input signal $x(t)$ and then, using formula (5.1), to find the dynamic error. This is difficult to do, however, since the operator of the measuring instrument is usually not known accurately enough.

Sometimes the problem can be solved without knowing the operator of the instrument at all. Suppose that we have an instrument and a record of some process realized with its help. We now disconnect the instrument from the process being studied and connect to the instrument a circuit in which we can control an analogous process. An example of such a device is a standard-signal generator (if an electric measuring instrument is employed). Next we record a signal at the input of the instrument such that at the output we obtain a process that is identical to the initially recorded process. When the records are identical enough, the input processes are also identical. Therefore we have found the input signal, and comparison of it (after multiplication by K in accordance with Eq. 5.1) with the given record can produce an estimate of the dynamic error.

In measurement practice, efforts are always made to use measuring instruments whose output signals would be quite close (so as to conform to the goals of the measurement) in form to the input signal. But in those cases when such measuring instruments are not available and existing measuring instruments must be used, in spite of the distortions created by them the reconstruction of the form (keeping the parameters unchanged) of the input signal becomes an important method for increasing measurement accuracy. It should be noted, however, that reconstruction of the form of the input signal presents great difficulties, associated with the fact that this problem is a so-called improperly posed problem (in the terminology of J. Hadamard), i.e., in this problem the solution does not depend continuously on the initial data. This means that when there are small errors in the specification of the dynamic characteristic of the measuring instrument and the reading of the values of the output signal, the error in determining the input signal can be so large that the solution obtained is physically meaningless.

Physically the main idea of the problem of improper formulation in application to reconstruction of the form of the input signal consists of the following. Ultimately the spectral composition of the output signal of a measuring instrument always decreases in intensity as the frequency increases. The amplitude–frequency response of a measuring instrument (which, naturally, is a stable system) at high frequencies also approaches the frequency axis. Thus it is required to find based on two functions with decreasing spectra a third function (the input signal) that provides a unique relation between them. At low and medium (for the given functions) frequencies, where the intensity of the spectra is high, the signal sought can be determined quite reliably, and the unavoidable errors in the initial data and the computational procedures operate in the normal fashion, i.e., they distort the solution without destroying its physical meaning. At high frequencies the intensities of the spectra drop to such an extent that their effect on the solution is comparable to that of errors in the initial data. The effect of these errors can be so large that the true solution is completely suppressed. Time-domain distortions usually have the form of rapidly oscillating functions, whose amplitude is often several orders of magnitude greater than the true solution.

Methods for solving improperly posed problems (methods of regularization) are under active development in mathematics, mathematical physics, geophysics, and the theory of automatic control. A list of the most important publications on this question relevant to metrology is given in Ref. 32.

The essential feature of methods of regularization consists of filtering out the distortions based on *a priori* information about the true solution. The main question is how to establish the optimal degree of filtering in order to filter out noise without distorting the true solution. Different methods of regularization require different volumes and forms of *a priori* information.

Dynamic errors most often are estimated when the choice of a recording instrument is being made. The problem is solved in the following way.

The worst form of input process is chosen and expressed analytically. One of the complete dynamic characteristics of the recording instrument is assumed to be known. Then it is possible to compute the corresponding output process. A superposition of these output and input processes gives the dynamic error of the expected measurement. If the absolute value of dynamic error lies within the permissible limits, then this recording instrument can be used for the measurement.

But it is inconvenient to work with an error as a function of time. For this reason, efforts are usually made to describe the dynamic error, when recording data, by a parameter that assumes a single value for the entire function. Most often, the error having the maximum modulus or its standard deviation is used.

It should be noted that the computational scheme presented above can be modified for different measurement problems. Thus a shift of the output signal in time relative to the input signal is often possible. In this case the signals can be artificially arranged so as to minimize the norm of the error.

In spite of the difficulty of estimating dynamic errors, the dynamic error is an elementary error. In those cases when the dynamic error is represented by its components, these components are regarded as elementary measurement errors.

We shall use the symbols presented in Table 1.1 to designate elementary errors. If an elementary error has both systematic and random components, we shall designate it with the symbol used for the dominant component.

5.4. Method for calculating the errors of single measurements

Once the errors of a single measurement have been analyzed, we have an estimate of the limits of all elementary errors of the measurement. We now proceed to the problem of synthesis. First of all, we single out absolutely constant errors, if they exist, and write out estimates of their limits H_f:

$$|h_f| \leqslant H_f \quad \text{or} \quad H_{fl} \leqslant h_f \leqslant H_{fr},$$

where $f = 1,\dots,k$ and k is rarely greater than 2.

The remaining elementary errors are conditionally constant:

$$|\zeta_i| \leqslant \theta_i, \quad i=1,\dots,n.$$

Above we modeled conditionally constant errors by a random quantity with a uniform probability distribution. For direct measurements, in the overwhelming majority of the cases, elementary errors can be assumed to be independent of one another. Starting from this we calculate the limiting value of the sum of all conditionally constant errors. For this we use the formula (3.13):

$$\theta = k \sqrt{\sum_{i=1}^{n} \theta_i^2}. \tag{5.3}$$

With a confidence probability $\alpha = 0.99$ and $n \leqslant 4$ it could turn out that $\theta > \Sigma_{i=1}^{n} \theta_i$. But it is obvious that this cannot happen. In this case it is possible to take

$$\theta = \sum_{i=1}^{n} \theta_i. \tag{5.4}$$

It would, of course, be more correct to take a more accurate value of the coefficient k from the curves presented in Fig. 3.3.

There arises, however, the question of how well founded the choice $\alpha = 0.99$ is. In most cases this limit does not correspond to the reliability of the initial data, and the limit $\alpha = 0.95$ is more appropriate. For $\alpha = 0.95$ formula (5.3) assumes the form (see page 70)

$$\theta = 1.1 \sqrt{\sum_{i=1}^{n} \theta_i^2}. \tag{5.5}$$

In this case $\theta < \Sigma_{i=1}^{n} \theta_i$. We shall show this by investigating the last inequality. First, let $n = 2$ and consider the inequality $1.1\sqrt{\theta_1^2 + \theta_2^2} < (\theta_1 + \theta_2)$. It is not difficult to verify that the inequality holds if $\theta_1/\theta_2 > 0.11$. This condition corresponds to practice, since an elementary error that is about ten times smaller than any other elementary error can be neglected.

Consider now the three terms $\theta_3 > \theta_2 > \theta_1$. Introducing $T = \theta_3 + \theta_2$, we obtain the identity

$$1.1\sqrt{T^2 + \theta_1^2 - 2\theta_3\theta_2} < (T + \theta_1).$$

The term $2\theta_3\theta_2 > 0$, and conditions of the inequality will be stronger if this term is dropped. Then, corresponding to the case we have just studied with two terms, we obtain

$$\frac{\theta_1}{\theta_2 + \theta_3} > 0.11.$$

It is obvious that this inequality holds easier then for two components. On the whole, as the number of terms increases, the inequality is more easily satisfied.

It is interesting to note that from the inequalities

$$k \sqrt{\sum_{i=1}^{n} \theta_i^2} < \sum_{i=1}^{n} \theta_i$$

it follows that $k < \sqrt{n}$, if all terms are equal.

It could happen that m of the n conditionally constant errors have unsymmetric limits:

$$\theta_{jl} \leqslant \zeta_j \leqslant \theta_{jr}, \quad j = 1, \ldots, n$$

where θ_{jl} is the left-hand limit and θ_{jr} is the right-hand limit.

For calculations, unsymmetric limits must be represented as symmetric limits with a shift by a_j, where

$$a_j = \frac{\theta_{jl} + \theta_{jr}}{2}.$$

The limits of an interval that is symmetric relative to a_j are calculated according to the formula

$$\theta_j = \frac{\theta_{jr} - \theta_{jl}}{2}.$$

Next, the limits of the error must be calculated according to the following formulas instead of Eq. (5.3) or (5.5):

$$\theta_r = \sum_{j=1}^{m} a_j + k\sqrt{\sum_{i=1}^{n-m} \theta_i^2 + \sum_{j=1}^{m} \theta_j^2},$$

$$\theta_l = \sum_{j=1}^{m} a_j - k\sqrt{\sum_{i=1}^{n-m} \theta_i^a + \sum_{j=1}^{m} \theta_j^2}. \qquad (5.6)$$

The absolutely constant elementary errors must now be taken into account. Since the probabilistic model is not appropriate for them, their limits must be summed arithmetically with the limits, calculating according to Eqs. (5.6), of the sum of the conditionally constant components θ_r and θ_l or θ:

$$\theta_r' = \sum_{f=1}^{k} H_{fr} + \theta_r,$$

$$\theta_l' = \sum_{f=1}^{k} H_{fl} - \theta_l. \qquad (5.7)$$

In the foregoing calculation the conditionally constant elementary errors were modeled by a random quantity with a uniform probability distribution. However, elementary errors, which appear in the resulting error after some transformation, are encountered. An example is the mismatch error in radioelectronic measurements. The elementary error here is the phase shift $\Delta\varphi_f$. As in the case of other elementary errors, for this error the limits of the phase shift are estimated, and it is assumed that the phase shift is uniformly distributed within these limits. But the error in the result contains not $\Delta\varphi_f$, but rather $\cos \Delta\varphi_f$. When $\Delta\varphi_f$ is distributed uniformly the quantity $\cos \Delta\varphi_f$ has the so-called arccosine distribution.

When transformed elementary errors are present their composition with the other errors must be constructed according to adopted mathematical methods. The universal numerical method described in Sec. 3.6 is convenient.

The distribution of the sum of elementary errors, for each of which a uniform distribution is adopted, can be regarded as a normal distribution, if $n > 4$ and they all have approximately the same limits. If, however, $n \leqslant 4$ or the limits of the elementary errors are substantially different, then the composition of these distributions must also be constructed.

We now return to the case when the elementary error has unsymmetric limits. The transformation of these limits to a symmetric form with a shift by a_j creates the temptation to introduce into the measurement result a correction corresponding to the shift a_j. One must be decisively cautioned against doing this: Information about errors is too unreliable to use for correcting the result of a measurement.

In the calculations performed above, it was assumed that the elementary errors are independent of one another. In some cases this assumption is not justified. An example is the case when several measuring devices, connected into a measuring system or forming a measurement channel, are used to perform a measurement and some influence quantity exceeds the limits of reference conditions for these instruments. In this case the measuring devices will acquire additional errors and they could be dependent.

From n elementary errors it is possible to single out $2m$ dependent errors. Then Eq. (5.3) must be transformed into the following form:

$$\theta = k\sqrt{\sum_{i=1}^{n-m} \theta_i^2 + \sum_{\mu=1}^{m} (\theta_{\mu_1} + \rho_{12}\theta_{\mu_2})^2}, \qquad (5.8)$$

where θ_{μ_1} and θ_{μ_2} are the moduli of the limits of pairs of dependent elementary errors, and the coefficient ρ_{12} is equal to $+1$ if the additional error ϑ_{μ_2} $(-\theta_{\mu_2} \leqslant \vartheta_{\mu_2} \leqslant \theta_{\mu_2})$ has the same sign as the error ϑ_{μ_1} $(-\theta_{\mu_1} \leqslant \vartheta_{\mu_1} \leqslant \theta_{\mu_1})$ and -1 if these errors have different signs.

The method presented above for calculating errors is equally suitable for *a priori* and *a posteriori* estimation, since at the synthesis stage there is no difference between these cases.

In conclusion we shall discuss the formula

$$\theta = \sqrt{\sum_{i=1}^{n} \theta_i^2},$$

which is often used in practice. This formula is obtained under the assumption that the elementary errors have a normal distribution and their limiting values were calculated for one and the same confidence limit. Let σ_i be the standard deviation of the ith elementary error and $\theta_i = z_\alpha \sigma_i$, where z_α is the quantile coefficient, determined according to a normal distribution and one and the same confidence limit for all i. It is obvious that

$$\sigma^2 = \sum_{i=1}^{n} \sigma_i^2.$$

We multiply both sides of this equality by z_α^2:

$$z_\alpha^2 \sigma^2 = \sum_{i=1}^{n} z_\alpha^2 \sigma_i^2.$$

But $\theta = z_\alpha \sigma$ and $\theta_i = z_\alpha \sigma_i$. From here we obtain the formula

$$\theta = \sqrt{\sum_{i=1}^{n} \theta_i^2}.$$

However, the limits of elementary errors are not estimated by probabilistic methods, and a probability cannot be assigned to them. Moreover, there are no grounds for using a normal distribution as the mathematical model of elementary errors.

The formula presented above can be interpreted differently, namely, as a particular case of formula (5.3) with $k = 1$.* The value $k = 1$ corresponds to a confi-

*S. G. Rabinovich, in: The 6th International Conference of the Israel Society for Quality Assurance, Tel-Aviv, November, 1986, Papers and abstracts, 4.3.4, pp. 1–4.

dence probability of 0.916. This explains the commonly held opinion that this formula somewhat underestimates the error, i.e., that this formula is not reliable enough.

On the other hand, this formula is widely used in practice. Its wide dissemination can be regarded as an indirect but practical confirmation of the fact that a uniformly distributed random quantity can be used as a model of conditionally constant elementary errors.

Sometimes the elementary components are summed according to the formula

$$\theta = \sum_{i=1}^{n} \theta_i,$$

Such summation, however, means that all elementary errors are assumed to be absolutely constant. This situation is rare. If, however, it is agreed that conditionally constant errors are also present, then the arithmetic summation means that all elementary errors simultaneously assume their limiting values, and with the same sign. This coincidence is very unlikely. Although this formula satisfies the principle of estimating the upper limit of errors, it is used less and less, and primarily only for obtaining a rough estimate of the error or in the extreme case, as done above, in formula (5.4) simply to eliminate an erroneous estimate.

5.5. Example: Calculation of errors in voltage measurements performed with a pointer-type voltmeter

We shall study several examples of the application of a class 1.0 pointer-type dc voltmeter with the following characteristics:

(i) The upper limits of measurement ranges are 3, 7.5, 15, 30, etc., up to 300 V.

(ii) The scale of the instrument has 75 graduations and starts at the 0 marker.

(iii) The limits of permissible intrinsic error are $\pm 1.0\%$ of a span (it is a fiducial error).

(iv) Full deflection of the pointer corresponds to a current of 15×10^{-6} A $\pm 1\%$.

(v) Under the reference conditions the temperature is equal to $+20 \pm 5$ °C and the measurements are performed with the instrument positioned horizontally. In this case we shall ignore all other influence quantities; we shall assume that they are identical to their normal reference values.

Additional errors. A deviation of the temperature from the reference value causes the indications of the instrument to change by not more than $\pm 1.0\%$ for each 10 °C change in temperature. Inclination of the instrument by 5° from the horizontal position changes the indications by not more than $\pm 1\%$ of the measurement limit.

5.5.1. A priori estimation

Suppose that some piece of equipment is to be monitored by measuring the voltage on several resistors. The equivalent output resistance (the source resistance) of the equipment in one case is equal to about 10 kΩ and in all other cases does not exceed 1 kΩ. The temperature of the medium can change from + 10 to + 25 °C. The slope relative to the horizontal position does not exceed 5°.

We are required to estimate the measurement error. The errors must be expressed in the form of relative errors.

Prior to the measurements the value of the measured quantity is unknown. It will supposedly be less than 3 V. The overlapping of limits in the voltmeter is equal to $3/7.5 = 0.4$ and $7.5/15 = 0.5$, after which these numbers repeat. Thus if the indication of the instrument drops below 0.4–0.5 of the upper limit of measurement, then the range of measurement must be switched. Developing this point of view, we shall assume that if the measured voltage is less than $0.4 \times 3\ \text{V} = 1.2\ \text{V}$, then a different voltmeter must be used.

In the range 1.2–3 V the largest relative error will occur when a voltage of the order of 1.2 V is being measured. The error will have to be estimated for this worst case.

The sources of error are as follows:

(1) the intrinsic error of the voltmeter,

(2) the reading error,

(3) the temperature error,

(4) the error introduced by the inclination of the instrument, and

(5) the error owing to the limited input resistance of the voltmeter.

We shall estimate all these errors.

(1) Intrinsic error: Its limits will be

$$\theta_{\text{in}} = \pm 1\% \times \frac{1}{0.4} = \pm 2.5\%, \quad |\theta_{\text{in}}| = 2.5\%.$$

(2) Reading error: This error does not exceed 0.25 of a graduation. When measuring 1.2 V at the limit 3 V this gives

$$\theta_\alpha = \pm 0.25 \times \frac{3 \times 100}{75 \times 1.2} = \pm 0.83\%, \quad |\theta_\alpha| = 0.83\%.$$

(3) Additional temperature error: The maximum deviation of the temperature from the normal value is $(20 - 5) - 10 = 5\ ^\circ\text{C}$. For this reason

$$\theta_T = \pm 1\% \times \frac{5}{10} = \pm 0.5\%, \quad |\theta_T| = 0.5\%.$$

(4) The additional error introduced by the 5° inclination of the instrument when measuring 1.2 V will be

$$\theta_l = \pm 1\% \times \frac{3}{1.2} = \pm 2.5\%, \quad |\theta_l| = 2.5\%.$$

(5) The errors $\{\varepsilon_R\}$ owing to the limited input resistance of the voltmeter are as follows. The input resistance of the voltmeter at the limit 3 V is

$$R_V = \frac{3}{1.5 \times 10^{-5}} = 2 \times 10^5\ \Omega.$$

The worst case occurs with the outside resistance $R'_{\text{or}} = 10\ \text{k}\Omega$.

The indications of the voltmeter correspond to the voltage on its terminals. This voltage U is less than the emf E in the circuit:

$$U=\frac{R_V}{R_V+R_{or}}E.$$

The error is

$$\varepsilon_R=\frac{U-E}{E}=\frac{-R_{or}}{R_V+R_{or}}.$$

For $R'_{or}=10$ kΩ

$$\varepsilon'_R=\frac{-10\times 10^3}{10\times 10^3+2\times 10^5}\times 100=-4.8\%.$$

If the outside resistance is 1 kΩ, then $\varepsilon''_R = -0.5\%$.

The errors $\{\varepsilon_R\}$ are absolutely constant for each unit being monitored. The remaining errors are conditionally constant.

Let us now add all conditionally constant errors. We shall use Eq. (5.5) and we shall assume that $\alpha=0.95$:

$$\delta=1.1\sqrt{2.5^2+0.83^2+0.5^2+0.25^2}=4\%.$$

We now take into account the absolutely constant error. Its limits are

$$H_{Rl}=-4.8\%,\quad H_{Rr}=-0.5\%,$$

but they are not known accurately enough in order to eliminate them by introducing the correction. Therefore, in accordance with (5.7) we obtain

$$\delta_r=-0.5+4=+3.5\%,\quad \delta_l=-4.8-4.0=-8.8\%.$$

Thus the error of the planned measurement will not exceed $\sim 10\%$.

5.5.2. *A posteriori estimation*

We shall now estimate the measurement error in the example examined above, assuming that the measurement has already been made. The significant difference from the foregoing case is that now we have an estimate of the measured quantity. Let the indication of the voltmeter in the case $R'_{or}=10$ kΩ be 62.3 graduations. Hence the voltmeter indicated

$$U=62.3\,\frac{3}{75}=2.492\text{ V}.$$

Suppose we found out that $R'_{or}=10$ kΩ$\pm 0.5\%$. The error ε'_R was calculated above: $\varepsilon'_R = -4.8\%$. Now we can introduce the correction C'_R:

$$C'_R=+4.8\times 10^{-2}\times 2.492=+0.120\text{ V}.$$

Taking the corrections into account we obtain

$$U'=U+C'_R=2.612\text{ V}.$$

The errors of the corrections are determined by the errors of the available values of the resistances R_V and R_{or}. We shall establish the relation between them.

$$C'_R = -\varepsilon'_R U = \frac{R_{\rm or}}{R_{\rm or}+R_V} U = \frac{R_{\rm or}}{R_{\rm or}+R_V} \times \frac{R_V}{R_{\rm or}+R_V} E = \frac{R_{\rm or}/R_V}{(1+R_{\rm or}/R_V)^2} E.$$

To simplify the notation, let $x = R_{\rm or}/R_V$. Then

$$C'_R = \frac{x}{(1+x)^2} E.$$

We now construct the differential relations:

$$dx = \frac{1}{R_V} dR_{\rm or} - \frac{R_{\rm or}}{R_V^2} dR_V = x \left(\frac{dR_{\rm or}}{R_{\rm or}} - \frac{dR_V}{R_V} \right),$$

$$dC'_R = E \left(\frac{dx}{(1+x)^2} - \frac{2x(1+x)dx}{(1+x)^4} \right) = E \frac{1-x}{(1+x)^3} dx,$$

$$dC'_R = E \frac{x(1-x)}{(1+x)^3} \left(\frac{dR_{\rm or}}{R_{\rm or}} - \frac{dR_V}{R_V} \right).$$

In the relative form, transforming from differentials to increments, we obtain

$$\varepsilon_c = \frac{\Delta C'_R}{C'_R} = \frac{1-x}{1+x} \left(\frac{\Delta R_{\rm or}}{R_{\rm or}} - \frac{\Delta R_V}{R_V} \right).$$

Since $\Delta R_{\rm or}$ and ΔR_V are independent, we shall regard each component of error of the correction as an elementary error of measurement. Obviously, both these components are conditionally constant:

$$\theta_{C1} = \left(\frac{1-x}{1+x} \right) \theta_{R_{\rm or}}, \quad \theta_{C2} = \left(\frac{1-x}{1+x} \right) \theta_{R_v}.$$

The limits of error of the internal resistance of the voltmeter are determined by the voltmeter class and are equal to $\pm 1\%$. Therefore, because $x = 5 \times 10^{-2}$

$$|\theta_{C2}| = \left(\frac{1-x}{1+x} \right) 1\% = 0.9 \times 1\% = 0.9\%.$$

The limits of the error in determining the input resistance of our apparatus (the outside resistance for voltmeter) are equal to $\pm 0.5\%$. Therefore

$$|\theta_{C1}| = \left(\frac{1-x}{1+x} \right) 0.5\% = 0.9 \times 0.5\% = 0.45\%.$$

The limits of the remaining errors are as follows:

$$|\theta_{\rm in}| = 1\% \times 75/62 = 1.2\%.$$

$$|\theta_\alpha| = \frac{0.25 \times 100}{62} = 0.4\%,$$

$$|\theta_T| = 0.5\%,$$

$$|\theta_l| = 1\% \times 75/62 = 1.2\%.$$

All these elementary errors can be assumed to be conditionally constant. According to formula (5.5), for $\alpha = 0.95$ we obtain

$$\delta = 1.1\sqrt{0.9^2 + 0.45^2 + 1.2^2 + 0.4^2 + 0.5^2 + 1.2^2} = 2.3\%.$$

When the result of the measurement is written in accordance with its error only three significant figures can be retained:

$$\tilde{U}' = 2.61\ \text{V};\quad \delta = \pm 2.3\%\quad (\alpha = 0.95),$$

$$U'_{0.95} = 2.61\text{V} \pm 2.3\%.$$

5.5.3. An accurate a posteriori estimation

The largest elementary errors were θ_{C2}, θ_{in}, and θ_l. How can they be reduced? The first two can be reduced by taking into account the individual properties of the voltmeter. If the voltmeter has a fresh table of corrections, then this can be done. Assume that at the limit 3 V on marker 60 the correction is equal to $+0.3$ graduations, while at marker 70 it is equal to $+0.2$ graduations. It can then be assumed that the correction to the indication at 62.3 graduations is also equal to $+0.3$ graduations. Therefore

$$C_{in} = +0.3 \times \frac{3}{75} = +0.012\ \text{V}.$$

Taking this correction into account, the voltmeter gives

$$U'' = 2.492 + 0.012 = 2.504\ \text{V}.$$

We shall assume that the limits of error in determining the correction, i.e., the calibration errors, are known and are equal to $\pm 0.2\%$. Converting to the indication of the instrument, we obtain $|\theta'_{in}| = 0.2 \times 75/62 = 0.24\%$.

With this correction we have eliminated the systematic component of the error of the voltmeter. The random component, however, remains and it must be taken into account. The dead zone, according to the indicating electric measurement instruments, can reach a value coinciding with the class designation of the instrument. In our case this is 1% of 3 V. The random error does not exceed half the dead zone. Thus the limits of random error are equal to

$$|\psi| = 0.5 \times 1\% \times \frac{75}{62} = 0.6\%.$$

The distribution of the random error in our case, when its limits have been estimated, can be assumed to uniform, as also the distributions of other conditionally constant elementary errors.

The input resistance of the voltmeter can be measured. Assume that this has been done and $R_V = 201.7\ \text{k}\Omega \pm 0.2\%$. Then

$$\varepsilon_R = -\frac{10 \times 10^3 \times 100}{(10 + 201.7) \times 10^3} = -4.72\%.$$

The correction will be

$$C_R = +4.72 \times 10^{-2} \times 2.504 = +0.118 \text{ V}.$$

Taking the correction C_R into account, we obtain

$$U'' = 2.504 + 0.118 = 2.622 \text{ V}.$$

The limits of error owing to the fact that the input resistance of the voltmeter is not known exactly will now become smaller:

$$|\theta'_{C2}| = 0.9 \times 0.2\% = 0.18\%. \quad |\theta_{C1}| = 0.45\%.$$

The error θ_l can be reduced by taking greater care in positioning the instrument horizontally. Assume that the deviation from the horizontal position does not exceed $\pm 2°$. Then

$$|\theta'_l| = 1 \times 2/5 \times 75/62 = 0.48\%.$$

The temperature error and the reading error will remain the same.

We now calculate the limits of measurement error again for $\alpha = 0.95$:

$$\delta = 1.1\sqrt{0.24^2 + 0.6^2 + 0.18^2 + 0.45^2 + 0.48^2 + 0.5^2 + 0.4^2} = 1.3\%.$$

We now write the result of the measurement as follows:

$$\tilde{U}'' = 2.62 \text{ V}, \quad \delta = \pm 1.3\% \quad (\alpha = 0.95), \quad \text{or } U''_{0.95} = 2.62 \text{ V} \pm 1.3\%.$$

The example examined above shows clearly how the measurement uncertainty decreases as one transfers from *a priori* to *a posteriori* estimation and then from approximate to accurate error estimation.

5.6. Method for calculating the errors of multiple measurements

The mathematical-statistical methods studied in Chap. 4 form the theoretical foundation for estimating measured quantities and their errors in the case of multiple direct measurements. These methods can also be regarded as practical methods, if the systematic components of the measurement error are negligible compared with the random or quasirandom component.

In the general case both the systematic and random components of the error must be estimated. The random error can be estimated only *a posteriori*; the systematic error, however, can also be estimated *a priori*.

Consider first the case when the measurements are repeated in order to reduce the random errors. Having n single measurements, we obtain $\{x_i\}$, $i = 1,\ldots,n$, where $x_i = A + \zeta_i$ and $\zeta_i = \psi_i + \vartheta_i$, i.e., the error has both random and systematic components.

By repeating the measurements we obtain information about the random error. Information about the systematic error cannot be extracted from the measurements themselves. In order to estimate this error it is necessary to know the properties of the measuring instrument employed, the method of measurement, and the conditions under which the measurements are performed.

Assume that the systematic error of the result of each observation (single measurement) is known. Then, introducing the corrections $C_i = -\vartheta_i$, we obtain a group of corrected measurement results

$$x_i = A + \psi_i.$$

Our problem is to find the estimate $A = f(x_i)$. A mathematically well-founded solution, which is unbiased, consistent, and effective, can be found if the form of the distribution of x_i is known. Measurement errors can often be assumed to have a normal distribution. The measurement results also have the same distribution. In principle it is possible to check whether the data obtained conformed to a normal distribution (see the methods presented in Chap. 4). Admittedly, this requires a significant number of measurements; in practice enough measurements to make such a check are very rarely performed, and it is usually simply assumed that the distribution is normal.

For a normal distribution, as shown in Sec. 4.2, the arithmetic mean is the optimal estimate of the center of the distribution A. As noted above, the arithmetic mean of the measurements is an unbiased, consistent, and effective estimate of the true value of the measured quantity only if the observations have a normal distribution. Irrespective of the form of the distribution of the measurement errors, however, the arithmetic mean has two important properties.

(1) The sum of the deviations from the arithmetic mean is equal to 0. Let $x_1,...,x_n$ be a group of observational results whose arithmetic mean is $\bar{x}$. We construct the differences $x_i - \bar{x}$ for all $i = 1,...,n$ and find their sum:

$$\sum_{i=1}^{n} (x_i - \bar{x}) = \sum_{i=1}^{n} x_i - \sum_{i=1}^{n} \bar{x}.$$

Since $\Sigma_{i=1}^{n} x_i = n\bar{x}$, $\Sigma_{i=1}^{n} \bar{x} = n\bar{x}$,

$$\sum_{i=1}^{n} (x_i - \bar{x}) = 0.$$

This property of the arithmetic mean can be used to check the calculations.

(2) The sum of the squares of the deviations from the arithmetic mean is minimum. Consider the function

$$Q = \sum_{i=1}^{n} (x_i - \tilde{A})^2.$$

We shall find A so as to minimize Q. For this we find

$$\frac{dQ}{d\tilde{A}} = -2 \sum_{i=1}^{n} (x_i - \tilde{A})$$

and set $dQ/d\tilde{A} = 0$, hence we obtain

$$\sum_{i=1}^{n} (x_i - \tilde{A}) = 0, \quad \sum_{i=1}^{n} x_i = n\tilde{A}, \quad \text{and} \quad \tilde{A} = \bar{x} = \frac{\sum_{i=1}^{n} x_i}{n}.$$

Since $dQ/d\tilde{A} < 0$ if $\tilde{A}_1 < \bar{x}$ and $dQ/d\tilde{A} > 0$ if $\tilde{A}_2 > \bar{x}$, for $\tilde{A} = \bar{x}$ we have the minimum of Q.

Although the sum of the squares of the deviations from the arithmetic mean is minimum, this only means that in the class of estimates that are a linear function of the measurement results, the arithmetic mean is the most effective estimate of the measured quantity. This estimate becomes absolutely effective if the errors are distributed normally. For other distributions, as pointed out in Chap. 3, there exist estimates that are more effective than the arithmetic mean. Obviously, these estimates are no longer a linear function of the measurement results.

Thus, for the estimate of the measured quantity we have

$$\tilde{A}=\frac{\sum_{i=1}^{n} x_i}{n}. \tag{5.9}$$

Because of random errors the measurement results also are random quantities; if another series of measurements is performed, then the new arithmetic mean obtained will differ somewhat from the previously found estimate.

The spread of the arithmetic means is characterized either by the variance of the arithmetic means or by the standard deviation. In accordance with Eq. (4.5) the standard deviation of the arithmetic mean is estimated from the experimental data as follows:

$$S_{\bar{x}}=\sqrt{\frac{\sum_{i=1}^{n} (x_i-\bar{x})^2}{n(n-1)}}. \tag{5.10}$$

In addition, for A it is possible to construct the confidence interval, determining the confidence limits of the random error in the measurement results. The confidence interval is determined by the inequalities

$$\tilde{A}-\Psi_{\bar{x}}\leqslant A\leqslant\tilde{A}+\Psi_{\bar{x}},$$

where $\Psi_{\bar{x}} = t_q S_{\bar{x}}$, and t_q is the q percent point of Student's distribution and depends on the confidence probability α and the number of degrees of freedom $\nu = n - 1$ (see Table A.4).

Therefore the random error ψ with probability equal to the confidence probability α falls within the limits $\pm\ \Psi_{\bar{x}}$:

$$\Psi_{\bar{x}}=t_q S_{\bar{x}}. \tag{5.11}$$

As one can see from what was said above, the random errors and confidence limits of these errors can be estimated from the data obtained as a result of measurements.

The situation is different in the case of the systematic errors. The biasness, characterized by the systematic errors, of the result of a measurement can be estimated either with the help of more accurate means of measurement or based on indirect data, including data on the metrological properties of the measuring instrument employed to perform the measurements. The first case is pointless; the more accurate measurement would replace the less accurate measurement. The problem of estimating the systematic error would remain, except that now it would pertain to the more accurate result. For this reason the second case is the main one.

In the case of multiple measurements the most common variant is the one when the most important systematic errors are eliminated with the help of corrections.

Then the errors in determining the corresponding corrections must be taken into account together with the eliminated errors.

An important characteristic of multiple measurements is that the random components of the elementary errors are manifested in multiple measurements and contribute to the random error of the result. For this reason, when estimating elementary errors, it is desirable to neglect their random components. We shall return to this question at the end of the section.

By summing the elementary errors freed of random components, wc obtain the limits of the systematic error of the result of measurement. The method of summation is presented in Sec. 5.4.

Thus, we obtain an estimate $S_{\bar{x}}$ of the standard deviation of the random error of the result of measurement with a known number of measurements n and the estimate of the limits of the systematic component θ.

In some cases this separate estimate of the components of the error of the result of measurement is sufficient. This is the case, for example, if the result of measurement is to be used for calculations together with other data, for which separate estimates of the components of error are also known or if the result of measurement is to be compared with the results of other measurements, for which the error components are determined separately.

Very often, however, it is necessary to find the total error of a measurement, including both the random and systematic components. Not too long ago specialists on accurate measurements objected to this formulation of the problem. They said that systematic and random errors are of different nature, and for this reason they cannot be added. In 1965 I still listened to these assertions. However, most people disagreed with this. Indeed, in a completed measurement these components are physically indistinguishable, i.e., physically they add: By measuring A we obtain $\tilde{A}$. The difference $\tilde{A}$-A contains both systematic and random components. When analyzing the error, theoretically or experimentally, we decompose,this error into its components: systematic and random. Obviously, the inverse problem of adding the components is legitimate.

To solve the problem we shall take into account only the conditionally constant and random errors. Regarding both types of errors as random quantities, in order to combine them we must construct the composition of the corresponding distributions. Unfortunately, this is too difficult in practice. For this reason, the limit of total error is sometimes calculated according to the formula

$$\Delta=\theta+\Psi,$$

where θ is the limit of the systematic error and $\Psi = t_q S_{\bar{x}}$ is the confidence limit of the random error.

This formula is quite simple, but it is clear that it gives an obviously overestimated estimate. A more plausible estimate for Δ can be found by the following method.[48,51]

The estimate of the standard deviation of the total error of the result is

$$S_{\Sigma}= \sqrt{S_{\bar{x}}^2+S_{\vartheta}^2}. \tag{5.12}$$

Estimates for $S_{\bar{x}}$ and S_{ϑ} can be found using formulas (5.10) and (3.15). Given S_{Σ}, the limit of the error Δ of the result could be calculated from the formula

$$\Delta = t_\Sigma S_\Sigma, \tag{5.13}$$

if the coefficient t_Σ were known; unfortunately, this coefficient is unknown.

Since the initial data, i.e., the data on the components of the total error, are not known accurately, an approximate estimate of the coefficient t_Σ can be used. In Ref. 48 the following formula was proposed for making such an estimate:

$$t_\Sigma = \frac{\Psi_{\bar{x}} + \theta}{S_{\bar{x}} + S_\vartheta}.$$

This formula was constructed based on the following considerations. The coefficient t_q, determining the ratio of the confidence limit and the standard deviation of the random error, is determined by Student's distribution and is known. Given estimates for θ and S_ϑ, it can be assumed that the analogous coefficient for the systematic error

$$t_\vartheta = \theta / S_\vartheta$$

is known.

It is natural to assume that the coefficient sought t_Σ is some function of t_q and t_ϑ, corresponding to the same probability. The weighted mean of t_q and t_ϑ for the weights $S_\vartheta/(S_{\bar{x}} + S_\vartheta)$ and $S_{\bar{x}}/(S_{\bar{x}} + S_\vartheta)$ respectively, was taken for this function. This results in the proposed formula

$$t_\Sigma = \frac{t_q S_{\bar{x}} + t_\vartheta S_\vartheta}{S_{\bar{x}} + S_\vartheta} = \frac{\Psi_{\bar{x}} + \theta}{S_{\bar{x}} + S_\vartheta}. \tag{5.14}$$

In order to be able to use this formula, its accuracy must be estimated. The extreme cases are those when the systematic error has a normal or uniform distribution. The distribution of the random error of the arithmetic mean may be assumed to be asymptotically normal.

If both terms have a normal distribution, then $t_q = t_\vartheta$, and as follows from formula (5.14) $t_\Sigma = t_q$. Since the composition of normal distributions gives a normal distribution, the obtained value of t_Σ is exact.

For the second case the results of calculations based on the approximate formula (5.14) must be compared with the results obtained from the exactly constructed composition of normal and uniform distributions.

An expression for the distribution density of the sum of two unknown centered random quantities, one of which has a normal distribution and the other a uniform distribution, is known from the theory of probability:

$$f(z) = \frac{1}{2h} \int_{-h}^{h} \frac{1}{\sigma\sqrt{2\pi}} e^{-(z-y)^2/2\sigma^2} dy,$$

where h is equal to one-half the interval in which the random quantity Y is distributed uniformly.

Setting $\sigma = 1$ and transforming to the probability distribution function, we find

$$\underset{z>0}{F(z)} = 0.5 + \frac{1}{2h\sqrt{2\pi}} \int_0^z \int_{-h}^{h} e^{-(z-y)^2/2} dy\, dz.$$

TABLE 5.1. Characteristic quantiles of the composition of centered normal ($\sigma = 1$) and uniform ($h = \text{var}$) distributions.

h	0.50	1.0	2.0	3.0	4.0	5.0	6.0	8.0	10
$z_{0.95}$ (α=0.90)	1.71	1.90	2.49	3.22	4.00	4.81	5.65	7.34	9.10
$z_{0.975}$ (α=0.95)	2.04	2.25	2.90	3.67	4.49	5.34	6.22	8.00	9.81
$z_{0.995}$ (α=0.99)	2.68	2.94	3.66	4.49	5.36	6.26	7.17	9.02	10.9

The starting distributions are symmetric relative to 0. For this reason, the density of the resulting distribution also has this property. We must find the limit of the confidence interval corresponding to the probability α. For this it is sufficient to find either the quantile z_r of the level r or the quantile of the level $1-r$, since $|z_r| = |z_{1-r}|$. Since $\alpha = 1 - 2r$, $r = (1-\alpha)/2$. Obviously, $r < 0.5$ and $z_r < 0$.

Table 5.1 gives values of z_{1-r} calculated using the presented formula for $\alpha = 0.90$, 0.95, and 0.99.

The relative error introduced by the use of the approximate formula (5.14) will be

$$\delta = \frac{\Delta - z_{1-r}}{z_{1-r}}.$$

The comparison should be made with $S_\Sigma = \sigma_\Sigma$, since $S_{\bar{x}} = \sigma = 1$. In so doing,

$$\sigma_\Sigma = \sqrt{\sigma^2 + \frac{h^2}{3}} = \sigma\sqrt{1 + \frac{1}{3}\left(\frac{h}{\sigma}\right)^2}.$$

For this reason, introducing the coefficient $t_r = z_{1-r}/\sigma_\Sigma$, we obtain

$$\delta = \frac{t_\Sigma - t_r}{t_r}.$$

The coefficient t_r depends only on the configuration of the resulting distribution, i.e., on the ratio of h and σ, and not on their absolute values. For this reason, a series of values of this coefficient can be calculated from the data in Table 5.1. These values are presented in Table 5.2.

TABLE 5.2. Values of the coefficient t_r as a function of the parameters of the normal and uniform distributions.

h/σ	0.5	1	2	3	4	5	6	8	10
σ_Σ (σ=1)	1.04	1.15	1.53	2.00	2.52	3.06	3.51	4.72	5.85
$t_{0.95}$ (α=0.90)	1.65	1.64	1.63	1.61	1.59	1.58	1.57	1.56	1.55
$t_{0.975}$ (α=0.95)	1.96	1.95	1.90	1.84	1.78	1.75	1.72	1.69	1.67
$t_{0.995}$ (α=0.99)	2.57	2.54	2.40	2.24	2.13	2.05	1.99	1.91	1.86

TABLE 5.3. Values of the coefficient t_Σ as a function of the parameters of the normal and uniform distributions.

h/σ	0.5	1	2	3	4	5	6	8	10
$t_{1\Sigma}$ (α=0.90)	1.63	1.61	1.60	1.59	1.58	1.58	1.58	1.57	1.57
$t_{2\Sigma}$ (α=0.95)	1.89	1.84	1.79	1.76	1.74	1.73	1.72	1.70	1.69
$t_{3\Sigma}$ (α=0.99)	2.38	2.26	2.11	2.03	1.97	1.94	1.91	1.87	1.84

We shall now once again turn to the approximate formula (5.14). The limits of the confidence interval, which are determined based on the uniform distribution, give θ. Since $r=(h-\theta)/2h$ and $r=(1-\alpha)/2$,

$$\theta=(1-2r)h=\alpha h. \tag{5.15}$$

The limit of the confidence interval for a normal distribution with the same confidence probability will be

$$\Delta_{\bar{x}}=z_{(1+\alpha)/2}\sigma,$$

where $z_{(1+\alpha)/2}$ is the quantile of the standard normal distribution.

Formula (5.14) assumes the form

$$t_\Sigma=\frac{z_{(1+\alpha)/2}+\alpha\dfrac{h}{\sigma}}{1+\dfrac{1}{\sqrt{3}}\dfrac{h}{\sigma}}.$$

The values of t_Σ calculated for the same ratios h/σ and confidence probabilities as were used for calculating t_r are presented in Table 5.3. The errors δ calculated based on the data given in Tables 5.2 and 5.3 are summarized in Table 5.4.

Thus comparing the results of exact calculations with the results of calculations performed using the approximate formula (5.14) shows that the errors due to the use of the approximate formula are in all cases negative and their absolute magnitude does not exceed 12% (for $\alpha=0.99$). This shows that the formula (5.14) can be used.

It should also be noticed that the error under study decreases as the distribution of the systematic errors approaches the normal distribution.

TABLE 5.4. Deviations of the coefficient t_Σ from t_r (in %).

h/σ	0.5	1	2	3	4	5	6	8	10
δ_1 (α=0.90)	1.2	1.9	1.8	1.1	0.6	0	0.8	0.6	1.2
δ_2 (α=0.95)	3.6	5.5	5.7	4.1	2.2	1.3	0	0.5	1.0
δ_3 (α=0.99)	7.4	11	12	9.4	7.3	5.5	4.0	2.2	1.1

The scheme presented above for estimating the error of a measurement, containing both random and systematic components, is a general scheme. In different particular problems it can be substantially modified.

We also discuss the question of when the systematic or random component of the error can be neglected.

Figure 5.2 shows plots, constructed based on the data in Table 5.1, of $z_{1-r}(h)$, corresponding to the composition of normal and uniform distributions and to the uniform distributions. Comparing the curves 1 with the values of z_{1-r} of a normal distribution (the points for $h=0$) we can find the error introduced by neglecting the systematic component. If it is assumed that this error does not exceed 15%, then we obtain the limiting values of the ratio h/σ. Depending on the confidence probability these ratios are as follows:

α	0.90	0.95	0.99
h/σ	1.2	1.1	1.1 .

If the random component is neglected, then the error arising is determined by the difference of the ordinates of the exact curve 1 and the straight line 2 for fixed h. For the same 15% error, we obtain the condition under which the random component can be neglected:

α	0.90	0.95	0.99
h/σ	3	4	7 .

It is obvious that when any component is neglected, the overall error decreases.

Thus if $h/\sigma < 1$, then the systematic error can be confidently neglected, and if $h/\sigma > 7$ then the random component can be neglected.

Admittedly, we do not know the exact values of the parameters h and σ; we know only their estimates θ and S. For this reason, in order to be rigorous the upper limit of the confidence interval for σ should be used instead of σ when determining whether or not the random component of the error can be neglected, and the lower limit should be used when determining whether or not the systematic component can be neglected.

It should be noted that in general the number of measurements should be chosen so that the random error of the arithmetic mean would be negligible compared with the systematic error.

In summing random and systematic errors we neglected the absolutely constant systematic component of the error. If this error is present, then its limits must be added arithmetically to the obtained estimate.

The method studied above pertains to measurements whose errors are estimated exactly. For measurements whose errors are estimated approximately the properties of the measuring instruments employed are taken into account based on the specifications.

We must now make a remark regarding the problem of taking into account the intrinsic error of instruments in the case of multiple measurements. The point of doing this is to avoid taking into account the random component of this error twice. The possibility of such an overestimation is concealed in the fact that, on the one hand, the spread in the results of single measurements reflects all sources of random error, including also the random component of the intrinsic error of instruments,

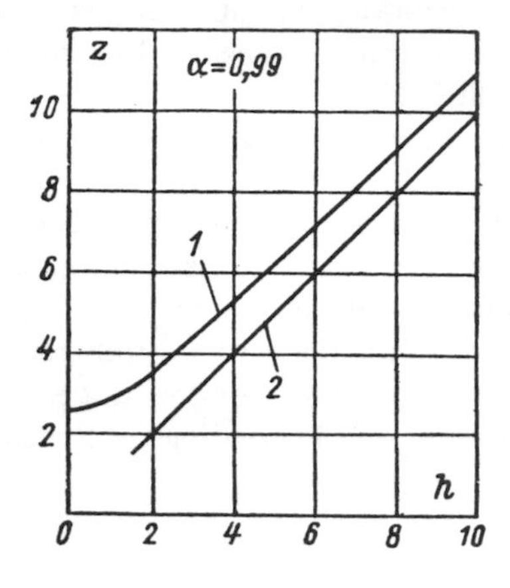

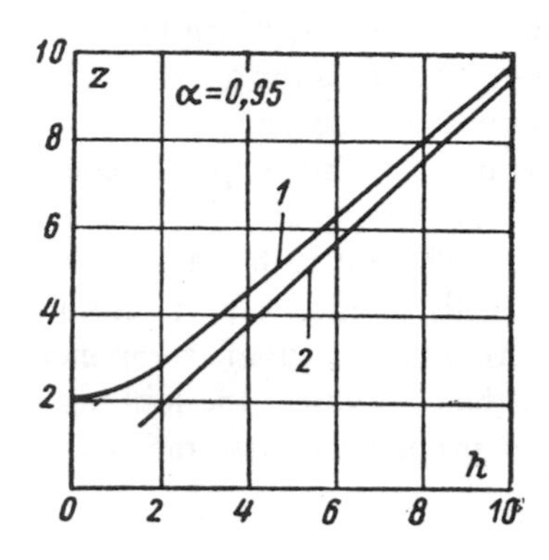

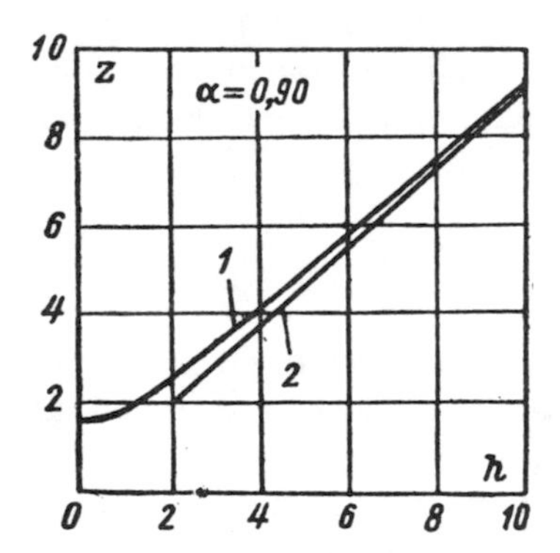

FIG. 5.2. Quantiles of the levels 0.99, 0.95, and 0.90 for a composition of the normal and uniform distributions (curves 1) and for uniform distribution (straight lines 2).

while on the other hand, when estimating the elementary component contributed by the intrinsic error of the instrument, this component also enters the calculation as part of the intrinsic error.

To avoid double counting the random component of the intrinsic error of the instrument, it must be eliminated from the intrinsic error when estimating the limits of the corresponding elementary error; it is more difficult to remove it from the spreads in the results of single measurements or the parameter characterizing them.

In the case when digital instruments are employed, the problem can be solved comparatively simply. For this, it is sufficient to drop in the calculation the second term $\pm(b+q)$ in the formula or, if the error is normalized according to formula (2.2), replace δ by $\delta' = c - d$.

In the case of analog instruments this problem must be solved depending on the properties of the particular instruments. For electric measuring instruments, for example, the random component of the intrinsic error is normalized together with the intrinsic error by prescribing the permissible limits of dead zone. As pointed out above, the limit of random error is equal to one-half the dead zone.

When the corresponding elementary error of a multiple measurement is estimated, the limit of intrinsic error of the instrument must be reduced by precisely one-half the limit of dead zone.

Naturally, the elementary error need not be reduced, but rather this random component can be taken into account when calculating the variance or the standard deviation of the measurement result. For this, the variance of the random component must be calculated from the random component of the intrinsic error of the instrument, and then its value must be subtracted from the estimate of the variance of the measurement result.

We shall return to the problem of combining the systematic and random components of the measurement error. This problem is significantly simplified, if the systematic component has many components and it can be assumed that it has a normal distribution. For this, in practice, it is sufficient that the systematic component consist of five or more elementary errors with approximately the same limits. If the number of measurements exceeds 20, then when calculating the limits of the systematic and random components the quantile coefficient will be the same for the same confidence probability. We shall have $\Psi = tS_{\bar{x}}$ and $\theta = tS_{\vartheta}$. As before $S^2_{\Sigma} = S^2_{\vartheta} + S^2_{\bar{x}}$ or $t^2S^2_{\Sigma} = t^2S^2_{\vartheta} + t^2S^2_{\bar{x}}$. From here it follows that

$$tS_{\Sigma} = \sqrt{(tS_{\vartheta})^2 + (tS_{\bar{x}})^2}$$

or, since $\Delta = tS_{\Sigma}$,

$$\Delta = \sqrt{\theta^2 + \psi^2}. \tag{5.16}$$

Clearly, this error corresponds to the same confidence probability for which Ψ and θ were calculated.

Now we shall discuss the absolutely constant elementary errors. Their limits are added to the sum of the other errors using the same procedure as that studied above for single measurements:

$$\Delta' = \sum_{f=1}^{k} H_f + \Delta, \tag{5.17}$$

where $\{H_f\}$, $f = 1,...,k$, are the limits of the absolutely constant elementary errors.

We note that if some elementary errors have unsymmetric limits, then they are represented by symmetric limits after being shifted relative to the result of measurement by a_j. The calculations are performed by the same method as in the case of single measurements. We recall once again that the biases a_j cannot be compensated by introducing corrections: The error estimates are too unreliable to change based on the measurement result.

We shall now discuss multiple measurements with quasirandom errors. The construction of the confidence interval for this case was discussed above. We shall discuss how to combine quasirandom and systematic errors.

Suppose we have a set of measurements $\{x_i\}$, $i = 1,...,n$, and the confidence interval $[x_k, x_{n-k+1}]$ for the measured quantity A. From the length of the nonparametric confidence interval we find the corresponding confidence probability. The solution of these problems is presented in Chap. 4. In this manner we find the confidence limit of the quasirandom error $\Psi = \frac{1}{2}[x_k, x_{n-k+1}]$.

In addition, the estimate $S_{\bar{x}}$ is calculated in the standard manner. Then the coefficient similar to Student's quantile coefficient $t_e = \Psi / S_{\bar{x}}$ can be found.

Since the confidence probability for Ψ is calculated and not selected in our case, as usual, the limit of the systematic error θ must also be calculated for the same probability. Using the by now familiar scheme we find

$$S_{\Sigma} = \sqrt{S_{\bar{x}}^2 + S_{\vartheta}^2}, \quad t_{\Sigma} = \frac{\theta + \psi}{S_{\vartheta} + S_{\bar{x}}}, \text{ and } \quad \Delta = t_{\Sigma} S_{\Sigma}.$$

Another way to solve the problem is to construct the composition of the distributions of the quasirandom and systematic errors. The distribution of the quasirandom error is represented by a histogram, which is constructed from the experimental data. The distribution of the systematic error, however, is most often obtained in an analytic form, i.e., in the form of an equation. The latter equation, however, is easily transformed into a histogram, after which the problem of constructing the composition of two histograms is solved by the numerical method studied in Chap. 3.

Finally, we note that when measuring average quantities (which is precisely when quasirandom errors can appear) the systematic errors are often negligibly small, which in general eliminates the problem of combining errors.

5.7. Comparison of different methods for combining systematic and random errors

The foregoing method for combining systematic and random errors is not the only method.

(1) The U.S. National Bureau of Standards (now the National Institute of Standards and Technology) gives in Ref. 22 the formula

$$\Delta = \theta + \Psi, \tag{5.18}$$

where $\theta = \sqrt{\Sigma_{i=1}^{m} \theta_i^2}$, if $\{\theta_i\}$, $i = 1,...,m$, are independent systematic components, and $\theta = \Sigma_{i=1}^{m} \theta_i$, if they are dependent, and $\Psi = t_q S_{\bar{x}}$.

The formulas for combining elementary systematic errors have already been investigated above. We shall now discuss formula (5.18). First of all, Ref. 22 is a working document of the National Institute of Standards and Technology (NIST), an organization that is especially interested in the problems of checking and calibrating measuring instruments. Measurements in this case can have absolutely constant elementary errors more often than in other cases. If such components predominate, then formula (5.18) is justified. But this method of calculating errors cannot be extended to arbitrary measurements, since in most cases it results in overestimation of the errors.

(2) The standard Ref. 2 and the manual Ref. 13 preceding it give two different formulas for calculating the total error of a measurement (uncertainties) with confidence probabilities of 0.95 and 0.99:

$$\Delta_{0.99} = \theta + t_{0.95} S_{\bar{x}}, \quad \Delta_{0.95} = \sqrt{\theta^2 + (t_{0.95} S_{\bar{x}})^2}.$$

The coefficient $t_{0.95}$ is chosen according to Student's distribution in both cases for the confidence probability 0.95.

The first formula is correct under certain conditions; it was analyzed above. The second formula appears to be very strange. The first term in this formula is not related with probabilistic relations, while the second formula was calculated for a confidence probability of 0.95. It is not entirely clear why the arithmetic sum of these terms is given a confidence probability of 0.99.

(3) The International Bureau of Weights and Measures (BIPM) in 1981 recommended an expression for measurement uncertainties.[30] This recommendation is developed in the projects ISO/TAG4/WG3, and method that is placed in the Fourth Draft of Guide to the Expression of Uncertainty in Measurement* was reflected in Ref. 19.

The elementary systematic errors are regarded as uniformly distributed random quantities. However, the limit of their sum is calculated using the formula $\theta = \sqrt{\Sigma_{i=1}^{n}\theta_i^2}$, i.e., without using the indicated model.

The systematic and random errors are combined using a formula that is essentially the same as Eq. (5.13). The only difference lies in the coefficient t_Σ. Here the coefficient is found from the Student's distribution corresponding to the selected confidence probability and the effective degree of freedom ν_{eff}. The following formula is given to calculate ν_{eff}:

$$\frac{S_\Sigma^4}{\nu_{\text{eff}}} = \frac{S_{\bar{x}}^4}{\nu} + \sum_{i=1}^{m} \left(\frac{\theta_i^2}{3}\right)^2.$$

It is assumed here that the random component of uncertainty has a degree of freedom $\nu = n - 1$ and each component of the systematic error has a degree of freedom equal to one. However, the notion of a degree of freedom is not applicable to random variables with a fully defined distribution function. Therefore it is impossible to assume that a quantity with uniform distribution within given limits has a degree of freedom equal to one (or to any other finite number). Thus the formula under discussion is not mathematically grounded.

We shall compare all methods enumerated above for summing the systematic and random errors in two numerical examples.

Suppose that as a result of some measurement the following indicators of its errors were obtained:

$$S_{\bar{x}} = 1, \quad n = 16, \quad \theta = 3.$$

Suppose also that the random errors have a normal distribution and the systematic errors have a uniform distribution. Then for the exact solution we can take the confidence limits presented in Table 5.1. As usual, we shall take $\alpha_1 = 0.95$ and $\alpha_2 = 0.99$. Then

$$\Delta_{T\ 0.99} = 4.49, \quad \Delta_{T\ 0.95} = 3.67.$$

Here there is an error: We assumed that $S_{\bar{x}} = \sigma_{\bar{x}}$. But for $n = 16$ this error is not very significant, and we shall neglect it.

We shall present the computational results obtained using all of the methods examined above.

---

*"Guide to the Expression of Uncertainty in Measurement; Fourth Draft." July 13, 1990. ISO/TAG 4/WG3.

TABLE 5.5. Errors of different methods of uncertainty calculation for example with $\theta = 3$, $\delta_x = 1$, $\nu = 16$.

Method of computation	$(\Delta_i - \Delta_T)/\Delta_T \times 100\%$	
	$\alpha = 0.99$	$d = 0.95$
1	32	39
2	14	0.3
3	340	132
4	3	0.8

(1) Reference 22 NBS (NIST). The coefficients of Student's distribution with $\nu = n - 1 = 15$ and the indicated values of the confidence probabilities will be as follows:

$$t_{0.99}(15) = 2.95, \quad t_{0.95}(15) = 2.13,$$

$$\Psi_{0.99} = 2.95 \times 1 = 2.95, \quad \Psi_{0.95} = 2.13 \times 1 = 2.13.$$

Therefore, $\Delta_{1,0.99} = 3 + 2.95 = 5.95$ and $\Delta_{1,0.95} = 3 + 2.13 = 5.13$.

(2) Reference 2. We shall make use of the calculations that were just performed:

$$\Delta_{2,0.99} = 3 + 2.13 \times 1 = 5.13, \quad \Delta_{2,0.95} = \sqrt{3^2 + (2.13)^2} = 3.68.$$

(3) Reference 19.

$$S_{\vartheta}^2 = 9/3 = 3, \quad S_{\vartheta} = 1.73,$$

$$S_{\Sigma}^2 = 1 + 3 = 4, \quad S_{\Sigma} = 2.$$

We shall calculate the effective number of degrees of freedom:

$$\frac{4^2}{\nu_{\text{eff}}} = \frac{1}{15} + 3^2, \quad \frac{16}{\nu_{\text{eff}}} = 9.07, \quad \text{and } \nu_{\text{eff}} = 2.$$

Next, we find from Student's distribution $t_{3,0.99} = 9.9$ and $t_{3,0.95} = 4.3$. Correspondingly, we obtain

$$\Delta_{3,0.99} = 9.9 \times 2 = 19.8, \quad \Delta_{3,0.95} = 4.3 \times 2 = 8.6.$$

(4) The formulas (5.12)–(5.14) give $S_{\vartheta} = 1.73$ and $S_{\Sigma} = 2.0$.

$$t_{\Sigma,\,0.99} = \frac{2.95 \times 1 + 0.99 \times 3}{1 + 1.73} = \frac{5.92}{2.73} = 2.17,$$

$$t_{\Sigma,\,0.95} = \frac{2.13 \times 1 + 0.95 \times 3}{1 + 1.73} = \frac{4.98}{2.73} = 1.82,$$

$$\Delta_{\Sigma,\,0.99} = 2.17 \times 2 = 4.34, \quad \Delta_{\Sigma,\,0.95} = 1.82 \times 2 = 3.64.$$

We shall compare the estimated errors with the exact values initially presented for the corresponding confidence intervals. The results are summarized in Table 5.5.

TABLE 5.6. Errors of different methods of uncertainty calculation for example with $\theta = 0.5$, $\delta_x = 1$, $\eta = 16$.

	$(\Delta_i - \Delta_T)/\Delta_T \times 100\%$	
Method of computation	$\alpha = 0.99$	$d = 0.95$
1	29	30
2	2	7
3	13	8
4	4	3

The errors for the case $\theta = 0.5$ and the previous values $S_{\bar{x}} = 1$ and $n = 16$ were calculated analogously. The results are presented in Table 5.6.

The examples presented show the following.

(a) As expected, the method of Ref. 19 cannot be used when the systematic error is significant.**

(b) The method from the standard Ref. 2, irrespective of the remarks made above, gave in both examples satisfactory results.

(c) The NBS (NIST) method, as expected, gave in the examples studied estimates that were too high.

(d) The formulas (5.12)–(5.14) gave good results in both examples.

Examples are not, of course, proof, but they nonetheless illustrate well the consideration stated above.

5.8. Essential aspects of the estimation of measurement errors when the number of measurements is small

In practice, measurements are often performed with a very small number of observations, for example, two or three observations. This is not enough for statistical analysis, and such measurements cannot be called multiple measurements. At the same time, they are also not single measurements.

What is the point of this seemingly strange choice of the number of measurements? Analysis shows that this is done in order to check the suitability of the model selected for the object of study. For example, when the diameter of a shaft is being measured, it is measured at several locations along the shaft and, in addition, in different directions. If the conditions under which the measurements are performed are sufficiently stable and the model (cylinder) corresponds to the object (shaft), then the differences between the single-measurement results should be small, and in any case less than twice the error of a single measurement. If the difference between the measurements is large, then it is pointless to use the selected

**The shortcomings of this method were discussed in the report "The U.S.A. and the U.S.S.R. standards for Measurement Uncertainty" given by S. Rabinovich at the Measurement Science Conference in Anaheim, CA, January 31 and February 1, 1991.

measurement method. If the difference between the measurements is small, then there arises the question of what should be regarded as the result of a measurement and what is its error.

In this case, in principle the result of any single measurement can be used as the result of the measurement. But since the measurements have already been performed, it would still be nice to use them somehow. For this reason, the arithmetic mean of the observational results is used as the result of measurement. In this respect, this case is no different from the case of multiple measurements. However, the situation is different with regard to the measurement error. Here two cases must be borne in mind.

(1) The difference between the observational results is insignificant, i.e., three or more times less than the limit of error. This means that the model corresponds well to the object and that the random component of the error of the observational results is small. As regards systematic errors, since the conditions under which the observations are performed are constant and the measuring instruments are not changed, the systematic error will be the same for all measurements. For this reason, in this case the measurement error is also the error of a single measurement.

(2) The difference between the observational results is significant. Let us see how the measurement error can be estimated in this case.

The difference with the largest modulus in a sample consisting of n elements is called in mathematical statistics the range R_n. The distribution function of the range is tabulated for normally distributed random quantities. Assuming that the random error of a single measurement has a normal distribution, we can write

$$P\{R_n \leqslant a\} = \alpha,$$

where a is the limit chosen for R_n. Next, assume that we have three single measurements and that $\alpha = 0.95$.

From the table presented for example in Ref. 34 we find, for $\alpha = 0.95$ and $n = 3$, that $a = 3.3\sigma$ or $\sigma = a/3.3$.

Let $a = b\Delta$, where Δ is the limit of error of a single measurement. But for $\alpha = 0.95$ the limit of random error can be estimated as

$$\Psi_1 = 1.96\sigma = \frac{1.96}{3.3} a = 0.594 b\Delta.$$

Let the systematic error consist of more than four components. Then

$$\Delta = \sqrt{\theta^2 + \Psi_1^2}, \quad \theta = \sqrt{\Delta^2 - \Psi_1^2} = \Delta\sqrt{1 - 0.353 b^2}.$$

The radicand must be greater than 0. Then the maximum value of the coefficient is $b_m = 1.7$.

We shall now consider the error of the arithmetic mean. The standard deviation in this case will be $\sigma_{\bar{x}} = \sigma/\sqrt{n} = \sigma/\sqrt{3}$. From here we find the limiting value of the random error:

$$\Psi_{\bar{x}} = t_{0.95}\sigma_{\bar{x}} = \frac{\Psi_1}{\sqrt{3}} = \frac{0.594}{\sqrt{3}} b\Delta.$$

The limiting value of the systematic error has already been estimated:

$$\Theta=\Delta\sqrt{1-0.353b^2}.$$

For the total measurement error, we obtain from here the expression

$$\Delta_{\bar{x}}=\sqrt{\theta^2+\Psi_{\bar{x}}^2}=\Delta\sqrt{(1-0.353b^2)+\left(\frac{0.594b}{\sqrt{3}}\right)^2}=\Delta\sqrt{1-0.235b^2}.$$

Now we can estimate the decrease in the error of the arithmetic mean compared with the error of a single measurement with $n=3$ measurements: $\mu = \Delta\bar{x}/\Delta = \sqrt{1-0.235b^2}$

$$b=0.5,\quad \mu=0.97,$$

$$b=1.0,\quad \mu=0.87,$$

$$b=1.5,\quad \mu=0.69,$$

$$b=1.7,\quad \mu=0.57.$$

The case $b=1.7$ means that the entire error of a single measurement is determined by the random component. In this case the measurement should be designed as a multiple measurement.

For $b=1.0$ we have $\mu=0.87$, i.e., the error in the measurement result is equal to the error in a single measurement.

Thus, if we assume that the range, i.e., the largest difference between the measurements, for $n=3$ can reach the limit of permissible error of a single measurement, then the error of the arithmetic mean will be approximated the same as the error of a single measurement. In addition, requiring that the limit of error of a single measurement be the smallest fraction of the range will not significantly reduce the estimate of the error of the arithmetic mean.

5.9. General plan for estimating measurement errors

The purpose of this section is to give, without getting into details, an overall plan for estimating the errors of direct measurements. This should help the reader concentrate on the essential points of each step in the solution of the problem.

1. Analyze the initial data.

1.1. Study the measurement problem. For this, one must first get an idea of the object whose parameter is being measured, the purpose of the measurement, and the required measurement accuracy. In connection with these questions it is necessary to determine a model of the object and try to check that the parameter to be measured (the measured quantity) corresponds to the required measurement accuracy. Next, it is necessary to write out the physical quantities characterizing the surrounding environment and affecting the size of the measured parameter, to estimate their nominal values and range of variation, and to determine how these quantities must be measured, if the measurement is being planned, or were measured, if the measurement has already been performed.

1.2. Establish which of the metrological properties of the measuring instruments chosen for the measurement or already employed in performing the measurement are important for the given measurement.

2. Prepare the data for the calculations.

2.1. Compile a list of all possible elementary errors in the given measurement.

2.2. Estimate the limits of all elementary errors. Express them in the form of absolute or relative errors and scale to the input of the measuring apparatus, instrument, or channel.

2.3. Determine whether or not it is useful to introduce corrections and the possibility of obtaining point estimates of the corresponding elementary errors necessary for this. This question must first be resolved for the dominant elementary errors. Determine the corrections to be made. Estimate the limits of inaccuracy each correction and add them to the list of elementary errors.

2.4. Check the independence of the elementary errors. If the errors ε_1 and ε_2 owing to different causes depend on some third physical quantity, then these errors will be dependent. To eliminate this dependence it is often sufficient to introduce a new elementary error that reflects the effect of this third quantity on the result of measurement. After this, instead of ε_1 and ε_2 we shall have new elementary errors ε_1' and ε_2', which can now be regarded as being independent.

2.5. Divide all elementary errors into conditionally and absolutely constant errors, and single out those errors whose limits differ in absolute magnitude, i.e., are unsymmetric relative to the result of measurement. If the measurement is multiple, then it is necessary to determine whether or not its error is purely random or quasirandom. Estimate the confidence limits of this error.

2.6. Estimate the quantities necessary for calculating the additional errors. For this it is desirable to measure these quantities; this is necessary if one intends to introduce the corresponding corrections. In the case when the limiting value of the additional error is estimated, it is usually sufficient to have the limiting value of the influential quantity.

2.7. Estimate the possible change in the intrinsic error of the instruments over the time period since they were calibrated. If there are grounds for assuming that the intrinsic error could have exceeded permissible values, then such instruments must be rechecked prior to performing the measurement and, if necessary, regulated or recalibrated.

3. Calculate the result of measurement.

3.1. In the case of single measurements the result of a measurement is often obtained directly from the indication of the measuring instrument, and no calculations are required for this. Sometimes, however, the indication of an instrument in units of the scale graduations must be multiplied by the scale factor, corrections must be introduced, and other nonspecific calculations must be performed. In the case of multiple measurements, the arithmetic mean is usually taken as the result of measurement. However, a different algorithm, determined by the definition of the measured quantity, can also be used. The corrections, if they are the same for all single measurements, can be introduced in the arithmetic mean and not in the result of each individual measurement.

3.2. *A priori* estimation of error is usually made for the least favorable case. If multiple measurement is planned, then the possible value of the standard deviation is taken based on recommendations of experts. The methods for performing the calculations were presented in this chapter.

3.3. *A posteriori* estimation of error is performed using the methods presented in this chapter.

3.4. The form in which the results of measurement are presented and their errors were presented in Chap. 1.

Chapter 6

Indirect measurements

6.1. Indirect measurements and calculations

Indirect measurements are measurements in which the value of the unknown quantity sought is calculated using matched measurements of other quantities related with the measured quantity by some known relation. We shall call these other quantities *measured arguments* or, briefly, arguments.

The difference between indirect measurements and calculations, if only with the use of results of measurements or other data, which are known with limited accuracy, was pointed out in Chap. 1. The main difference is that in the case of indirect measurements the values of the arguments are found based on specially performed measurements that are matched with one another. This does not happen in the case of calculations. However, this distinction is only of physical meaning and is reflected in the manner in which the value of the quantity of interest can be determined. This distinction has no bearing on the calculations of the error in the result obtained. For this reason the methods studied in this chapter will be applicable to both indirect measurements and calculations.

In keeping with this remark, in what follows we shall, for brevity, speak only about indirect measurements.

It should also be noted that the methods used to estimate errors of indirect measurements are often necessary to estimate the errors in direct measurements. This happens when several different measuring instruments are connected into a single network, forming the measurement channel. An example will be studied in this chapter.

6.2. Classification of indirect measurements

The measured quantity (whose true value is A) is related to the measured arguments A_i $(i=1,\dots,m)$ by a function that can usually be resolved relative to A, i.e., it can be represented in the form

$$A=f(A_1,\dots,A_m). \tag{6.1}$$

The cases of an implicit dependence between A and A_i are atypical.

We shall distinguish, according to the form of the functional dependence (6.1), indirect measurements with a linear dependence between the measured quantity and measured arguments, indirect measurements with a nonlinear dependence between these quantities, and indirect measurements in which the dependence between the quantities is of a mixed type. In the case of a linear dependence Eq. (6.1) has the form

$$A = \sum_{i=1}^{m} b_i A_i,$$

where b_i is a constant coefficient of the ith argument A_i, and m is the number of terms.

We shall call indirect measurements linear when the quantities are related linearly and nonlinear when they are related nonlinearly.

In the general case, Eq. (6.1) for nonlinear indirect measurements can be represented as a product of several functions:

$$A = \prod_{i=1}^{m} f_i(A_i).$$

In the particular case, $A = f(A_i)$.

In the case when the dependence between the quantities is of the mixed type, Eq. (6.1) assumes the form

$$A = \prod_{i=1}^{m} f_i(A_i) + \cdots + \prod_{l=1}^{r} f_l(A_l).$$

If methods for analyzing the results of observations for linear and nonlinear indirect measurements are known, then the analogous problem for the case with a relation of a mixed type reduces in an elementary fashion to the two preceding cases. For this reason, this form of the indirect measurements need not be specially studied.

Indirect measurements, just like direct measurements, are divided into static and dynamic. Static indirect measurements can be very different depending on the properties of the measured arguments. If the measured arguments can be regarded as being constant in time, then the indirectly measured quantity is also constant, i.e., we have the usual static situation.

However, the measured quantity can also be constant when the arguments vary. For example, suppose we are measuring the resistance of a resistor by the ammeter and voltmeter method, and the voltage of the source changes in time according to the conditions of the measurement. Although the measured arguments change, the measured quantity remains unchanged.

To obtain the correct result in the case under study the arguments must be measured with instruments such that the arguments do not change significantly over the time interval during which the indications of the instruments settle down.

Indirect measurements are also possible, in principle, when the measured arguments and the indirectly measured quantity itself change in time. Indirect measurements, for which the measuring instruments or some of them are in the dynamic regime, in accordance with the general definition of dynamic measurements, must be regarded as dynamic.

In addition, single and multiple indirect measurements must be distinguished.

A specific technique for performing indirect measurements is to measure the arguments simultaneously. This makes it possible to substitute simultaneously the obtained values of the arguments into the relation that relates them to the measured quantity, and to obtain in this manner an instantaneous value of the measured quantity corresponding to the time at which the arguments are measured. The collection of such values is in no way different from the collection of instantaneous values of the quantity obtained with direct measurements.

It is natural to process the collections of instantaneous values so obtained in the same way as collections of data obtained with direct measurements.

It is desirable to reduce indirect measurements to direct measurements not only in the case of dynamic measurements, but also—in the case when the measured quantity and the measured arguments are related by a complicated nonlinear relation—with static indirect measurements. The necessary condition for realizing this technique is that the arguments must be measured in a matched fashion, for example, simultaneously.

We call the method of indirect measurement in which a group of values of the measured quantity is obtained and the result of the indirect measurement is found by analyzing the group as a group of observations obtained by means of direct measurements the method of reduction.

6.3. Error owing to incomplete correspondence between the model and the object of study

The equation expressing the relation between A and $\{A_i\}$, $i = 1,...,m$, is the definition of the quantity to be measured indirectly and is closely related with the model of the object of study. It is often difficult to give a clear interpretation of this model. In many cases this is possible only for each measured argument separately.

Problems from the field of lineal–angular measurements are, as always, clearest. For example, suppose that we are required to measure the area of a plot of land that is depicted by a rectangle on a sketch. Here the rectangle is the model of the object. Its area is $S_m = ab$, where a and b are the lengths of the sides of the rectangle.

The discrepancies between the model and the object can in this case be expressed by the fact that the angle between the sides will not be exactly 90°, that the opposite sides of the section will not be precisely identical, and that the lines bounding the area will not be strictly straight. Each discrepancy can be estimated quantitatively and then the error introduced by it can be calculated. It is usually obvious beforehand which source of error will be most important.

Suppose that in our example the most important source of error is that the angle between adjoining sides differs from 90° by α, as shown in Figs. 6.1. Then the area of the plot would have to be calculated according to the formula $S_t = ab \cos \alpha$.

Therefore the error owing to the threshold discrepancy in this case will be

$$\Delta_m = S_m - S_t = ab(1 - \cos \alpha), \quad \delta_m = \frac{\Delta_m}{S_t} = \frac{1 - \cos \alpha}{\cos \alpha}.$$

The admissable angle α_a must be estimated from the required accuracy in determining the area of the plot of land. If $\alpha \geqslant \alpha_a$, then the model must be redefined

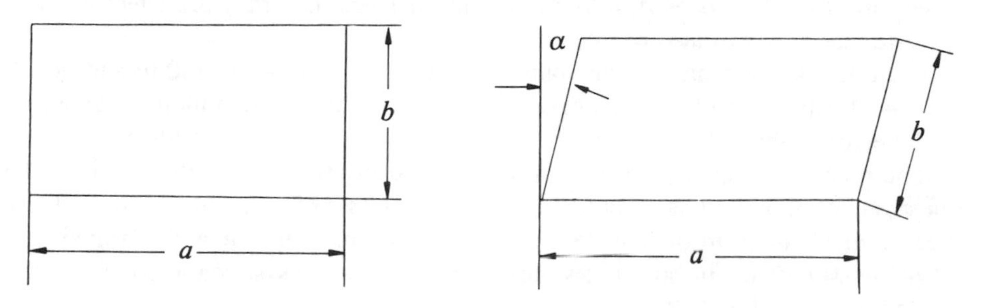

FIG. 6.1. Rectangle and parallelogram as a model of the plot of land.

and the measured quantity must be defined differently. Correspondingly, we shall obtain a different formula for calculating the measured area.

6.4. Linear indirect measurements

The linear functional relation between the measured quantity A and the measured arguments A_i can be expressed in the general form by the formula

$$A = \sum_{i=1}^{m} b_i A_i, \tag{6.2}$$

where b_i is a constant coefficient for the ith argument.

In some cases the constant coefficients are known exactly. For example, the inaccuracy can be determined by the spacing in a table from which the value of some coefficient is taken. In such cases this coefficient must be regarded as the measured argument and the linear indirect measurement becomes nonlinear.

If a random quantity Y, related linearly with the random quantities X_i,

$$Y = \sum_{i=1}^{m} b_i X_i,$$

is used, then

$$M[Y] = \sum_{i=1}^{m} b_i M[X_i].$$

For this reason, given estimates of the arguments A_i, it is natural to take for the estimate of the measured quantity A

$$\tilde{A} = \sum_{i=1}^{m} b_i \tilde{A}_i, \tag{6.3}$$

since if the estimates of $\tilde{A}_i$ are unbiased and consistent, in this case we obtain an unbiased and consistent estimate of $\tilde{A}$. In addition,

$$D[Y] = \sum_{i=1}^{m} b_i^2 D[X_i].$$

Therefore if the estimates $\tilde{A}_i$ have minimum variance, i.e., they are effective, then the estimate of $\tilde{A}$ will also be effective.

Each obtained estimate $\tilde{A}_i$ has some fixed error $\zeta_i = \tilde{A}_i - A_i$, and in addition $\zeta_i = \vartheta_i + \psi_i$, where ζ_i and ψ_i are realizations of the systematic and random components of the error, respectively.

The random component of the error in measuring the argument A_i varies from one group of observations to another if the measurement is performed as a multiple measurement, or from one single measurement to another if it is performed as a single measurement. In both cases, however, because the systematic component is singled out, $M[\psi_i] = 0$.

Under the conditions of a specific measurement, the realization of the systematic error remains unchanged. It can be different when the same argument is measured using different samples of the measuring instrument.

The error in the result of an indirect measurement is made up of the errors in the measurements of the arguments. Substituting into Eq. (6.3) the estimates of A_i we obtain

$$\tilde{A} = \sum_{i=1}^{m} b_i(A_i + \zeta_i).$$

From here we find the error in the indirect measurement

$$\zeta = \tilde{A} - A = \tilde{A} - \sum_{i=1}^{m} b_i A_i = \sum_{i=1}^{m} b_i \zeta_i,$$

i.e.,

$$\zeta = \sum_{i=1}^{m} b_i \vartheta_i + \sum_{i=1}^{m} b_i \psi_i. \tag{6.4}$$

Thus, when indirect measurements are performed both the limits of the systematic error in the result, as for direct measurements, and the random errors must be found by summing the components.

We shall first study random errors in the result of an indirect measurement. The realizations of the systematic components of the errors in estimates of all A_i must be assumed to be unchanged. We shall find the variance of the random error

$$D[\zeta] = D\left[\sum_{i=1}^{m} b_i \zeta_i\right] = \sum_{i=1}^{m} b_i^2 D[\psi_i].$$

We note that $D[\zeta] = D[\psi] = D[\tilde{A}]$.

This relation is valid in the case when the errors in the arguments are independent. If, however, the errors in the arguments are correlated with one another, then

$$D[\zeta] = D[\tilde{A}] = \sum_{i=1}^{m} b_i^2 D[\psi_i] + 2 \sum_{k \neq l}^{m} b_k b_l \rho_{kl} \sqrt{D[\psi_k] D[\psi_l]},$$

where ρ_{kl} is the correlation coefficient between the random errors in the measurements of the kth and lth arguments.

The relations obtained between the variances of the quantities can also be used to estimate the variances. Thus we obtain

$$S^2 = \sum_{i=1}^{m} b_i^2 S_i^2 + 2 \sum_{k \neq l}^{m} \tilde{\rho}_{kl} b_k b_l S_k S_l, \tag{6.5}$$

where S^2 is an estimate of the variance of the result of measurement, i.e., $S^2 = S^2(\tilde{A})$, S_i^2 is the estimate of the variance of the result of measurement of the ith argument, i.e., $S_i^2 = S^2(\tilde{A}_i)$, and

$$\tilde{\rho}_{kl} = \frac{\sum_{j=1}^{n} (x_{kj} - \tilde{A}_k)(x_{lj} - \tilde{A}_l)}{n^2 S_k S_l}$$

is an estimate of the correlation coefficient between the errors in the measurements of the arguments A_k and A_l.

A correlation between the errors in the measurements of the arguments most often arises in those cases when the measurements are performed simultaneously and the changes in the influence quantities (the air temperature, the supply voltage, etc.), though they are in themselves permissible, affect the observational results. If, however, the arguments are measured at a different time and they are measured using different measuring instruments, then there are no grounds for expecting that a correlation will appear between the errors in these measurements. In these cases, in formula (6.5) the second term can be neglected, and it assumes the form

$$S^2(\tilde{A}) = \sum_{i=1}^{m} b_i^2 S_i^2(\tilde{A}_i). \tag{6.6}$$

The spread in the observations owing to the random error in the measurement of each argument can usually be regarded as a normally distributed random quantity. But even if one of these distributions must be assumed to be different from a normal distribution, the distribution of the arithmetic mean, in practice, can still be regarded as being normal. The random error in the result of an indirect measurement, determined by adding the random errors in the measurements of the arguments, can be regarded with even greater justification to be a normally distributed random quantity. This makes it possible to find the confidence interval for the true value of the measured quantity A.

If the number of observations performed in a measurement of each argument is greater than 25–30, then the confidence limit of the random error Ψ in the result of an indirect measurement will be

$$\Psi = z_{(1+\alpha)/2} S(\tilde{A}), \tag{6.7}$$

where $z_{(1+\alpha)/2}$ is the quantile of the normalized normal distribution corresponding to the chosen confidence probability α.

A difficulty arises when the number of observations is less than 25–30. In this case, in principle, one could use Student's distribution, but it is not known how to determine the number of degrees of freedom.

The problem does not have an exact solution. An approximate estimate of the number of degrees of freedom, called the effective estimate, can be found using formula (6.8), based on the well-known Welch–Satterthwaite formula,

$$\nu_{\text{eff}}=\frac{\left(\sum_{i=1}^{m} b_i^2 S_i^2(\tilde{A}_i)\right)^2}{\sum_{i=1}^{m} \frac{b_i^4 S_i^4(\tilde{A}_i)}{\nu_i}}. \quad (6.8)$$

Given ν_{eff} and the confidence probability, from Student's distribution we can find t_q and therefore the confidence limit Ψ. Just as in the case of application of formula (6.7), we have

$$P\{|\tilde{A}-A|\leqslant\Psi\}=\alpha.$$

We shall now consider the problem of estimating the systematic components of the error. It can be assumed that the constant and determined components of the systematic errors in the measurements of the arguments have been excluded by introducing corrections. But their residuals remain.

We shall divide the residuals of systematic components into conditionally and absolutely constant components. We shall first study the conditionally constant components. For each such component the limit of the possible values is found and, as usual, a uniformly distributed random quantity will be used as the model.

Correspondingly, to calculate the limits of the systematic error in the measurement result $\tilde{A}$, it is necessary to use different formulas that follow in a natural manner from the material presented in Chap. 5.

If the components ϑ_i of the total error ϑ are assumed to be uniformly distributed within their limits θ_i, then

$$\theta=k\sqrt{\sum_{i=1}^{m} b_i^2\theta_i^2}. \quad (6.9)$$

The values of k are presented on p. 70. For $m\leqslant 3$ and $\alpha=0.99$ it is also necessary to estimate the algebraic sum $\Sigma_{i=1}^{m} b_i\theta_i$, and if this sum is less than the value of θ calculated using Eq. (6.9), then the limit of the systematic error must be taken as

$$\theta=\sum_{i=1}^{m} b_i\theta_i. \quad (6.10)$$

If all components of the total error ϑ can be assumed to be normally distributed (and this is justified when they are all formed by a large number of terms) and all limits θ_i are calculated for the same confidence probability, then

$$\theta=\sqrt{\sum_{i=1}^{m} b_i^2\theta_i^2}. \quad (6.11)$$

In the intermediate cases the solution can be obtained based on the following considerations. Let k terms have a normal distribution and l terms a uniform distribution. The estimate of the variance of the sum of the first terms is

$$S_k^2=\sum_{i=1}^{k} b_i^2\left(\frac{\theta_i}{z_{(1+\alpha)/2}}\right)^2,$$

where $z_{(1+\alpha)/2}$ is the quantile of the normal distribution corresponding to the probability α.

An estimate of the variance of the sum of the second terms is

$$S_l^2=\sum_{j=1}^{l} b_j^2\frac{\theta_j^2}{3}.$$

An estimate of the standard deviation of the systematic component is

$$S(\vartheta)=\sqrt{S_k^2+S_l^2},$$

and

$$\theta=t_\Sigma S(\vartheta). \tag{6.12}$$

We find the coefficient t_Σ from the transformed formula (5.14)

$$t_\Sigma=\frac{\theta_k+\theta_l}{S_k+S_l},$$

where θ_k and θ_l are the limits of the sum of the k and l terms of the error calculated using formulas (6.9) and (6.11), respectively. Both limits must be calculated for the same confidence probability. The obtained confidence limit of the systematic component of the error of the result of an indirect measurement will also correspond to this value of the confidence probability.

The random and conditionally constant components of the error of an indirect measurement are summed using the formulas presented in Chap. 5. The absolutely constant components must then be added to this sum. For this, their limiting values, multiplied by the corresponding coefficients b_i, must be added arithmetically with the limiting values of the sum of the random and conditionally constant components. We note once more that often there are no absolutely constant elementary errors.

If, for some reason, the corrections to the results of measurements of the arguments must be used to refine the direct result of an indirect measurement, then in accordance with formula (6.4) this must be done using the relation

$$C=\sum_{i=1}^{m} b_iC_i,$$

where C and C_i are the corrections to the results of the measurements $\tilde{A}$ and $\tilde{A}_i$.

6.5. Calculation of the error of a compound resistor

We shall study the case when twelve resistors with three different nominal resistances are connected in series:

$$R_\Sigma=2R_1+4R_2+6R_3.$$

This equation is a particular case of the dependence (6.2) with $m=3$, $b_1=2$, $b_2=4$, and $b_3=6$.

For resistors each having a nominal resistance R_n the limits of permissible errors are known:

R_i	R_1	R_2	R_3
R_n,Ω	100.00	10.00	1.00
θ_i,Ω	0.03	0.02	0.01 .

For the case at hand, we shall assume that the distribution of the actual resistance over the collection of resistors having the same nominal resistance is normal and truncated at the probability $\alpha = 0.98$.

The nominal resistance of our group of resistors, according to the relation (6.2), is equal to

$$R_\Sigma = 2\times 100 + 4\times 10 + 6\times 1 = 246\ \Omega.$$

We shall find the confidence limits of the error of this group of resistors for $\alpha = 0.98$ using formula (6.11):

$$\theta = \sqrt{\sum_{i=1}^{3} b_i^2\theta_i^2} = \sqrt{2^2\times 0.03^2 + 4^2\times 0.02^2 + 6^2\times 0.01^2} = 0.11\ \Omega.$$

Rounding off, we obtain $\theta = 0.1\ \Omega$. Finally, taking into account the required number of significant figures, we can write

$$R_\Sigma = 246.0 \pm 0.1\ \Omega, \quad \alpha = 0.98.$$

If it must be assumed that the real resistances of the resistors are distributed uniformly, then the confidence limits θ must be calculated using formula (6.9), in which case we would obtain for the same confidence probability $\alpha = 0.98$

$$\theta' = k\theta = 1.3\times 0.11 = 0.14\ \Omega.$$

The difference between θ' and θ for many cases is significant.

We shall consider the variant when m resistors with the same nominal resistance and the same tolerance are connected in series:

$$R_\Sigma = mR_N, \quad \theta_R = \text{const}.$$

For example, $R_N = 100\ \Omega$, $\theta_{R\%} = 0.5\%$, and $m = 10$. What is the error of a compound resistor?

The problem is not as simple as it looks, since its solution depends on the technology employed to fabricate the resistors. Suppose that resistors fabricated at different times using different equipment are connected together. In this case, their errors are independent, and the error of each resistor can be regarded as a realization of a uniformly distributed random quantity. Then, according to formula (6.9), we shall have

$$\theta = k\sqrt{\sum_{i=1}^{m} \theta_R^2} = k\theta_R\sqrt{m}.$$

It is convenient to transform this formula so that it would contain the relative errors. For this, we shall divide both sides of the equation by $R_\Sigma = mR_N$ and write $100 \times \theta/R_\Sigma = \theta_\%$, $100 \times \theta_R/R_N = \theta_{R\%}$. Then we obtain

$$\theta_\% = (k/\sqrt{m})\theta_{R\%}.$$

In the case at hand, we must focus on a high confidence probability. Let $\alpha = 0.99$. Then $k = 1.4$ and

$$\theta_{\%} = 1.4/\sqrt{10} \times 0.5 = 0.2\%,$$

i.e., the accuracy of a compound resistor is higher than that of a single resistor. However, the increase in accuracy is limited by the accuracy of the measuring instrument used in checking the resistance of the resistors.

Sometimes a compound resistor can be made up from resistors whose resistance was adjusted individually by the same operator with the help of the same measuring instrument. This is most often possible when accurate resistors especially are fabricated. The actual errors in all the resistors must become approximately equal to one another, and since they are all, essentially, systematic, the relative error of the compound resistor will become the same as that of the separate resistors.

Expanding on this example, we can move on to the case when some additive quantity is measured in several identical applications of the same measuring instrument, for example, measurement of the length of a flat body with the help of a short ruler. It is obvious from the foregoing discussion that the relative systematic component of the error of such a length measurement will be equal to the relative systematic error of the ruler. The random component, however, must be estimated for a concrete measurement.

6.6. Nonlinear indirect measurements

In the case of nonlinear indirect measurements the relation between the measured quantity A and the measured arguments A_i can be very diverse. But even in the simplest cases, the accurate estimation of the measured quantity based on the estimates of the measured arguments presents great difficulties. This is because for random quantities X and Y

$$M[f(X,Y)] \neq f(MX,MY).$$

Although $M[f(X,Y)]$ can be found for many particular functions $f(X,Y)$, such a solution in most cases is too complicated for practical use. For this reason, the measured quantity is most often estimated by the value obtained by substituting estimates of the arguments into formula (6.1):

$$\tilde{A} = \prod_{i=1}^{n} f_i(\tilde{A}_i).$$

This procedure results in a definite uncertainty in the result. An exception are those cases when it corresponds to the definition itself of the measured quantity.

The errors in indirect nonlinear measurements are usually estimated based on expansion of the function (6.1) in a Taylor series. The Taylor series for functions of two variables x and y is written, as is well known, as follows:

$$f(x+h,y+k) = f(x,y) + \left(\frac{\partial}{\partial x}h + \frac{\partial}{\partial y}k\right)f(x,y) + \frac{1}{2!}\left(\frac{\partial}{\partial x}h + \frac{\partial}{\partial y}k\right)^2 f(x,y) + \cdots$$
$$+ \frac{1}{n!}\left(\frac{\partial}{\partial x}h + \frac{\partial}{\partial y}k\right)^n f(x,y) + R_{n+1},$$

where the remainder is

$$R_{n+1}=\frac{1}{(n+1)!}\left(\frac{\partial}{\partial x}h+\frac{\partial}{\partial y}k\right)^{n+1}f(x+v_1h,y+v_2k),\quad 0<v_{1,2}<1.$$

For our problem, we shall set $x = A_1$, $y = A_2$, $x + h = \tilde{A}_1$, and $y + k = \tilde{A}_2$. Therefore

$$h=\zeta_1=\tilde{A}_1-A_1,\quad k=\zeta_2=\tilde{A}_2-A_2,$$

where ζ_1 and ζ_2 are the errors in the results of measurements of the first and second arguments.

To obtain a linear dependence the series must be truncated at $n = 1$. Then we obtain

$$f(\tilde{A}_1,\tilde{A}_2)=f(A_1,A_2)+\left(\frac{\partial}{\partial A_1}\zeta_1+\frac{\partial}{\partial A_2}\zeta_2\right)f(A_1,A_2)+R_2, \qquad (6.13)$$

$$R_2=\frac{1}{2}\left(\frac{\partial}{\partial A_1}\zeta_1+\frac{\partial}{\partial A_2}\zeta_2\right)^2 f(A_1+v_1\zeta_1,A_2+v_2\zeta_2). \qquad (6.14)$$

The partial derivatives are customarily called influence coefficients. We introduce for them the following notation:

$$W_i=\frac{\partial f}{\partial A_i},\quad i=1,\ldots,m.$$

We note that in Eq. (6.13) the influence coefficients are calculated at a point with the coordinates A_1 and A_2, and in principle they do not have errors, but only in principle.

We shall investigate the relation obtained. We shall first investigate the mathematical expectation

$$M[\,f(\tilde{A}_1,\tilde{A}_2)\,]=A+M[\,W_1\zeta_1+W_2\zeta_2]+MR_2.$$

Estimates of the arguments have both systematic and random errors, but since the systematic components are separated out, it can be assumed with accuracy sufficient for each specific measurement that

$$M\zeta_1=M\zeta_2=0.$$

Then

$$M[\,W_1\zeta_1+W_2\zeta_2]=0.$$

The remainder is estimated using formula (6.14). But the true values of the arguments are unknown, and the second derivatives must be calculated at a point with the coordinates $\tilde{A}_1$ and $\tilde{A}_2$, i.e., it must be assumed that $v_1 = v_2 = 1$. The values of the derivatives become random quantities, and because of this it is impossible to obtain a general expression for the mathematical expectation of the remainder, and it must be estimated. The errors in the results of measurements of the arguments for calculating this estimate are assumed to be equal to their practical limiting values. Then, using formula (6.14), we find

$$\tilde{R}_2=\frac{1}{2}\left(\frac{\partial^2 f}{\partial A_1^2}\right)_{\tilde{A}_1,\tilde{A}_2}\Delta_1^2+\frac{1}{2}\left(\frac{\partial^2 f}{\partial A_2^2}\right)_{\tilde{A}_1,\tilde{A}_2}\Delta_2^2+\left(\frac{\partial^2 f}{\partial A_1\partial A_2}\right)_{\tilde{A}_1,\tilde{A}_2}\Delta_1\Delta_2.$$

If the estimate of the remainder obtained in this manner is so much smaller than the other errors in the result of measurement that it can be neglected, then $M[f(\tilde{A}_1,\tilde{A}_2)] \approx A$ and the measured quantity can be estimated using the formula

$$\tilde{A}=f(\tilde{A}_1,\tilde{A}_2,\ldots,\tilde{A}_m), \tag{6.15}$$

and the estimate can be assumed to be unbiased, and it is also consistent if consistent estimates of the arguments were employed.

The remainder can be reduced by retaining one additional term in the Taylor series expansion. In this case, instead of Eq. (6.13) we obtain (for m arguments)

$$f(\tilde{A}_1,\ldots,\tilde{A}_m)=f(A_1,\ldots,A_m)+\left(\frac{\partial}{\partial A_1}\zeta_1+\cdots+\frac{\partial}{\partial A_m}\zeta_m\right)f(A_1,\ldots,A_m)$$

$$+\frac{1}{2}\left(\frac{\partial}{\partial A_1}\zeta_1+\cdots+\frac{\partial}{\partial A_m}\zeta_m\right)^2 f(A_1,\ldots,A_m)+R_3, \tag{6.16}$$

$$R_3=\frac{1}{1\times2\times3}\left(\frac{\partial}{\partial A_1}\zeta_1+\cdots+\frac{\partial}{\partial A_m}\zeta_m\right)^3\times f(A_1+\nu_1\zeta_1,\ldots,A_m+\nu_m\zeta_m). \tag{6.17}$$

Since ζ_i are centered random quantities and $D\zeta_i\neq0$, the mathematical expectation of the quadratic term is different from 0:

$$M\left[\frac{1}{2}\left(\frac{\partial}{\partial A_1}\zeta_1+\cdots+\frac{\partial}{\partial A_m}\zeta_m\right)^2 f(A_1,\ldots,A_m)\right]$$

$$=M\left[\frac{1}{2}\sum_{i=1}^{m}\frac{\partial^2 f}{\partial A_i^2}\zeta_i^2+\sum_{l\neq k}^{m}\frac{\partial^2 f}{\partial A_l\,\partial A_k}\zeta_l\,\zeta_k\right]$$

$$=\frac{1}{2}\sum_{i=1}^{m}\left(\frac{\partial^2 f}{\partial A_i^2}\right)D\zeta_i+\sum_{l\neq k}^{m}\rho_{lk}\frac{\partial^2 f}{\partial A_l\,\partial A_k}\sqrt{D\zeta_l\,D\zeta_k}. \tag{6.18}$$

Therefore, a correction calculated according to the formula

$$\tilde{C}=-\frac{1}{2}\sum_{i=1}^{m}\frac{\partial^2 f}{\partial A_i^2}S_i^2-\sum_{l\neq k}^{m}\tilde{\rho}_{lk}\frac{\partial^2 f}{\partial A_l\,\partial A_k}S_l\,S_k \tag{6.19}$$

can be introduced into the estimate obtained using the formula (6.15).

The partial derivatives are calculated by replacing the true values of the arguments by their estimates.

Thus in the more general form the formula for estimating $\tilde{A}$ assumes the form

$$\tilde{A}=f(\tilde{A}_1,\ldots,\tilde{A}_m)+\tilde{C}. \tag{6.20}$$

Whether or not corrections should be introduced is determined by the ratio of $\tilde{C}$ and the errors of $\tilde{A}$. It should also be noted that in most cases the conditions under which the errors in the measurements of the arguments are independent, noted in

Sec. 6.2, are satisfied, and it is sufficient to retain only the first term in formula (6.19) in order to calculate the correction.

To determine whether or not the remainder R_3 can be neglected, the arguments made above for the remainder R_2 should be applied.

We shall now estimate the errors. In the general form

$$\zeta=\tilde{A}-A.$$

We shall use relation (6.13) and, bearing in mind the fact that the possibility of neglecting the remainder R_2 is determined separately, we obtain

$$\zeta=\left(\frac{\partial}{\partial A_1}\zeta_1+\cdots+\frac{\partial}{\partial A_m}\zeta_m\right)f(A_1,\ldots,A_m)$$

or

$$\zeta=\sum_{i=1}^{m} W_i\zeta_i. \tag{6.21}$$

For the exact values of the influence coefficients relation (6.21) is identical to relation (6.4) for linear indirect measurements. For this reason, the parameters of random, unexcluded systematic errors and the total error of the result of measurement can be estimated using formulas (6.5)–(6.12).

In particular, when the errors of the measurements of the arguments are not related, we obtain in accordance with formula (6.6)

$$S(\tilde{A})=\sqrt{W_1^2S_1^2+\cdots+W_m^2S_m^2}. \tag{6.22}$$

It should be borne in mind that S_i^2 in formula (6.22) is an estimate of the variance of the result of measurement $\tilde{A}_i$ of the argument and not of the results of single measurements.

The influence coefficients are most often estimated by substituting into the expression for the partial derivatives the estimates obtained $\tilde{A}_i$. Therefore instead of the influence coefficients themselves we obtain only their estimates. In addition, sometimes the influence coefficients are determined experimentally. In either case, they are determined with some error.

The error in determining the influence coefficients is another source of error of nonlinear indirect measurements. This error can be avoided, if the dependence (6.1) has the form

$$A=A_1^kA_2^l\cdots A_m^n. \tag{6.23}$$

In this case the influence coefficients are determined by the expressions

$$W_1=\frac{\partial A}{\partial A_1}=kA_1^{k-1}A_2^l\cdots A_m^n,$$

$$W_2=\frac{\partial A}{\partial A_2}=A_1^klA_2^{l-1}\cdots A_m^n,$$

$$\cdots$$

$$W_m = \frac{\partial A}{\partial A_m} = A_1^k A_2^l \cdots n A_m^{n-1}.$$

We estimate the measured quantity using formula (6.15), and the absolute error is determined by formula (6.21). We shall now transfer from the absolute error to the relative error:

$$\varepsilon = \frac{\tilde{A} - A}{A} = \frac{k A_1^{k-1} A_2^l \cdots A_m^n}{A} \zeta_1 + \frac{l A_1^k A_2^{l-1} \cdots A_m^n}{A} \zeta_2 + \cdots + \frac{A_1^k A_2^l \cdots n A_m^{n-1}}{A} \zeta_m.$$

Substituting formula (6.23) for A we obtain

$$\varepsilon = k \frac{\zeta_1}{A_1} + l \frac{\zeta_2}{A_2} + \cdots + n \frac{\zeta_m}{A_m}. \qquad (6.24)$$

The influence coefficients for the relative errors in the measurements of the arguments are equal to the powers of the corresponding arguments. The latter are known exactly *a priori*, so that the error noted above does not arise. This is another advantage of expressing the measurement errors in the form of relative errors.

Using formula (6.24) instead of formula (6.21), it is possible to estimate, as examined above, the total error in an indirect measurement as well as its components. In particular, the standard deviation of the result of measurement can be estimated using the formula

$$S_{\%}(\tilde{A}) = \sqrt{k^2 S_{\%}^2(\tilde{A}_1) + \cdots + n^2 S_{\%}^2(\tilde{A}_m)}. \qquad (6.25)$$

The difference lies in the fact that instead of the absolute errors in the measurements of the arguments, their relative errors appear in all formulas.

The practical technique for finding the influence coefficients when expressing the errors in the form of relative errors consist of the following. First, the logarithm of the dependence under study, for example, Eq. (6.23), is taken and then the result is differentiated. In our case we obtain

$$\ln A = k \ln A_1 + l \ln A_2 + \cdots + n \ln A_m,$$

$$\frac{dA}{A} = k \frac{dA_1}{A_1} + l \frac{dA_2}{A_2} + \cdots + n \frac{dA_m}{A_m}.$$

Then, assuming that the measurement errors are small and replacing the differentials by increments, we obtain formula (6.24). The error resulting from this substitution, naturally, is the same as that made in using the Taylor series expansion. It is determined by the sum of the products of the errors in pairs, triplets, etc., i.e., by second- and higher-order infinitesimals.

It is important to note one other practical recommendation. In some nonlinear dependences one or another argument can appear several times and the expression cannot be simplified. In calculating the influence coefficients in this case, after expanding in a Taylor series, the terms with similar argument must be reduced. The reduction of similar terms must be done retaining the signs that were obtained upon differentiation. The numerical calculations can be performed, using, depending on the circumstances, the signs of the terms only when the final formula, i.e., the formula in which each argument (more accurately, the error in the measurement of each argument) appears only once, has been obtained, and only then. They must be retained when calculating the corrections; in calculating the errors, when the errors of the measurements of the arguments are assumed to be random quantities, the summation is performed by probabilistic methods, and these signs are usually unimportant.

The error owing to the uncertainty of the influence coefficients can be estimated. For this, we shall find the relation between W_i and its estimate $\widetilde{W}_i$, using the Taylor series expansion. It is obvious that

$$\widetilde{W}_i = W_i + \sum_{j=1}^{m} \frac{\partial W_i}{\partial A_j} \zeta_j, \quad j \neq i.$$

Here ζ_j is the error in the measurement of the jth argument. According to (6.21)

$$\zeta = \sum_{i=1}^{m} \left(\widetilde{W}_i - \sum_{j=1}^{m} \frac{\partial W_i}{\partial A_j} \zeta_j \right) \zeta_i.$$

From here,

$$\zeta = \sum_{i=1}^{m} \widetilde{W}_i \zeta_i - 2 \sum_{j \neq i}^{m} \frac{\partial^2 f}{\partial A_i \, \partial A_j} \zeta_j \zeta_i .$$

if $\widetilde{\zeta} = \Sigma_{i=1}^{m} \widetilde{W}_i \zeta_i$, is used instead of ζ, then we obtain the error

$$\Delta\zeta = \widetilde{\zeta} - \zeta = 2 \sum_{j \neq i}^{m} \frac{\partial^2 f}{\partial A_i \, \partial A_j} \zeta_j \zeta_i.$$

Substituting into this formula the practical limiting errors in the estimates of the arguments, it is possible to estimate the error owing to the uncertainty in the influence coefficients and decide whether or not this error can be neglected. If this error is found to be too large compared with the permissible error for the given measurement and it is impossible to improve the estimates of the influence coefficients, then, obviously, the Taylor series expansion cannot be used, and the plan of the entire measurement must be correspondingly reexamined.

The mathematical expectation and variance of the error owing to the uncertainty of the influence coefficients are equal to

$$M[\Delta\zeta] = 2 \sum_{i \neq j}^{m} \frac{\partial^2 f}{\partial A_i \, \partial A_j} M\vartheta_i \, M\vartheta_j,$$

$$D[\Delta\zeta] = 4 \sum_{i \neq j}^{m} \left(\frac{\partial^2 f}{\partial A_i \, \partial A_j} \right)^2 D\zeta_i \, D\zeta_j.$$

These expressions were derived under the assumption that the errors in the measurements of the arguments are independent; they are approximate owing to the fact that the second derivatives were assumed to be determinate quantities. To solve the problem more exactly, they must be assumed to be random quantities.

In the overwhelming majority of the cases linearization is permissible, and for this reason it is the main method used for obtaining a result and estimating the errors in the case of nonlinear indirect measurements.

When linearizing, the arguments are estimated by the arithmetic means of the results of the measurements. It is well known, however, that in the general case $M[f(X,...,Z)]\neq f(MX,...,MZ)$. For this reason the estimate found in this manner for the measured quantity is biased. Although this biasness enters into the estimate of the error of an indirect measurement, it is undesirable.

An unbiased estimate of the measured quantity can be obtained by using the method of reduction. For this, as pointed in Sec. 6.2, the observations performed when making measurements of the arguments must be matched. One way to obtain matched observations is to perform them simultaneously. However, this can result in the appearance of a correlation between the errors of the measurements of the arguments. Taking this correlation into account makes it more difficult to process the observational results, and for this reason it is desirable to design the experiment so that there still would be no correlation between the errors of the measurements of the arguments. The existence of a correlation can be checked by statistical methods.

The method of reduction makes it possible to find an estimate of the measured quantity and its random error without resorting to a Taylor series expansion. For systematic errors this problem can be solved by the method of sorting of variants—a numerical method for constructing the distribution of estimates of the measured quantity from distributions of estimates of the arguments. As noted above, the latter distributions are assumed to be uniform. The method of sorting is described in Sec. 3.6.

One would think that the method of sorting can also be used to find an estimate of an indirectly measured quantity. In reality, from the histograms of observations, obtained in measurements of the arguments, one can find an approximation to the corresponding distribution function of the possible values of the measured quantity and from this distribution function the most probable estimate of the measured quantity. Unfortunately, in this case the random error of the result cannot be estimated.

A different method for solving the problem is proposed in Ref. 40.

To explain the essence of this method, we shall examine the particular case when the measured quantity is a function of two arguments that can be represented as a product of two functions:

$$A=f(A_1,A_2)=f_1(A_1)f_2(A_2). \qquad (6.26)$$

The true value of each measured argument is a determinate quantity. Because of random errors appearing in the measurements, however, the results of observations, performed when measuring the arguments, must be regarded as random quantities. Since it is known that in the general case

$$M[f(X)]\neq f(MX),$$

the estimates $\tilde{A}_i$ cannot be inserted directly into formula (6.26). It is necessary to transfer from the random quantities x_{1i} and x_{2j}—the results of the observations—to the random quantities y_{1i} and y_{2j} according to the formulas

$$y_{1i}=f_1(x_{1i}), \quad y_{2j}=f_2(x_{2j}).$$

If the errors in the measurements of the arguments are independent, then the random quantities Y_1 and Y_2 are also independent. Then, since

$$M[f(X_1,X_2)]=M[f_1(X_1)f_2(X_2)]=M[f_1(X_1)]M[f_2(X_2)],$$

as the estimate of the measured quantity we must take

$$\widetilde{A}=\widetilde{y}_1\widetilde{y}_2, \tag{6.27}$$

where

$$y_1=M[f_1(X_1)], \quad y_2=M[f_2(X_2)],$$

and $\widetilde{y}_1$ and $\widetilde{y}_2$ are their estimates.

Each group of observations y_{1i} and y_{2i} is processed as a group of observations made when performing direct measurements, as a result of which we obtain $\widetilde{y}_1$ and $\widetilde{y}_2$ and estimates of their standard deviations $S(\widetilde{y}_1)$ and $S(\widetilde{y}_2)$. The former, according to Eq. (6.27), determine the estimate of the measured quantity and the latter determine the estimate of its standard deviation.

To solve the second problem, we shall first present a formula for the variance of the product of two independent random quantities $Y = Y_1Y_2$:

$$\begin{aligned} D[Y]&=M[Y^2]-(MY)^2 \\ &=M[Y_1^2Y_2^2]-(M[Y_1Y_2])^2 \\ &=M[Y_1^2]M[Y_2^2]-(MY_1MY_2)^2, \end{aligned}$$

$$M[Y_i^2]=DY_i+(MY_i)^2.$$

For this reason, we obtain

$$DY=[DY_1+(MY_1)^2][DY_2+(MY_2)^2]-(MY_1MY_2)^2.$$

Therefore, for our case we can write

$$D\widetilde{A}=(D\widetilde{y}_1+\widetilde{y}_1^2)(D\widetilde{y}_2+\widetilde{y}_2^2)-(\widetilde{y}_1\widetilde{y})^2=D\widetilde{y}_1D\widetilde{y}_2+D\widetilde{y}_1\widetilde{y}_2^2+D\widetilde{y}_2\widetilde{y}_1^2.$$

Usually $D\widetilde{y}_1D\widetilde{y}_2 \ll \widetilde{y}_2^2D\widetilde{y}_1 + \widetilde{y}_1^2D\widetilde{y}_2$, and for this reason it can be assumed that

$$D\widetilde{A}=\widetilde{y}_2^2D\widetilde{y}_1+\widetilde{y}_1^2D\widetilde{y}_2.$$

The relative form is convenient:

$$\frac{D\widetilde{A}}{\widetilde{A}^2}=\frac{D\widetilde{y}_1}{\widetilde{y}_1^2}+\frac{D\widetilde{y}_2}{\widetilde{y}_2^2}.$$

Transferring from variances to their estimates, we obtain

$$\frac{S^2(\widetilde{A})}{\widetilde{A}^2}=\frac{S^2(\widetilde{y}_1)}{\widetilde{y}_1^2}+\frac{S^2(\widetilde{y}_2)}{\widetilde{y}_2^2}.$$

Correspondingly,

$$\frac{S(\widetilde{A})}{\widetilde{A}}=\sqrt{\frac{S^2(\widetilde{y}_1)}{\widetilde{y}_1^2}+\frac{S^2(\widetilde{y}_2)}{\widetilde{y}_2^2}}.$$

If there are more than two cofactors, then the relations presented will assume the following form:

$$\tilde{A} = \prod_{i=1}^{m} \tilde{y}_i,$$

$$\frac{S^2(\tilde{A})}{\tilde{A}^2} = \sum_{i=1}^{m} \frac{S^2(\tilde{y}_i)}{\tilde{y}_i^2},$$

$$\frac{S(\tilde{A})}{\tilde{A}} = \sqrt{\sum_{i=1}^{m} \frac{S^2(\tilde{y}_i)}{\tilde{y}_i^2}}.$$

Confidence intervals can also be constructed for the products of quantities. The procedure is based on constructing the confidence intervals for measured arguments; this is done by the standard methods. Next, the limits of the intervals are transformed into the limits of the intervals of the quantities y_i, and by combining them the confidence intervals are obtained for the measured quantity. The probability that the measured quantity A will fall within a given interval is calculated as the product of the confidence probabilities of the corresponding intervals for the arguments.

It must be underscored that the main method for estimating the errors of nonlinear indirect measurements is expansion in a Taylor series. As a result the nonlinear dependence is transformed into a linear dependence, and everything said above for linear indirect measurements, single and multiple, becomes applicable to the nonlinear indirect measurement.

In the case of single measurements with large errors linearization may be inadmissable. Then it is necessary to use the method of *sorting of variants*, which was described in Sec. 3.6.

In application to indirect measurements, this method consists of the following.

The form of the error distribution for each argument must be chosen based on the available information about the errors of the measurements of the arguments. If only the limits of the measurement error of some argument are known, then, as always, we shall assume that the distribution of these errors is uniform. If it is known, however, that the limits of some error are the confidence limits calculated by summing several components, then the distribution of the resulting error can be assumed to be normal.

After the distribution of the measurement error of each argument has been chosen, it is necessary to transfer from the continuous distributions to their discrete distribution. For this, the entire interval between the limits is divided into several different sections. Often three to five sections are sufficient. Then the values of the arguments corresponding to the center of each section (separate values) are calculated. The probability of each such separate value is assumed to be equal to the area under the probability distribution curve of the error on the corresponding section. Substituting into the equation relating the quantity and the separate values of the arguments we obtain the separate value of the measured quantity. The probability of this value is equal to the product of the probabilities of the separate values of the arguments substituted into the formula.

From the separate values obtained for the measured quantity and the corresponding probabilities the distribution function of these values can be constructed, and selecting a confidence probability, the confidence limits of the error in the result of measurement can be found from this distribution function.

In conclusion we call attention to the fact that at the last stage it is necessary to construct the probability distribution function and not the probability density, since the integral function smooths out while the differential function accentuates the inaccuracies in the initial data and in the solution.

6.7. Calculation of the error of the measurement of the density of a solid body

The accurate measurement of the density of a solid body can serve as an example of a multiple nonlinear indirect measurement.

The density of a solid body is given by the formula

$$\rho = m/V,$$

where m is the mass of the body and V is the volume of the body. In the experiment considered, the mass of the body was measured by methods of precise weighing using a collection of standard weights whose uncertainty did not exceed 0.01 mg. The volume of the body was determined by the method of hydrostatic weighing using the same set of weights.

The results of measurements are presented in Table 6.1 in the first and fourth columns.

The difference between the observational results is explained by the random error of the balances. As follows from the data presented, this error is so much larger than the systematic errors due to the uncertainties in the masses of the weights that these errors can be neglected.

TABLE 6.1. The results of measurements of the density of a solid body and data from initial processing.

Mass of body $m_i \times 10^3$ (kg)	$(m_i - \bar{m})$ $\times 10^7$ (kg)	$(m_i - \bar{m})^2$ $\times 10^{14}$ (kg^2)	Volume of body $V_i \times 10^6$ (m^3)	$(V_i - \bar{V})$ $\times 10^{10}$ (m^3)	$(V_i - \bar{V})^2$ $\times 10^{20}$ (m^6)
1	2	3	4	5	6
252.9119	−1	1	195.3799	+1	1
252.9133	+13	169	195.3830	+32	1024
252.9151	+31	961	195.3790	−8	64
252.9130	+10	100	195.3819	+21	441
252.9109	−11	121	195.3795	−3	9
252.9094	−26	676	195.3788	−10	100
252.9113	−7	49	195.3792	−6	36
252.9115	−5	25	195.3794	−4	16
252.9119	−1	1	195.3791	−7	49
252.9115	−5	25	195.3791	−7	49
252.9118	−2	4	195.3794	−4	16

Since the mass of the solid body and its volume are constants, to estimate the density of the solid, the mass and volume of the solid must be estimated with the required accuracy and their ratio must be formed. For this, we find the average values of the observational results and estimates of the standard deviations for the groups of measurements:

$$\bar{m}=252.9120\times10^{-3}\ \text{kg},\quad \bar{V}=195.3798\times10^{-6}\ \text{m}^3,$$

$$S^2(m_i)=\frac{1}{n_1-1}\sum_{i=1}^{n_1}(m_i-\bar{m})^2=\frac{2132\times10^{-14}}{10}=213.2\times10^{-14}\ \text{kg}^2,$$

$$S^2(V_i)=\frac{1}{n_2-1}\sum_{i=1}^{n_2}(V_i-\bar{V})^2=\frac{1805\times10^{-20}}{10}=180.5\times10^{-20}\ \text{m}^6.$$

The estimates of the variances, in the relative form, are equal to

$$S_r^2(m_i)=\frac{213\times10^{-14}}{(252.9\times10^{-3})^2}=3.32\times10^{-11},$$

$$S_r^2(V_i)=\frac{180\times10^{-20}}{(195.4\times10^{-6})^2}=4.74\times10^{-11}.$$

The estimate of the measured quantity is

$$\tilde{\rho}=\frac{\bar{m}}{\bar{V}}=\frac{252.9120\times10^{-3}}{195.3798\times10^{-6}}=1.294\,463\times10^3\ \text{kg/m}^3.$$

To estimate the error of the obtained result it is most convenient to use the method of linearization, but it is necessary to check that this method is admissible. For this, it is necessary to estimate the remainder R_2 according to formula (6.14):

$$R_2=\frac{1}{2}\left[\frac{\partial^2\rho}{\partial m^2}(\Delta m)^2+\frac{\partial^2\rho}{\partial V^2}(\Delta V)^2+2\frac{\partial^2\rho}{\partial m\partial V}\Delta m\Delta V\right],$$

$$\frac{\partial\rho}{\partial m}=\frac{1}{V},\quad \frac{\partial^2\rho}{\partial m^2}=0,$$

$$\frac{\partial\rho}{\partial V}=-\frac{m}{V^2},\quad \frac{\partial^2\rho}{\partial V^2}=\frac{2m}{V^3},$$

$$\frac{\partial\rho}{\partial m\partial V}=-\frac{1}{V^2}.$$

We shall calculate the partial derivatives at the point with the coordinates $\bar{m}$ and $\bar{V}$, so that the errors Δm and ΔV are relatively insignificant. We obtain

$$R_2=\frac{\bar{m}}{\bar{V}^3}(\Delta V)^2-\frac{1}{\bar{V}^2}\Delta m\Delta V=\frac{\bar{m}}{\bar{V}}\left(\frac{\Delta V}{\bar{V}}\right)^2-\frac{\bar{m}}{\bar{V}}\frac{\Delta m}{\bar{m}}\frac{\Delta V}{\bar{V}}=\tilde{\rho}\frac{\Delta V}{\bar{V}}\left(\frac{\Delta V}{\bar{V}}-\frac{\Delta m}{\bar{m}}\right)$$

or

$$\frac{R_2}{\tilde{\rho}}=\frac{\Delta V}{\bar{V}}\left(\frac{\Delta V}{\bar{V}}-\frac{\Delta m}{\bar{m}}\right).$$

For ΔV and Δm we take the largest deviations from the average values observed in the experiment:

$$\Delta V = 32 \times 10^{-10}\ \mathrm{m}^3, \quad \Delta m = 31 \times 10^{-7}\ \mathrm{kg}.$$

The relative errors are equal to

$$\frac{\Delta V}{\bar{V}} = \frac{32 \times 20^{-10}}{195.4 \times 10^{-6}} = 1.64 \times 10^{-5},$$

$$\frac{\Delta m}{\bar{m}} = \frac{31 \times 10^{-7}}{252.9 \times 10^{-3}} = 1.22 \times 10^{-5}.$$

Since the errors are random, the sign of the error should not be specified. We obtain

$$R_2/\tilde{\rho} = 1.64 \times 10^{-5}(1.64 + 1.22) \times 10^{-5} = 4.7 \times 10^{-10}.$$

This error is so much smaller than the errors associated with $\Delta V/\bar{V}$ and $\Delta m/\bar{m}$ it is obvious that linearization is possible.

We shall estimate the standard deviation in the relative form using formula (6.25)

$$S_r(\tilde{\rho}) = \sqrt{\frac{S_r^2(m_i)}{n_1 \bar{m}^2} + \frac{S_r^2(V_i)}{n_2 \bar{V}^2}} = \sqrt{\frac{3.32 \times 10^{-11} + 4.74 \times 10^{-11}}{11}} = 2.7 \times 10^{-6}.$$

Here the influence coefficients are $k = 1$ and $l = -1$. In terms of percentages, $S_{\%}(\tilde{\rho}) = 2.7 \times 10^{-4}\%$.

In units of density we obtain

$$S(\tilde{\rho}) = 2.7 \times 10^{-6} \times 1.294 \times 10^{3} = 3.5 \times 10^{-3}\ \mathrm{kg/m}^3.$$

We shall now find the confidence limits of the error in the result. Based on formula (6.24) we can write

$$\psi_{r\rho} = \psi_{rm} - \psi_{rV}.$$

Since the errors are random, the minus sign need not be included. We shall find the confidence limits of the components. We shall take the confidence probability $\alpha = 0.95$. In addition, we have $\nu = 10$. Then we can find from Student's distribution $t_q = 2.23$.

We shall also calculate in the form of relative errors the confidence limits of the component errors:

$$\Psi_{m\%} = 100 t_q \frac{S_r(\bar{m})}{\sqrt{n_1}} = 100 \times 2.23 \sqrt{\frac{3.32 \times 10^{-11}}{11}} = 3.88 \times 10^{-4}\%,$$

$$\Psi_{V\%} = 100 t_q \frac{S_r(\bar{V})}{\sqrt{n_2}} = 100 \times 2.23 \sqrt{\frac{4.74 \times 10^{-11}}{11}} = 4.64 \times 10^{-4}\%.$$

Our case meets the requirement discussed above in connection with the formula $\theta = \sqrt{\Sigma_{i=1}^{n} \theta_i^2}$ (p. 126). Therefore we can find the value

$$\Psi_{\rho\%} = \sqrt{\Psi_{m\%}^2 + \Psi_{V\%}^2} = 10^{-4}\sqrt{3.88^2 + 4.64^2} = 6.0 \times 10^{-4}\%.$$

So,

$$\Psi_{p\%}=6\times10^{-4}\%.$$

It is interesting to calculate the confidence limits for the same confidence probability, using an approximate estimate of the number of degrees of freedom. We shall use the formula (6.8), transforming it, bearing in mind the fact that in our case $n_1=n_2=n=11$ and $b_1=b_2=1$:

$$\nu_{\text{eff}}=\frac{\left(\dfrac{S_r^2(m_i)}{n}+\dfrac{S_r^2(V_i)}{n}\right)^2}{\dfrac{S_r^4(m_i)}{n^2\nu}+\dfrac{S_r^4(V_i)}{n^2\nu}}=\frac{(3.32+4.74)^2\times10^{-22}}{\left(\dfrac{3.32^2}{10}+\dfrac{4.74^2}{10}\right)\times10^{-22}}=\frac{64.96}{33.49}\times10=19.$$

For $\nu_{\text{eff}}=19$ and $\alpha=0.95$ we find from a table of Student's distribution $t_q=2.10$. From here

$$\Psi_{p\%}=2.1\times2.7\times10^{-6}\times100=5.7\times10^4\%.$$

The value obtained is quite close to the result calculated by a different method.

6.8. Calculation of the error of the measurement of ionization current by the compensation method

Accurate measurements of weak currents, generated, for example, by γ rays from standards of unit radium mass, are performed by the compensation method using an electrometer. The measured current I is defined by the expression

$$I=CU/\tau,$$

where C is the capacitance of the capacitor, with whose help the ionization current is compensated, U is the initial voltage on the capacitor, and τ is the compensation time.

We shall examine the measurement of ionization current on the special apparatus described in Ref. 38. A capacitor, whose capacitance $C=4006.3$ pF is known to within 0.005%, is employed. The voltage on the capacitor is established with the help of a class 0.1 voltmeter with a measurement range of 0–15 V. The time is measured with a timer whose scale is divided into tenths of a second.

The measurement is performed by making repeated observations. Each time the same indication of the voltmeter $U=7$ V is established and the compensation time is measured. The results of 27 observations are given in the first column of Table 6.2.

The largest difference between the obtained values of the compensation time is equal to 0.4 s, i.e., the deviations from the average reach 0.25%. What can explain this spread? Obviously, the systematic errors of the measuring instrument employed here have nothing to do with the error. We shall estimate the random error of the instruments and their role.

According to the norms, the dead zone of an electric measuring instrument must not exceed the limit of the intrinsic error permissible for it. This limit in indicating a voltage of 7 V is $\delta U=0.1\times(15/7)=0.21\%$.

TABLE 6.2. The results of measurements of the ionization current and data from initial processing.

τ_i (s)	$I_i \times 10^{10}$ (A)	$(I_i - \bar{I}) \times 10^{14}$ (A)	$(I_i - \bar{I})^2 \times 10^{28}$ (A^2)
74.4	3.7694	7	49
74.6	3.7593	−94	8836
74.3	3.7745	58	3364
74.6	3.7593	−94	8836
74.4	3.7694	7	49
74.4	3.7694	7	49
74.4	3.7694	7	49
74.4	3.7694	7	49
74.4	3.7694	7	49
74.3	3.7745	58	3364
74.5	3.7643	−44	1936
74.4	3.7694	7	49
74.5	3.7643	−44	1936
74.4	3.7694	7	49
74.6	3.7593	−94	8836
74.2	3.7705	18	324
74.5	3.7643	−44	1936
74.3	3.7745	58	3364
74.4	3.7694	7	49
74.4	3.7694	7	49
74.5	3.7643	−44	1936
74.5	3.7643	−44	1936
74.3	3.7745	58	3364
74.3	3.7745	58	3364
74.3	3.7745	58	3364
74.4	3.7694	7	49
74.5	3.7643	−44	1936

For this reason, when a voltage of 7 V is set on the voltmeter, voltages that can differ from the average by not more than one-half the dead zone, i.e., 0.1%, will be obtained on the capacitor.

The timer has virtually no random error.

The compensation time is a function of the voltage on the capacitor, and the spread in the voltage on the capacitor is accompanied by the same spread in the compensation time. The obtained spread of 0.25% is larger than expected. Therefore the observed phenomenon should have a different reason.

In the experiment under study, the background current could be the reason. This current adds to the measured current and is indistinguishable from it. But it is known that the background current can be assumed to be a stationary process over a time interval shorter than that required to perform a measurement. This makes it possible to eliminate the background current by measuring during the experiment the average background current and subtracting it from the value obtained for the ionization current. For this, however, the ionization current must also be measured as an average current.

Essentially, in this manner the model of the phenomenon was redefined and a new definition of the specific measured quantity was given.

The starting dependence is nonlinear. For this reason, to solve the problem we shall use the method of reduction; for each value of τ_i ($i = 1,...,27$) we find the corresponding current I_i and then calculate, for the entire group of values, the average value, giving an estimate of the measured quantity.

The values of I_i are presented in the second column of Table 6.2. The average value $\overline{I}' = 3.7687 \times 10^{-10}$ A.

We shall now estimate the errors. First we shall find an estimation of the standard deviation of the measurement result. Since

$$\sum_{i=1}^{27} (I_i - \overline{I})^2 = 59\,171 \times 10^{-28},$$

we have

$$S(\overline{I}) = \sqrt{\frac{\sum_{i=1}^{27} (I_i - \overline{I})^2}{27 \times 26}} = 9.2 \times 10^{-14} \text{ A}.$$

In the relative form, $S_{\%}(\overline{I}) = 0.027\%$.

We shall now estimate the systematic errors. The influence coefficients (in the relative form) V_C, V_U, and V_τ of the errors of the estimates of the arguments C, U, and τ are equal to $V_C = 1$, $V_U = 1$, and $V_\tau = -1$. Therefore the existing systematic errors are related by the relation $\vartheta_{I\%} = \vartheta_{C\%} + \vartheta_{U\%} - \vartheta_{\tau\%}$.

For each error we estimate its limit $|\vartheta_i| \leqslant \theta_i$.

The limit of the total error of the voltmeter (neglecting the sign) is equal to 0.21%. Since the voltmeter also has a random error, we shall take $\theta_U = 0.15\%$. For the capacitor, $\theta_C = 0.05\%$ is given. The limit of the systematic error of the timer is equal to the value of one graduation, i.e., $\theta_\tau = 0.1 \times 100/74 = 0.135\%$.

Turning to formula (6.9) and setting $\alpha = 0.95$, we obtain

$$\theta_{I\%} = k\sqrt{\theta_C^2 + \theta_U^2 + \theta_\tau^2}$$
$$= 1.1\sqrt{(5 \times 10^{-2})^2 + (13.5 \times 10^{-2})^2 + (15 \times 10^{-2})^2} = 0.23\%.$$

The average background current is usually equal to $(0.5\text{–}1) \times 10^{-12}$ A. It can usually be measured to within 5%. With respect to the measured ionization current this error is equal to 0.013%, and it can obviously be neglected.

If during the measurement the average background current was equal to $\overline{I}_b = 0.75 \times 10^{-12}$ A, then

$$\tilde{I} = \overline{I} - \overline{I}_b = 3.7612 \times 10^{-10} \text{ A}.$$

So

$$\tilde{I} = 3.7612 \times 10^{-10} \text{ A}, \quad S(\tilde{I}) = 9 \times 10^{-14} \text{ A} \quad (n = 27),$$

$$\theta_I = 8.7 \times 10^{-13} \text{ A} \quad (\alpha = 0.95).$$

Since $\theta/S > 7$, the random error can be neglected. After rounding off, we obtain finally

$$I = (3.761 \pm 0.009) \times 10^{-10} \text{ A} \quad (\alpha = 0.95).$$

We note that the limits of the unexcluded systematic error of the result could also have been estimated by the method of sorting of variants.

6.9. Calculation of the error of indirect measurement of power at high frequency

As an example of a single indirect measurement we shall study the measurement of the power generated by a high-frequency current in a resistor according to the formula $P = I^2R$, where P is the power measured, I is the effective current, and R is the active resistance of the resistor.

Measurements of the current and resistance give estimates of their values $\tilde{I}$ and $\tilde{R}$ and the limits of the relative errors $\delta I = 0.5\%$ and $\delta R = 1\%$.

It is assumed that the starting equation can be linearized. We shall prove this assumption:

$$\tilde{P}=P+\frac{\partial f}{\partial I}\zeta_I+\frac{\partial f}{\partial R}\zeta_R+R_2,$$

$$R_2=\frac{1}{2}\left(\frac{\partial^2 f}{\partial I^2}\Delta I^2+\frac{\partial^2 f}{\partial R^2}\Delta R^2+2\frac{\partial^2 f}{\partial I\partial R}\Delta I\Delta R\right),$$

$$\frac{\partial f}{\partial I}=2IR,\quad \frac{\partial f}{\partial R}=I^2,\quad \frac{\partial^2 f}{\partial I^2}=2R,\quad \frac{\partial^2 f}{\partial R^2}=0,\quad \frac{\partial^2 f}{\partial I\partial R}=2I.$$

Using the relative form of the errors

$$\varepsilon_P=\frac{\tilde{P}-P}{P}=2\frac{\zeta_I}{I}+\frac{\zeta_R}{R}+r_2=2\varepsilon_I+\varepsilon_R+r_2,$$

where

$$r_2=\frac{R_2}{P}=\left(\frac{\Delta I}{I}\right)^2+2\frac{\Delta I}{I}\frac{\Delta R}{R}=\delta I^2+2\delta I\delta R.$$

It is obvious that $r_2 \ll (2\varepsilon_I+\varepsilon_R)$, i.e., linearization is permissible, and the error in the result can be estimated according to the formula

$$\varepsilon_P=2\varepsilon_I+\varepsilon_R.$$

We know the errors δI and δR, and in addition it is assumed that $|\varepsilon_I| \leqslant \delta I$ and $|\varepsilon_R| \leqslant \delta R$. The limit of the error of the result must be found differently depending on what about the errors δI and δR are known. If these errors are determined only by the properties of the instruments employed, then several situations are possible. We shall study the most typical ones.

(1) The measurement is performed under reference conditions, the measuring instruments have been recently checked, and the working standards were at least five times more accurate than the instruments. In this case, it can be assumed that the limits of error of the measurements of the current and resistance δI and δR are reliable. Assuming that within the estimated limits the actual errors ε_I and ε_R are distributed uniformly over the set of instruments, the limit of error of the measurement result can be found as the confidence error according to the formula

$$\delta P_1 = k\sqrt{4(\delta I)^2 + (\delta R)^2}.$$

For $\alpha = 0.95$, $k = 1.1$ and we obtain

$$\delta P_1 = 1.1\sqrt{4 \times 0.25 + 1} = 1.5\%.$$

(2) The measurement is performed under reference conditions, but the measuring instruments were checked comparatively a long time ago or the working standards were only two or three times more accurate. Under these conditions it cannot be assumed that the actual errors of measurement of the current and resistance do not exceed the limits of error δI and δR obtained based on the ratings of the instruments employed. Since in our case there are only two elementary errors and they have not been accurately estimated, we shall add them arithmetically. This gives

$$\delta P_2 = 2\delta I + \delta R = 2 \times 0.5 + 1 = 2\%.$$

(3) The measurement is performed under the normal operating conditions appropriate for the instruments employed and the estimates of the errors δI and δR are determined based on a number of components. For this reason, these estimates can be regarded as uncertainties, and the errors themselves can be assumed to have a normal distribution.

Then the uncertainty of the result of measurement can be found from the formula

$$\delta P_3 = \sqrt{4(\delta I)^2 + (\delta R)^2} = 1.4\%.$$

This estimate is practically just as reliable as the estimates of its components.

6.10. Calculation of the error of measurement of voltage with the help of a potentiometer and a voltage divider

The measurement of voltage with the help of a potentiometer is a direct measurement. However, when the errors of the potentiometers and the errors of the standard cell are standardized separately, and also when measurements are performed using a voltage divider, the error of the result of such a measurement is estimated by methods that are specifically designed for indirect measurements, and this is why this example is studied in Chap. 6, which is devoted to indirect measurements.

We shall study the case of single measurements with accurate estimation of errors, for example, measurement of voltage with the help of a class 0.005 P309 potentiometer, a class 0.005 standard cell, and a class 0.005 P35 voltage divider. These instruments were manufactured in the former USSR.

It is well known that when working with such potentiometers at first the potentiometerical current I_p is adjusted in the circuit with accurate resistors so that the voltage drop on the section of the circuit with the resistance R_{sc} would balance the emf of the standard cell U_{sc}. In this case

$$I_p = U_{sc}/R_{sc}.$$

Next, the standard cell is disconnected and the measured voltage U_p is connected to the potentiometer circuit. By switching the potentiometer a fraction of the resistors of the potentiometer is introduced into the comparison circuit such that the voltage drop on their resistance R_p would compensate U_p, i.e., $U_p = I_pR_p$. Then

$$U_p=\frac{R_p}{R_{sc}} U_{sc},$$

and knowing the emf of the standard cell and the ratio R_p/R_{sc}, we find U_p.

The indications of the potentiometer are proportional to R_p, but the error of the potentiometer is determined not by the errors of the resistances R_p and R_{sc}, but rather by the error of the ratio R_p/R_{sc}. The uncertainty associated with the operations of comparing the voltages can be neglected, since the smoothness of the unit controlling the potentiometer and the sensitivity of the zero indicator were designed so that this condition would be satisfied.

The potentiometer has six decades and a builtin self-balancing amplifier. The limit of permissible error as a function of the measured voltage U is calculated using the formula (given in the manufacturer's documentation)

$$\Delta U=\pm(50U+0.04)\times 10^{-6} \text{ V}.$$

The error of the potentiometer does not exceed the indicated limits if the ambient air temperature ranges from + 15 to + 30 °C and differs by not more than 2.5 °C from the temperature at which the measuring resistors of the potentiometer were adjusted (the P309 potentiometer has builtin calibration and adjusting systems).

The emf of the class 0.005 standard cell can be determined with an error of ± 10 μV. The effect of the temperature is taken into account with the help of a well-known formula, which describes quite accurately the temperature dependence of the emf.

Assume that in a measurement of one and the same voltage, performed using a voltage divider whose voltage division ratio was set equal to 1:10, the following potentiometer indications were obtained:

$$x_1=1.256\,316 \text{ V}, \quad x_2=1.256\,321 \text{ V}, \quad x_3=1.256\,318 \text{ V}.$$

The limit of permissible error of the potentiometer itself in this case is

$$\Delta U=\pm 50\times 1.26\times 10^{-6}=\pm 63 \ \mu\text{V}.$$

For this reason the difference of 5 μV between the results of the three observations presented above can be regarded as resulting from the random error of the measurement, whose magnitude is acceptable. In the calculation, therefore, any of the results obtained or their average value can be used.

In the process of adjusting the measuring resistors, which is donc prior to the measurement, the corrections of the higher-order decades were estimated. We shall introduce them into the indications of the potentiometer.

Let the correction for indication "12" of the decade "$\times$100 mV" equal $+ 15 \times 10^{-6}$ V, and the correction of the indication "5" of the decade "$\times$10 mV" equal $- 3 \times 10^{-6}$ V. The corrections for the other decades are so small that they are no longer of interest. Each correction is determined with an error of $\pm 5 \times 10^{-8}$ V. The error of the potentiometer corresponding to the indications of the remaining decades that are 0.0063...V falls within the limits determined in accordance with the formula given above and are equal to

$$\Delta U = \pm(50 \times 0.0063 + 0.04) \times 10^{-6} = \pm 0.4 \times 10^{-6} \text{ V}.$$

In addition, it is necessary to take into account the possible change in the air temperature in the room. If this change falls within permissible limits, then according to the specifications of the potentiometer the error can change approximately by 1/4 of the permissible limit, i.e., by 16 μV.

We shall take for the result the average value of the observations performed, correcting it by the amount $C = (15 - 3) \times 10^{-6} = 12 \times 10^{-6}\,\mu$V:

$$U_p = \bar{x} = 1.256\,318 + 0.000\,012 = 1.256\,330 \text{ V}.$$

The errors of the potentiometer itself, which enter into this result, are

$$\theta_1 = \pm 16 \times 10^{-6} \text{ V}, \quad \theta_2 = \pm 0.4 \times 10^{-6} \text{ V}.$$

The error in determining the corrections and the error θ_2 can be neglected.

Thus the limits of error of the potentiometer itself are equal to θ_1:

$$\theta_p = \theta_1 = \pm 16 \times 10^{-6} \text{ V}.$$

Next, we must estimate the errors owing to the standard cell and the voltage divider. Assuming that the division coefficient of the voltage divider is equal to K_d, the measured voltage is determined from formula $U_x = K_d U_p$, where

$$U_p = \frac{R_p}{R_{sc}} U_{sc},$$

and for this reason we can write

$$U_x = K_d \frac{R_p}{R_{sc}} U_{sc}. \tag{6.28}$$

The error of the voltage divider can reach $5 \times 10^{-3}\%$. But the real division coefficient of the divider can be found and taken into account. This is precisely what we must do in the case at hand. In the given measurement $K_d = 10.0003$ and the error in determining K_d falls within the range $\pm 2 \times 10^{-3}\%$.

The emf of the standard cell is taken into account with the help of the special decades of the potentiometer. The discrepancy between the real value of the emf of the standard cell and the value exhibited on the potentiometer falls within the limits of error in determining the emf of the standard cell ($\pm 10\,\mu$V).

We estimate the measured voltage U_x according to the formula

$$\tilde{U}_x = K_d U_p = 10.0003 \times 1.256\,330 = 12.563\,68 \text{ V}.$$

To estimate the measurement error, we shall use the standard trick. First, we shall take the logarithm of expression (6.28). Then we find the differentials of both sides of the equation and, neglecting errors that are second-order infinitesimals, we replace the differentials by the increments. This gives

$$\frac{\Delta U_x}{U_x} = \frac{\Delta K_d}{K_d} + \frac{\Delta(R_p/R_{sc})}{R_p/R_{sc}} + \frac{\Delta U_{sc}}{U_{sc}}.$$

For the terms we have only estimates of the limits, and not the values of the errors themselves. For this reason, we shall estimate the limits of the measurement error. For this, we can use the formula (5.3). First, all components must be represented in the form of relative errors. The relative error of the potentiometer itself, more accurately, its limits in percent, will be

$$\theta_{p\%}=\frac{100\theta_p}{U_p}=\frac{100\Delta(R_p/R_{sc})}{R_p/R_{sc}}=\pm\frac{100\times 16\times 10^{-6}}{1.26}=\pm 1.3\times 10^{-3}\%.$$

The limits of the relative error of the voltage divider were estimated directly as $\theta_{K\%} = \pm 2 \times 10^{-3}\%$. The limits of error in determining the emf of the standard cell in the form of a relative error will be

$$\theta_{sc\%}=\pm\frac{100\times 10\times 10^{-6}}{1.018}=\pm 1\times 10^{-3}\%.$$

We now find the limit of the measurement error according to formula (5.3):

$$\theta_{\%}=k\sqrt{1.3^2+2^2+1^2}\times 10^{-3}=k\times 2.6\times 10^{-3}\%.$$

Let $\alpha = 0.95$. Then $k = 1.1$ and

$$\theta_{\%}=1.1\times 2.6\times 10^{-3}=2.9\times 10^{-3}\approx 3\times 10^{-3}\%.$$

Finally, we must check the number of significant figures in the result of measurement. For this, we shall put the limits $\theta_{\%}$ in the form of absolute errors

$$\theta=\pm 2.9\times 10^{-3}\times 10^{-2}\times 12.6=\pm 37\times 10^{-5}\ \text{V}.$$

Since the measurement is accurate, the error in the result of measurement is expressed by two significant figures and there are no extra figures in the result obtained. The final result is

$$U_x=12.563\,68\pm 0.000\,37\ \text{V}\quad(\alpha=0.95).$$

If the measurement were performed with approximate estimation of the errors, then the errors of all components would have to be set equal to $5 \times 10^{-3}\%$ and the limit of the measurement error would be

$$\theta_{\%}=1.1\times 10^{-3}\sqrt{3\times 5^2}=0.01\%.$$

Then $\theta' = \pm 0.0013$ V and the result of measurement would have to be written with fewer significant figures:

$$U_x=12.5637\pm 0.0013\ \text{V}\quad(\alpha=0.95).$$

Here two significant figures are retained in the numerical value of the measurement error because the value of the most significant digit is less than 3.

Chapter 7

Simultaneous and combined measurements

7.1. General remarks about the method of least squares

Simultaneous and combined measurements, as pointed out in Chap. 1, are usually performed so that the number of equations relating the measured quantities is larger than the number of the latter. Because of measurement errors, even when the exact relation between the measured quantities is known, it is impossible to find values of the unknowns such that all equations would be satisfied. Under these conditions, the estimated values of the unknowns are found with the help of the method of least squares.

The method of least squares is a widely employed computational technique that makes it possible to eliminate the nonuniqueness of experimental data. This method is easily implemented with the help of computers and good least-squares software is available.

There is an extensive literature on the method of least squares, and it has been well studied. It is known that this method does not always give results that satisfy the criteria of optimality of estimation theory. This method gives unbiased estimates of unknown quantities, which have minimum variances, only if errors of measurements are normally distributed.

The indicated conditions are rarely satisfied in practice. The method of least squares is nonetheless widely employed. This is because in general it is simple, and the biasness of the estimates obtained is usually not very significant. In addition, it must be acknowledged that so far there is simply no other universal method. More grounded methods—methods of confluent analysis—have been developed only for the simplest cases.[54]

We shall discuss the method of least squares, since it is the main computational method used for simultaneous and combined measurements and in order to use this method knowingly it is necessary to know its basic ideas.

An example of simultaneous measurements is finding the parameters of the equation

$$R=R_{20}+a(t-20)+b(t-20)^2,$$

which expresses the temperature dependence of an accurate measuring resistor.

By measuring simultaneously R (the resistance of the resistor) and t (the temperature of the resistor) and by varying the temperature, we obtain several equations, from which it is necessary to find R_{20}—the resistance of the resistor at $t = 20\,^\circ\text{C}$—and the temperature coefficients a and b.

We can write the basic equation in the general form

$$F_0(A,B,C,...,x,y,z,...)=l, \tag{7.1}$$

where x, y, z, and l are known coefficients and directly measured quantities, and A, B, and C are the unknowns to be determined.

Substituting the experimentally obtained numerical values of x_i, y_i, and z_i into Eq. (7.1) we obtain a series of equations of the form

$$F_i(A,B,C,...,x_i,y_i,z_i)=l_i, \tag{7.2}$$

which contain only the unknown quantities A, B, and C to be found and numerical coefficients.

The quantities sought are found by solving the obtained equations simultaneously.

An example of a combined measurement is finding the capacitances of two capacitors from measurements of the capacitance of each one of them separately as well as when the capacitors are connected in parallel and in series. Each of these measurements is performed with one observation, but ultimately we shall have four equations for two unknowns:

$$C_1=x_1, \quad C_2=x_2, \quad C_1+C_2=x_3, \quad \frac{C_1C_2}{C_1+C_2}=x_4.$$

Substituting into these equations the experimentally found values of x_i, we obtain a system of equations analogous to Eqs. (7.2).

As we have already pointed out, the number of equations in the system (7.2) is usually greater than the number of unknowns, and because of measurement errors it is impossible to find values of the measured quantities such that all equations would be satisfied simultaneously, even if they are themselves known exactly. For this reason, Eqs. (7.2), in contrast to normal mathematical equations, are said to be conditional. When the values of the unknowns found by some method are substituted into the conditional equations (7.2), for the reasons mentioned we obtain

$$F_i(\tilde{A},\tilde{B},\tilde{C},...)-l_i=v_i\neq 0.$$

The quantities v_i are called residuals. The solution of the conditional equations that minimizes the sum of the squares of the residuals is generally recognized. This proposition was first published by Legendre and is called Legendre's principle. He implemented this principle by the method that is now called the method of least squares.

7.2. Measurements with linear equally accurate conditional equations

To simplify the formulas we shall consider the case of three unknowns. Let the system of conditional equations have the form

$$Ax_i+By_i+Cz_i=l_i \quad (i=1,...,n,\ n>3), \tag{7.3}$$

where A, B, and C are the unknowns to be determined and x_i, y_i, z_i, and l_i are the results of the ith series of measurements and known coefficients.

In the general case the number of unknowns $m<n$; if $m=n$, then the system of conditional equations can be solved uniquely, though the obtained results are burdened with errors.

If some estimates of the measured quantities $\tilde{A}$, $\tilde{B}$, and $\tilde{C}$ are substituted into Eq. (7.3), then we obtain the residuals

$$v_i=\tilde{A}x_i+\tilde{B}y_i+\tilde{C}z_i-l_i.$$

We shall find estimates of $\tilde{A}$, $\tilde{B}$, and $\tilde{C}$ from the conditions

$$Q=\sum_{i=1}^{n} v_i^2=\min.$$

In order for this condition to be satisfied it is necessary that

$$\frac{\partial Q}{\partial \tilde{A}}=\frac{\partial Q}{\partial \tilde{B}}=\frac{\partial Q}{\partial \tilde{C}}=0.$$

We shall find these particular derivatives and equate them to 0:

$$\frac{\partial Q}{\partial \tilde{A}}=2\sum_{i=1}^{n}(\tilde{A}x_i+\tilde{B}y_i+\tilde{C}z_i-l_i)x_i=0,$$

$$\frac{\partial Q}{\partial \tilde{B}}=2\sum_{i=1}^{n}(\tilde{A}x_i+\tilde{B}y_i+\tilde{C}z_i-l_i)y_i=0,$$

$$\frac{\partial Q}{\partial \tilde{C}}=2\sum_{i=1}^{n}(\tilde{A}x_i+\tilde{B}y_i+\tilde{C}z_i-l_i)z_i=0.$$

From here we obtain a system of so-called normal equations:

$$\tilde{A}\sum_{i=1}^{n}x_i^2+\tilde{B}\sum_{i=1}^{n}x_iy_i+\tilde{C}\sum_{i=1}^{n}x_iz_i=\sum_{i=1}^{n}x_il_i,$$

$$\tilde{A}\sum_{i=1}^{n}y_ix_i+\tilde{B}\sum_{i=1}^{n}y_i^2+\tilde{C}\sum_{i=1}^{n}y_iz_i=\sum_{i=1}^{n}y_il_i,$$

$$\tilde{A}\sum_{i=1}^{n}z_ix_i+\tilde{B}\sum_{i=1}^{n}z_iy_i+\tilde{C}\sum_{i=1}^{n}z_i^2=\sum_{i=1}^{n}z_il_i.$$

The normal equations are often written using Gauss's notation:

$$\sum_{i=1}^{n}x_i^2=[xx],\quad \sum_{i=1}^{n}x_iy_i=[xy],\ \text{etc.}$$

It is obvious that

$$\sum_{i=1}^{n}x_iy_i=\sum_{i=1}^{n}y_ix_i,\quad [xy]=[yx].$$

In Gauss's notation the normal equations assume the simpler form

$$[xx]\tilde{A}+[xy]\tilde{B}+[xz]\tilde{C}=[xl],$$
$$[xy]\tilde{A}+[yy]\tilde{B}+[yz]\tilde{C}=[yl],$$
$$[xz]\tilde{A}+[yz]\tilde{B}+[zz]\tilde{C}=[zl]. \tag{7.4}$$

We call attention to two obvious but important properties of the matrix of coefficients of the unknowns in the system of equations (7.4):

(1) The matrix of these coefficients is symmetric relative to the main diagonal.

(2) All elements on the main diagonal are positive.

These properties are general. They do not depend on the number of unknowns, but in this example they are shown in application to the case with three unknowns.

The number of normal equations is equal to the number of unknowns, and solving these equations by known methods we obtain estimates of the measured quantities. The solution can be written most compactly with the help of the determinants

$$\tilde{A}=\frac{D_x}{D}, \quad \tilde{B}=\frac{D_y}{D}, \quad \tilde{C}=\frac{D_z}{D}, \tag{7.5}$$

where

$$D=\begin{vmatrix} [xx] & [xy] & [xz] \\ [yx] & [yy] & [yz] \\ [zx] & [zy] & [zz] \end{vmatrix}.$$

The determinant D_x is obtained from the principal determinant D of the system by replacing the column with the coefficients of the unknown $\tilde{A}$ with the column of free terms:

$$D_x=\begin{vmatrix} [xl] & [xy] & [xz] \\ [yl] & [yy] & [yz] \\ [zl] & [zy] & [zz] \end{vmatrix}.$$

The determinants D_y and D_z are found analogously, i.e., by replacing the second and third columns, respectively, with the indicated column.

Now we must estimate the errors of the obtained results. The estimate of the variance of the conditional equations is calculated from the formula

$$S^2=\frac{\sum_{i=1}^{n} v_i^2}{n-m}, \tag{7.6}$$

where v_i is the residual of the ith conditional equation. Then the estimates of the variances of the values found for the unknowns can be calculated using the formulas

$$S^2(\tilde{A})=\frac{D_{11}}{D}S^2, \quad S^2(\tilde{B})=\frac{D_{22}}{D}S^2, \quad S^2(\tilde{C})=\frac{D_{33}}{D}S^2, \tag{7.7}$$

where D_{11}, D_{22}, and D_{33} are the algebraic complements of the elements $[xx]$, $[yy]$, and $[zz]$ of the determinant D, respectively (they are obtained by removing from the matrix of the determinant D the column and row whose intersection is the given element).

The confidence intervals for the true values of the measured quantities are constructed based on Student's distribution. In this case the number of degrees of freedom for all measured quantities is equal to $\nu = n - m$.[42]

7.3. Reduction of linear unequally accurate conditional equations to equally accurate conditional equations

In Sec. 7.2 we studied the case when all conditional equations had the same variance. Such conditional equations are said to be equally accurate. In practice there can be cases when the conditional equations have different variances. This usually happens if equations reflecting the measurements performed under different conditions are added to the system of equations. For example, if in calibrating a collection of weights, special measures are not taken, because of the different loading of the weights the weighing errors will be different for different combinations of weights. Correspondingly, the conditional equations will not be equally accurate either.

For unequally accurate conditional equations the most likely set of values of the unknowns $A,B,C,\ldots$ will be obtained if the expression

$$Q = \sum_{i=1}^{n} g_i v_i^2,$$

where g_i is the weight of the ith conditional equation, is minimized.

The introduction of weights is equivalent to multiplying the conditional equations by $\sqrt{g_i}$. Finally, the cofactors g_i will appear in the coefficients of the unknowns in the normal equations.

Thus the first equation of the system of normal equations (7.4) will assume the form

$$[gxx]\tilde{A} + [gxy]\tilde{B} + [gxz]\tilde{C} + [gxl] = 0.$$

All remaining equations will change analogously. Each coefficient in the equation is a sum of terms of the form

$$[gxy] = g_1 x_1 y_1 + g_2 x_2 y_2 + \cdots + g_n x_n y_n.$$

The weights of the conditional equations are found from the conditions

$$\sum_{i=1}^{n} g_i = 1,$$

$$g_1 : g_2 : \cdots : g_n = \frac{1}{\sigma_1^2} : \frac{1}{\sigma_2^2} : \cdots : \frac{1}{\sigma_n^2}.$$

Therefore, to solve the problem it is necessary to know the variance of the conditional equations. If the weights have been determined (or chosen), then after the transformations presented above the further solution of the problem proceeds in the manner described in Sec. 7.2, and finally we obtain estimates of the measured quantities and their rms deviations. However, the weights are usually determined approximately.

7.4. Linearization of nonlinear conditional equations

For a number of fundamental reasons, the method of least squares has been developed only for linear conditional equations. For this reason, nonlinear conditional equations must be put into a linear form.

The general method for doing this is based on the assumption that the incompatibility of the conditional equations is small, i.e., their residuals are small. Then, taking from the system of conditional equations as many equations as there are unknowns and solving them, we find the initial estimates of the unknowns A_0, B_0, and C_0. Next, assuming that

$$A=A_0+a, \quad B=B_0+b, \quad C=C_0+c,$$

and substituting these expressions into the conditional equations, we expand the conditional equations in series. Let

$$F_i(A,B,C)=l_i.$$

Then retaining only terms with the first powers of the corrections a, b, and c we obtain

$$f_i(A_0,B_0,C_0)-l_i+\left(\frac{\partial f_i}{\partial A}\right)_0 a+\left(\frac{\partial f_i}{\partial B}\right)_0 b+\left(\frac{\partial f_i}{\partial C}\right)_0 c=0.$$

We find the partial derivatives by differentiating the functions $f_i(A,B,C)$ with respect to A, B, and C, respectively, and then we substitute A_0, B_0, and C_0 into the obtained formulas and find their numerical values. In addition,

$$f_i(A_0,B_0,C_0)-l_i=\lambda_i\neq 0.$$

Thus we obtain a system of linear conditional equations for a, b, and c. The solution of this system gives their estimates and standard deviations. Then

$$\tilde{A}=A_0+\tilde{a}, \quad \tilde{B}=B_0+\tilde{b}, \quad C=C_0+\tilde{c}.$$

Since A_0, B_0, and C_0 are nonrandom quantities, $S^2(\tilde{A}) = S^2(\tilde{a})$, $S^2(\tilde{B}) = S^2(\tilde{b})$, etc.

In principle, once $\tilde{A}$, $\tilde{B}$, and $\tilde{C}$ have been obtained, the second approximation can be constructed, etc.

In addition to the foregoing method of linearization of the conditional equations, the method of substitutions is also employed. Thus if, for example, the conditional equation has the form

$$y_i=x_i \sin A+z_i e^{-2B},$$

where x, y, and z are directly measured quantities, and A and B must be determined, then the substitution

$$U=\sin A, \quad E=e^{-2B}$$

can be made.

Then we obtain the linear conditional equation

$$y_i = x_i U + z_i E.$$

The solution of these equations gives $\tilde{U}$ and $\tilde{E}$ and estimates of their variances which through their use we can then find the quantities A and B sought.

The method of substitutions is convenient, but it is not always applicable. In principle, one can imagine one other general method for solving a system of equations when the number of equations is greater than the number of unknowns. This method is as follows.

Take from the available conditional equations a group of equations such that their number is equal to the number of unknowns. Such a group gives a definite value for each of the unknowns.

Next, replacing in turn the equations in the group by each of the other equations that were not in the group, we obtain other values of the same unknowns. Irrespective of the method used to combine the equations, all possible combinations of equations must be sorted through and for each combination the values of the unknowns must be found. As a result of such calculations we obtain for each unknown a group of values that can be regarded as the group of observations obtained with direct measurements.

All values in the group are equivalent, but, unfortunately, they are not independent. This presents difficulties in estimating the variances of the values obtained for the unknowns.

7.5. Examples of the application of the method of least squares

The examples studied below are presented in order to demonstrate the computational technique as well as the physical meaning of the method, and for this reason they were chosen so that the calculations would be as simple as possible. The initial data for the examples are taken from Ref. 43.

Example 1. Determine the angles of a trihedral prism. Each angle is measured three times. The measurements of all angles are equally accurate. The results of all single measurements are as follows:

$$x_1 = 89^\circ\,55', \quad y_1 = 45^\circ\,5', \quad z_1 = 44^\circ\,57',$$

$$x_2 = 89^\circ\,59', \quad y_2 = 45^\circ\,6', \quad z_2 = 44^\circ\,55',$$

$$x_3 = 89^\circ\,57', \quad y_3 = 45^\circ\,5', \quad z_3 = 44^\circ\,58'.$$

If each angle is found as the arithmetic mean of the corresponding observations, then we obtain

$$A_0 = 89^\circ\,57', \quad B_0 = 45^\circ\,5.33', \quad C_0 = 44^\circ\,56.67'.$$

The sum of the angles of the triangle must satisfy the condition

$$A + B + C = 180^\circ.$$

We obtain 179° 59′. This discrepancy is the result of measurement errors. The values of A_0, B_0, and C_0 obtained must be changed so that the exactly known condition is satisfied.

The relations between the unknowns that must be satisfied exactly are called constraints.

In Ref. 43 the problem is solved by the method of least squares. However, in this case the method of least squares must be regarded only as a procedure leading to a unique answer. Its application here can be justified by the fact that the residual is equal to only 180° − 179° 59′ = 1′, so that the changes in the directly obtained values of the angles should be insignificant. Therefore these changes need not be found with high accuracy. We now proceed to the solution of the problem.

In order that the constraints be satisfied, we proceed as follows.

If we have n conditional equations, m unknowns, and k constraints, and $n > m - k$ and $m > k$, then k unknowns can be eliminated from the conditional equations by expressing these unknowns in terms of the remaining unknowns. Next, using the method of least squares we find the values of $m - k$ unknowns and the estimates of standard deviations of these estimates. We obtain the remaining k unknowns using the constraint equations. To find their standard deviations, strictly speaking, another cycle of calculations with the conditional equations, in which the previously excluded unknowns are retained and the other unknowns are excluded, must be performed. This repeated calculation is often not performed, based on the fact that any conclusion about the standard deviation of the previously excluded unknowns can be made using the estimate of the standard deviation of the other unknowns.

Let us return to our problem. To simplify the calculations we shall assume that

$$A = A_0 + a, \quad B = B_0 + b, \quad C = C_0 + c$$

and we shall find the values of the corrections a, b, and c.

The system of conditional equations transforms into the following system:

$$a_1 = -2', \quad b_1 = -0.33', \quad c_1 = +0.33',$$

$$a_2 = +2' \quad b_2 = +0.67', \quad c_2 = -1.67',$$

$$a_3 = 0, \quad b_3 = -0.33' \quad c_3 = +1.33'.$$

The constraint equation will assume the form

$$A_0 + a + B_0 + b + C_0 + c = 180°.$$

Therefore

$$a + b + c = 180° - 179°\,59' = 1'.$$

We exclude c from the conditional equations using the relation

$$c = 1' - a - b,$$

and in each equation we indicate both unknowns. We obtain the following system of conditional equations:

$$1\times\tilde{a}+0\times\tilde{b}=-2', \quad 0\times\tilde{a}+1\times\tilde{b}=-0.33', \quad 1\times\tilde{a}+1\times\tilde{b}=+0.67',$$
$$1\times\tilde{a}+0\times\tilde{b}=+2', \quad 0\times\tilde{a}+1\times\tilde{b}=+0.67', \quad 1\times\tilde{a}+1\times\tilde{b}=+2.67',$$
$$1\times\tilde{a}+0\times\tilde{b}=0, \quad 0\times\tilde{a}+1\times\tilde{b}=-0.33', \quad 1\times\tilde{a}+1\times\tilde{b}=-0.33'.$$

We now construct the system of normal equations. Its general form will be

$$[xx]\tilde{a}+[xy]\tilde{b}=[xl],$$
$$[xy]\tilde{a}+[yy]\tilde{b}=[yl],$$

Here we obtain:

$$[xx]=1+1+1+1+1+1=6,$$
$$[xy]=1+1+1=3,$$
$$[yy]=1+1+1+1+1+1=6,$$
$$[xl]=-2'+2'+0.67'+2.67'-0.33'=+3',$$
$$[yl]=-0.33'+0.67'-0.33'+0.67'+2.67'-0.33'=+3'.$$

Therefore the normal equations will assume the form

$$6\tilde{a}+3\tilde{b}=3', \quad 3\tilde{a}+6\tilde{b}=3'.$$

In accordance with the relations (7.5) we calculate

$$D=\begin{vmatrix} 6 & 3 \\ 3 & 6 \end{vmatrix}=36-9=27,$$

$$D_a=\begin{vmatrix} 3' & 3 \\ 3' & 6 \end{vmatrix}=18'-9'=9',$$

$$D_b=\begin{vmatrix} 6 & 3' \\ 3 & 3' \end{vmatrix}=18'-9'=9'$$

and we find

$$\tilde{a}=\tilde{b}=9'/27=0.33'.$$

Therefore $\tilde{c}=0.33'$ also.

Substituting the estimates obtained into the conditional equations, we calculate the residuals:

$$v_1=2.33', \quad v_4=0.67', \quad v_7=0,$$
$$v_2=-1.67', \quad v_5=-0.33', \quad v_8=-2',$$
$$v_3=0.33', \quad v_6=0.67', \quad v_9=1'.$$

From formula (7.6) we calculate an estimate of the variance of the equations

$S^2=\dfrac{\sum_{i=1}^{9} v_i^2}{9-2}=\dfrac{14.34}{7}=2.05$. Now $D_{11}=6$, $D_{22}=6$ and formulas (7.7) give

$$S^2(\tilde{a})=S^2(\tilde{b})=\frac{6}{27}\times 2.05=0.456,\quad S(\tilde{a})=S(\tilde{b})=0.675.$$

In view of the fact that the conditional equations are equally accurate and the estimates $\tilde{a}$, $\tilde{b}$, and $\tilde{c}$ are equal to one another the repeated calculations need not be performed, and we can write down immediately $S(\tilde{c})=0.675'$. Finally, we obtain $\tilde{A}=89°\ 57.33'$, $\tilde{B}=45°\ 5.67'$, $\tilde{C}=44°\ 57.00'$, and $S(\tilde{A})=S(\tilde{B})=S(\tilde{C})=0.68'$.

We construct the confidence interval for each angle based on Student's distribution. The number of degrees of freedom in this case is equal to $9-2=7$ and for $\alpha=0.95$ Student's coefficient $t_{0.95}=2.36$. Therefore $\Delta_{0.95}=2.36\times 0.68'=1.6'$. Thus, we obtain finally

$$A_{0.95}=89°\ 57.3'\pm 1.6',\quad B_{0.95}=45°\ 5.7'\pm 1.6',\quad C_{0.95}=44°\ 57.0'\pm 1.6'.$$

Example 2. We shall study the example, presented at the beginning of this chapter, of combined measurements of the capacitance of two capacitors. The results of the direct measurements are as follows:

$$x_1=0.2071\ \mu\text{F},\quad x_2=0.2056\ \mu\text{F},$$

$$x_1+x_2=0.4111\ \mu\text{F},\quad \frac{x_1x_2}{x_1+x_3}=0.1035\ \mu\text{F}.$$

The last equation is nonlinear. We expand it in a Taylor series, for which we first find the partial derivatives

$$\frac{\partial f}{\partial C_1}=\frac{C_2(C_1+C_2)-C_1C_2}{(C_1+C_2)^2}=\frac{C_2^2}{(C_1-C_2)^2}$$

and analogously

$$\frac{\partial f}{\partial C_2}=\frac{C_1^2}{(C_1+C_2)^2}\cdot$$

Since $C_1\approx x_1$ and $C_2\approx x_2$, we can write

$$C_1=0.2070+e_1,\quad C_2=0.2060+e_2.$$

We make the expansion for the point with the coordinates $C_{10}=0.2070$ and $C_{20}=0.2060$. We obtain

$$\frac{C_{10}C_{20}}{C_{10}+C_{20}}=0.103\ 25,$$

$$\frac{\partial f}{\partial C_1}=\frac{0.206^2}{(0.207+0.206)^2}=0.249,$$

$$\frac{\partial f}{\partial C_2}=\frac{0.207^2}{(0.207+0.206)^2}=0.251.$$

We find the conditional equations, setting $x_1=C_1$ and $x_2=C_2$:

$$1\times e_1+0\times e_2=0.0001,$$

$$0\times e_1+1\times e_2=-0.0004,$$

$$1\times e_1+1\times e_2=-0.0019,$$

$$0.249e_1+0.251e_2=0.000\,25.$$

We now calculate the coefficients of the normal equations

$$[xx]=2.062,\quad [xy]=1.0625,\quad [yy]=2.063,\quad [xl]=-0.001\,738,$$

$$[yl]=-0.002\,237.$$

The normal equations will be

$$2.062e_1+1.0625e_2=-0.001\,738,$$

$$1.0625e_1+2.063e_2=-0.002\,237.$$

We now find the unknowns e_1 and e_2. According to Eqs. (7.5) we calculate

$$D=\begin{vmatrix} 2.062 & 1.0625 \\ 1.0625 & 2.063 \end{vmatrix}=3.125,$$

$$D_x=\begin{vmatrix} -0.001\,738 & 1.0625 \\ -0.002\,237 & 2.063 \end{vmatrix}=-0.001\,22,$$

$$D_y=\begin{vmatrix} 2.062 & -0.001\,738 \\ 1.0625 & -0.002\,237 \end{vmatrix}=-0.002\,75.$$

From here we find

$$e_1=\frac{D_x}{D}=-0.000\,39,\quad e_2=\frac{D_y}{D}=-0.000\,88.$$

Therefore

$$\tilde{C}_1=0.2070-0.000\,39=0.206\,61\ \mu\text{F},$$

$$\tilde{C}_2=0.2060=0.000\,88=0.205\,12\ \mu\text{F}.$$

We find the residuals of the conditional equations by substituting the estimates obtained for the unknowns into the conditional equations:

$$v_1=0.000\,49,\quad v_3=-0.000\,63,$$

$$v_2=0.000\,58,\quad v_4=0.000\,48.$$

Now we can calculate from formula (7.6) an estimate of the variance of the conditional equations:

$$S^2=\frac{\sum_{i=1}^{4} v_i^2}{4-2}=\frac{120\times 10^{-8}}{2}=6\times 10^{-7}.$$

The algebraic complements of the determinant D will be $D_{11}=2.063$ and $D_{22}=2.062$. Since $D_{11}\approx D_{22}$,

$$S^2(\widetilde{C}_1)=S^2(\widetilde{C}_2)=\frac{D_{11}}{D}S^2=\frac{2.063}{3.125}\times 6\times 10^{-7}=4\times 10^{-7},$$

$$S(\widetilde{C}_1)=S(\widetilde{C}_2)=6.3\times 10^{-4}\ \mu\text{F}.$$

The method, examined above, for measuring the capacitances of the capacitors was apparently chosen in order to reduce somewhat the systematic error of the measurement, which is different at different points of the measurement range; to reduce the random component of the error it would be sufficient to measure each capacitance multiply.

7.6. Determination of the parameters in formulas from empirical data and construction of calibration curves

The purpose of almost any investigation in natural science is to find regularities in the phenomena in the material world, and measurements are the characteristic method that gives objective data for achieving this goal.

It is desirable to represent the regular correlations determined between physical quantities based on measurements in an analytic form, i.e., in the form of formulas. The initial form of the formulas is usually established based on an unformalized analysis of the collection of data obtained. One of the important prerequisites of analysis is the assumption that the dependence sought must be expressed by a smooth curve; physical laws usually correspond to smooth curves.

Once the form of the formula is chosen, its parameters are then found by an interpolation approximation of the empirical data by the formula obtained, and this is most often done by the method of least squares.

This problem is of great importance, and there are many mathematical and applied works that are devoted to it.

We shall discuss some aspects of the solution of this problem that are connected with the application of the method of least squares. The application of this method is based on the assumption that the criterion for the optimal choice of the parameter sought can be assumed to be that the sum of squares of the deviations of the empirical data from the curve obtained is minimized. This assumption is very often justified, but not always. For example, sometimes the curve must be drawn so that it exactly passes through all prescribed points. This is natural, if the coordinates of the points mentioned are given as exact coordinates. The problem is solved by the methods of the interpolation approximation, and in addition it is known that the degree of the interpolation polynomial will be only one less than the number of points mentioned.

Sometimes the maximum deviation of the experimental data from the curve is minimized. As we have pointed out, however, most often the sum of the squares of the indicated deviations is minimized by the method of least squares. For this purpose all values obtained for the quantities (in physically justified combinations) are substituted successively into the chosen formula. Ultimately a system of conditional equations, from which the normal equations are constructed, is obtained; the solution of these equations gives the values sought for the parameters.

Next, substituting the values obtained for the parameters into the conditional equations, the residuals of these equations can be found and the standard deviation of the conditional equations can be estimated from them.

It is significant that in this case the standard deviation of the conditional equations is determined not only by the measurement errors but also by the imperfect structure of the formula chosen to describe the dependence sought. For example, it is well known that the temperature dependence of the electric resistance of many metals is reminiscent of a parabola. In engineering, however, it is often found that some definition section of this dependence can be approximated by a linear function. The inaccuracy of the chosen formula, naturally, is reflected in the standard deviation of the conditional equations. Even if all experimental data were free of any errors, the standard deviation would still be different from 0. For this reason, in this case the standard deviation characterizes not only the error of the conditional equations, but also the fact that the empirical formula adopted does not correspond to the true relation between the quantities.

In connection with what we have said above, the estimates, obtained by the method described above, of the variances of the determined parameters of the empirical formulas become conditional in the sense that they characterize not only the random spread in the experimental data, as usual, but also the uncertainty of the approximation, which is nonrandom.

It should be noted that if the empirical formula can be assumed to be linear, then the parameters of this formula can also be determined by the method of correlation and regression analysis. They also make it possible to construct the confidence intervals for the parameters and the confidence zone for the approximating straight line.

Everything said above is completely relevant to the problem of constructing calibration curves of measuring transducers and instruments.

We shall discuss the problem of constructing linear calibration curves, most often encountered in practice.

Thus, the relation between a quantity y at the output of a transducer and the quantity x at the input of the transducer must be expressed by the dependence

$$y = a + bx. \tag{7.8}$$

When calibrating the transducer the values of $\{x_i\}$, $i = 1,\dots,n$, in the range $[x_{min}, x_{max}]$ are given and the corresponding values $\{y_i\}$ are found.

Relation (7.8) gives a system of conditional equations

$$\tilde{b}x_i + \tilde{a} = y_i + v_i.$$

The residuals v_i are determined by the relation

$$v_i = \tilde{b}x_i + \tilde{a} - y_i.$$

Following the least-squares scheme presented above, we obtain the system of normal equations,

$$\tilde{b}\sum_{i=1}^{n} x_i^2 + \tilde{a}\sum_{i=1}^{n} x_i = \sum_{i=1}^{n} x_i y_i, \quad \tilde{b}\sum_{i=1}^{n} x_i + n\tilde{a} = \sum_{i=1}^{n} y_i. \tag{7.9}$$

The principal determinant of the system (7.9) will be

$$D=\begin{vmatrix} \sum_{i=1}^{n} x_i^2 & \sum_{i=1}^{n} x_i \\ \sum_{i=1}^{n} x_i & n \end{vmatrix} = n \sum_{i=1}^{n} x_i^2 - \left(\sum_{i=1}^{n} x_i \right)^2 .$$

The determinant D_x is given by

$$D_x=\begin{vmatrix} \sum_{i=1}^{n} x_iy_i & \sum_{i=1}^{n} x_i \\ \sum_{i=1}^{n} y_i & n \end{vmatrix} = n \sum_{i=1}^{n} (x_iy_i) - \sum_{i=1}^{n} x_i \sum_{i=1}^{n} y_i .$$

From here we find an estimate of the coefficient b:

$$\tilde{b}=\frac{D_x}{D}=\frac{n \sum_{i=1}^{n} x_iy_i - \sum_{i=1}^{n} x_i \sum_{i=1}^{n} y_i}{n \sum_{i=1}^{n} x_i^2 - \left(\sum_{i=1}^{n} x_i \right)^2}=\frac{\sum_{i=1}^{n} x_iy_i - n\bar{x}\bar{y}}{\sum_{i=1}^{n} x_i^2 - n(\bar{x})^2} .$$

It is not difficult to show that

$$\sum_{i=1}^{n} x_iy_i - n\bar{x}\bar{y} = \sum_{i=1}^{n} (x_i-\bar{x})(y_i-\bar{y}) \tag{7.10}$$

and that

$$\sum_{i=1}^{n} x_i^2 - n\bar{x}^2 = \sum_{i=1}^{n} (x_i-\bar{x})^2 . \tag{7.11}$$

Then the expression for $\tilde{b}$ assumes the simpler form

$$\tilde{b}=\frac{\sum_{i=1}^{n} (x_i-\bar{x})(y_i-\bar{y})}{\sum_{i=1}^{n} (x_i-\bar{x})^2} . \tag{7.12}$$

The determinant D_y is given by

$$D_y=\begin{vmatrix} \sum_{i=1}^{n} x_i^2 & \sum_{l=1}^{n} x_iy_i \\ \sum_{i=1}^{n} x_i & \sum_{i=1}^{n} y_i \end{vmatrix} = n\bar{y} \sum_{i=1}^{n} x_i^2 \quad n\bar{x} \sum_{i=1}^{n} x_iy_i .$$

Therefore

$$\tilde{a}=\frac{D_y}{D}=\frac{n\bar{y} \sum_{i=1}^{n} x_i^2 - n\bar{x} \sum_{i=1}^{n} x_iy_i}{n \sum_{i=1}^{n} x_i^2 - n^2(\bar{x})^2} .$$

Using the identities (7.10) and (7.11) we put the estimate $\tilde{a}$ into the form

$$\tilde{a}=\bar{y}-\tilde{b}\bar{x}. \tag{7.13}$$

Relations (7.12) and (7.13) solve the problem, i.e., they determine the calibration curve

$$y=\tilde{a}+\tilde{b}x. \tag{7.14}$$

To evaluate the uncertainty of the calibration curve the characteristics of each specific problem must be carefully analyzed. We shall study two typical variants.

(1) From the experimental data and the obtained estimates $\tilde{a}$ and $\tilde{b}$ we find the residuals of the conditional equations

$$v_i=\tilde{a}+\tilde{b}\cdot x_i-y_i.$$

Next, according to the general scheme of the least-squares method, we calculate the estimate of variance of the conditional equations using formula (7.6)

$$S^2=\frac{\sum_{i=1}^{n} v_i^2}{n-2}$$

and estimates of the variances of $\tilde{a}$ and $\tilde{b}$ using formulas (7.7). After this we find the confidence limits for Δ_a and Δ_b, which essentially solves the problem. As pointed out above, the confidence limits are constructed based on Student's distribution with $n-2$ degrees of freedom in our case, since for us two parameters are being determined.

If the calibration curve is plotted, then it is not difficult to construct based on the points the confidence band for this curve. This band has the form shown in Fig. 7.1.

(2) Digressing from the mathematical solution, the limits of error of the calibration curve can be found based on the limits of measurement errors of x_i and y_i.

We refer the error $|\zeta_x| \leqslant \Delta_x$ to the output $\zeta_{yx} = \tilde{b}\zeta_x$. Correspondingly, we shall have $\Delta_{yx} = \tilde{b}\Delta_x$. Now both components can be combined and the problem can be solved:

$$\Delta_{cy}=k\sqrt{\Delta_y^2+b_x^2\Delta_x^2},$$

where Δ_{cy} is the limit of error of the calibration curve based on the value at the output, i.e., y.

In the same manner we can refer the error $|\zeta_y|\leqslant\Delta_y$ to the input and calculate Δ_{cx}, and we can use either Δ_{cx} or Δ_{cy}.

Both variants do not always give the complete solution. Thus the mathematical solution does not take into account the constants for the entire range of measurement errors of x and y, and in general a mathematical solution can be obtained if the conditions under which the least-squares method is applicable are satisfied. The calculation of the limits of errors of the calibration curve based on the limits of measurement errors Δ_x and Δ_y does not take into account some specific errors introduced by the transducer itself. However, knowing in principle the characteristics of a particular type of transducer the properties of the transducer can always be taken into account.

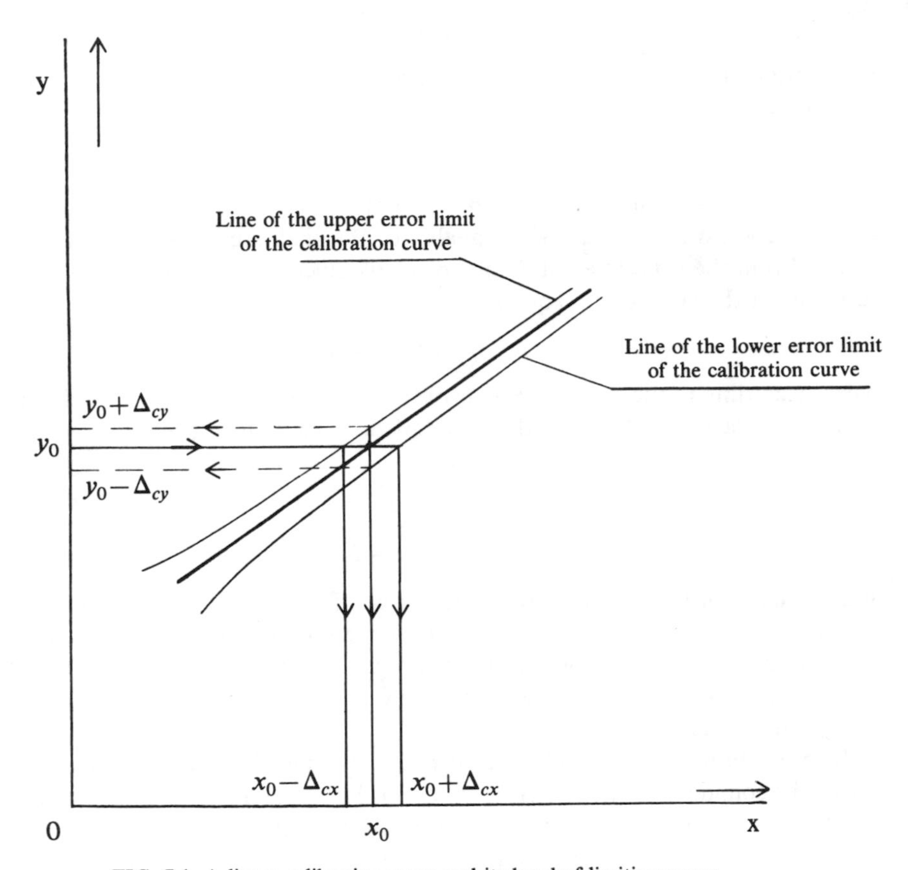

FIG. 7.1. A linear calibration curve and its band of limiting errors.

When working with measuring transducers the dependence $x=f(y)$ and not $y=\varphi(x)$ is required. Obviously, there is no difficulty in making the transformation

$$x=(y-\tilde{a})/\tilde{b}.$$

If the calibration curve is plotted, then there is no need to convert it; it can simply be used in the "reverse" direction, as shown in Fig. 7.1.

Figure 7.1 shows the possible form of the calibration curve of a measuring transducer, its limiting error band, and also the construction of the input signal x_0 based on the output signal y_0 and the error limits $\pm\Delta_{cx}$ of the value found.

The methods of confluent analysis in application to the problem of constructing linear dependences are reviewed in Ref. 54.

Chapter 8

Combining the results of measurements

8.1. Introductory remarks

Measurements of the same quantity are often performed in different laboratories and, therefore, under different conditions and occasionally even by different methods. Sometimes there arises the problem of combining the data obtained in order to find the most accurate estimate of the measured quantity.

In many cases, in the investigation of new phenomena, measurements of the corresponding quantities take a great deal of time. By collecting into groups measurements performed over a limited time, intermediate estimates of the measured quantity can be obtained in the course of the measurements. It is natural to find the final result of a measurement by combining the intermediate results.

The examples presented show that the problem of combining the results of measurements is of great significance for metrology. At the same time, it is important to distinguish situations in which one is justified in combining results from those in which one is not justified in doing so. It is pointless to combine results of measurements in which quantities in the essence of different dimension were measured.

It should be noted that when comparing results of measurements the data analysis is often performed based on the intuition of the experimenters without using formalized procedures. It is interesting that in the process, as a rule, the correct conclusions are drawn. On the one hand, this indicates that modern measuring instruments are of high quality and on the other hand that the experimenters, who by estimating the errors were able to determine all sources of error and exhibited reasonable care, were highly qualified.

8.2. Theoretical principles

The following problem has a mathematically rigorous solution. Consider L groups of measurements of the same quantity A. Estimates of the measured quantity $\bar{x}_1,...,\bar{x}_L$ were made from the measurements of each group, and in addition

$$M[\bar{x}_1]=\cdots=M[\bar{x}_L]=A.$$

The variances of the measurements in each group $\sigma_1^2,...,\sigma_L^2$ and the number of measurements in each group $n_1,...,n_L$ are known.

The problem is to find an estimate of the measured quantity based on data from all groups of measurements. This estimate is denoted as $\bar{\bar{x}}$ and is called the combined average or weighted mean.

We shall seek $\bar{\bar{x}}$ as a linear function of $\bar{x}_j$, i.e., as their weighted mean,

$$\bar{\bar{x}} = \sum_{t=1}^{L} g_j \bar{x}_j. \tag{8.1}$$

Therefore the problem reduces to finding the weights g_j. Since $M[\bar{x}_j] = A$ and $M[\bar{\bar{x}}] = A$, we obtain from Eq. (8.1)

$$M[\bar{\bar{x}}] = M\left[\sum_{j=1}^{L} g_j \bar{x}_j\right] = \sum_{j=1}^{L} g_j M[\bar{x}_j], \quad A = A \sum_{j=1}^{L} g_j.$$

Therefore,

$$\sum_{j=1}^{L} g_i = 1.$$

Next, we require that $\bar{\bar{x}}$ be an effective estimate of A, i.e., $D[\bar{\bar{x}}]$ must be minimum. For this, we find an expression for $D[\bar{\bar{x}}]$, using the formula

$$D[\bar{\bar{x}}] = D\left[\sum_{j=1}^{L} g_j \bar{x}_j\right] = \sum_{j=1}^{L} g_j^2 D[\bar{x}_j] = g_1^2 \sigma^2(\bar{x}_1) + g_2^2 \sigma^2(\bar{x}_2) + \cdots + g_L^2 \sigma^2(\bar{x}_L). \tag{8.2}$$

Using the condition $\Sigma_{j=1}^{L} g_j = 1$, we write $g_L = 1 - g_1 - g_2 - \cdots - g_{L-1}$. We shall now find the condition under which $D[\bar{\bar{x}}]$ has a minimum. For this we differentiate Eq. (8.2) with respect to g_j and equate the derivatives to 0. Since we have $L-1$ unknowns, we take $L-1$ derivatives:

$$2g_1 \sigma^2(\bar{x}_1) - 2(1 - g_1 - g_2 - \cdots - g_{L-1}) \sigma^2(\bar{x}_L) = 0,$$

$$2g_2 \sigma^2(\bar{x}_2) - 2(1 - g_1 - g_2 - \cdots - g_{L-1}) \sigma^2(\bar{x}_L) = 0,$$

$$\cdots$$

$$2g_{L-1} \sigma^2(\bar{x}_{L-1}) - 2(1 - g_1 - g_2 - \cdots - g_{L-1}) \sigma^2(\bar{x}_L) = 0.$$

Since the second term is identical in each equation, we obtain

$$g_1 \sigma^2(\bar{x}_1) = g_2 \sigma^2(\bar{x}_2) = \cdots = g_L \sigma^2(\bar{x}_L).$$

The transfer from g_{L-1} to g_L is made based on the fact that the elimination of g_L was not dictated by some fundamental considerations and instead of g_L a weighting coefficient with any number could have been taken.

Thus we have found a second condition that the weights of the arithmetic means of the groups of measurements must satisfy:

$$g_1 : g_2 : \cdots : g_L = \frac{1}{\sigma^2(\bar{x}_1)} : \frac{1}{\sigma^2(\bar{x}_2)} : \cdots : \frac{1}{\sigma^2(\bar{x}_L)}. \tag{8.3}$$

To find the weight g_j, it is necessary to know either the variances of the arithmetic means or the ratio of the variances. If we have all the variances $\sigma^2(\bar{x}_j)$, then we can set $g_i' = 1/\sigma^2(\bar{x}_j)$. We then obtain

$$g_j = \frac{g_j'}{\sum_{j=1}^{L} g_j'}. \tag{8.4}$$

Since the weights are nonrandom quantities, it is not difficult to determine the variance for $\bar{\bar{x}}$. According to relation (8.2) we have

$$D[\bar{\bar{x}}] = \sum_{j=1}^{L} g_j^2 D[\bar{x}_j] = \frac{\sum_{j=1}^{L} (g_j')^2 D[\bar{x}_j]}{\left(\sum_{j=1}^{L} g_j'\right)^2} = \frac{\sum_{j=1}^{L} \left(\frac{1}{\sigma^2(\bar{x}_j)}\right)^2 \sigma^2(\bar{x}_j)}{\left(\sum_{j=1}^{L} \frac{1}{\sigma^2(\bar{x}_j)}\right)^2} = \frac{1}{\sum_{j=1}^{L} \frac{1}{\sigma^2(\bar{x}_j)}}. \tag{8.5}$$

The relation (8.3) makes it possible to obtain the exact weights g_j if the variances $\sigma^2(\bar{x}_j)$ themselves are not known but only their ratios are known. In this case, having estimates of the variances of the arithmetic means of the groups instead of their values, an expression can be derived for the estimate of the variance of the weighted mean:

$$S^2(\bar{\bar{x}}) = \frac{1}{N-1}\left(\sum_{j=1}^{L} g_j \frac{n_j - 1}{n_j} S_j^2 + \sum_{j=1}^{L} g_j(\bar{x}_j - \bar{\bar{x}})^2\right). \tag{8.6}$$

The particular case when the variances of the measurements are the same for all groups but the number of observations in the groups is different is of interest. In this case we can set $g_j' = n_j$. Then the weights of the arithmetic means will be

$$g_j = n_j/N, \tag{8.7}$$

where $N = \sum_{j=1}^{L} n_j$, and the relation (8.6) will assume the form

$$S^2(\bar{\bar{x}}) = \frac{1}{N(N-1)}\left(\sum_{j=1}^{L} (n_j - 1)S_j^2 + \sum_{j=1}^{L} n_j(\bar{x}_j - \bar{\bar{x}})^2\right). \tag{8.8}$$

This result can also be obtained directly, combining the measurements of all groups into one large group of measurements.

The number of measurements in the combined group is $N = \sum_{j=1}^{L} n_j$.

If the measurements are collected according to groups, then the combined average will be

$$\bar{\bar{x}} = \frac{\sum_{j=1}^{L} \sum_{i=1}^{n_j} x_{ji}}{N}.$$

Let us expand the numerator. This gives

$$\bar{\bar{x}}=\frac{(x_{11}+x_{12}+\cdots+x_{1n_1})+(x_{21}+x_{22}+\cdots+x_{2n_2})+\cdots}{N}$$

$$=\frac{n_1\bar{x}_1+n_2\bar{x}_2+\cdots+n_L\bar{x}_L}{N}=\sum_{j=1}^{L} g_j\bar{x}_j,$$

where $g_j = n_j/N$ is the weight of the jth arithmetic mean.

The aggregate average $\bar{\bar{x}}$, for this reason, is also called the weighted mean. The estimate of standard deviation of the weighted mean can be estimated by regarding the weighted mean as the average of the large group of combined measurements:

$$S^2(\bar{\bar{x}})=\frac{\sum_{k=1}^{N}(x_k-\bar{\bar{x}})^2}{N(N-1)}.$$

We gather together the terms in the numerator

$$S^2(\bar{\bar{x}})=\frac{\sum_{j=1}^{L}\sum_{i=1}^{n_f}(x_{ji}-\bar{\bar{x}})^2}{N(N-1)}$$

and perform simple transformations of the numerator in order to simplify the calculations:

$$\sum_{j=1}^{L}\sum_{i=1}^{n_j}(x_{ji}-\bar{\bar{x}})^2=\sum_{j=1}^{L}\sum_{i=1}^{n_j}(x_{ji}-\bar{x}_j+\bar{x}_j-\bar{\bar{x}})^2$$

$$=\sum_{j=1}^{L}\sum_{i=1}^{n_j}(x_{ji}-\bar{x}_j)^2+2\sum_{j=1}^{L}\sum_{i=1}^{n_j}(x_{ji}-\bar{x}_j)(\bar{x}_j-\bar{\bar{x}})$$

$$+\sum_{j=1}^{L}\sum_{i=1}^{n_j}(\bar{x}_j-\bar{\bar{x}})^2.$$

The second term in the last expression is equal to 0, since by virtue of the properties of the arithmetic mean $\Sigma_{i=1}^{n_j}(x_{ji}-\bar{x}_j)=0$ and $\Sigma_{j=1}^{L}(\bar{x}_j-\bar{\bar{x}})=0$. For this reason,

$$S^2(\bar{\bar{x}})=\frac{1}{N(N-1)}\left(\sum_{j=1}^{L}\sum_{i=1}^{n_j}(x_{ij}-\bar{x}_j)^2+\sum_{j=1}^{L}\sum_{i=1}^{n_j}(\bar{x}_j-\bar{\bar{x}})^2\right).$$

Note that

$$\sum_{i=1}^{n_j}(x_{ji}-\bar{x}_j)^2=(n_j-1)S_j^2,\quad \sum_{i=1}(\bar{x}_j-\bar{\bar{x}})^2=n_j(\bar{x}_j-\bar{\bar{x}})^2.$$

Then, retaining the summation over groups, we obtain

$$S^2(\bar{\bar{x}})=\frac{1}{N(N-1)}\left[\sum_{j=1}^{L}(n_j-1)S_j^2+\sum_{j=1}^{L}n_j(\bar{x}_j-\bar{\bar{x}})^2\right].$$

The first term in the formula obtained characterizes the spread in the measurements in groups and the second term characterizes the spread of the arithmetic means of the groups.

8.3. Effect of the error of the weights on the error of the weighted mean

Looking at the general form of the formula determining the weighted mean, one would think, since the weights g_j and the weighted values of $\bar{x}_j$ appear in it symmetrically, that the weights must be found with the same accuracy as $\bar{x}_j$. In practice, however, the weights are usually expressed by numbers with one or two significant figures. How is the uncertainty of the weights reflected in the error of thc weighted mean?

The general expression for the weighted mean has the form

$$\bar{\bar{x}} = \sum_{j=1}^{L} g_j \bar{x}_j,$$

where $\bar{x}_j$ is the jth weighted value, g_j is the weight used for the jth value of the quantity, and $\bar{\bar{x}}$ is the weighted mean.

We shall regard weighted values as fixed, constant. In addition, as usual, we shall assume that $\Sigma_{j=1}^{L} g_j = 1$. This equality is also satisfied for the inaccurately determined weighting coefficients, i.e., for $\tilde{g}_j$. Therefore

$$\sum_{j=1}^{L} \Delta g_j = 0,$$

where Δg_j is the error in determining the coefficient g_j.

Assuming that the exact value of the weighted mean is y, we estimate the error of its estimate:

$$\Delta y = \sum_{j=1}^{L} \tilde{g}_j \bar{x}_j - \sum_{j=1}^{L} g_j \bar{x}_j = \sum_{j=1}^{L} \Delta g_j \bar{x}_j.$$

We shall express Δg_1 in terms of the other errors:

$$\Delta g_1 = -(\Delta g_2 + \cdots + \Delta g_L)$$

and substitute it into the expression for Δy:

$$\Delta y = (\bar{x}_2 - \bar{x}_1)\Delta g_2 + (\bar{x}_3 - \bar{x}_1)\Delta g_3 + \cdots + (\bar{x}_L - \bar{x}_1)\Delta g_L$$

or in the form of relative error

$$\frac{\Delta y}{y} = \frac{g_2(\bar{x}_2 - \bar{x}_1)\dfrac{\Delta g_2}{g_2} + \cdots + g_L(\bar{x}_L - \bar{x}_1)\dfrac{\Delta g_L}{g_L}}{\sum_{j=1}^{L} g_j \bar{x}_j}.$$

The errors of the weights $\Delta g_j/g_j$ are themselves unknown. But let us assume that we can estimate their limits. Let either these limits be equal in modulus or $\Delta g/g$ be the modulus of the limit of the largest error.

Replacing all relative errors $\Delta g_j/g_j$ by $\Delta g/g$, we obtain

$$\frac{\Delta y}{y} \leqslant \frac{\Delta g}{g} \frac{|g_2(\bar{x}_2 - \bar{x}_1) + g_3(\bar{x}_3 - \bar{x}_1) + \cdots + g_L(\bar{x}_L - \bar{x}_1)|}{\sum_{j=1}^{L} g_j \bar{x}_j}.$$

The numerator on the right-hand side of the inequality can be put into the following form:

$$g_2(\bar{x}_2 - \bar{x}_1) + g_3(\bar{x}_3 - \bar{x}_1) + \cdots + g_L(\bar{x}_L - \bar{x}_1)$$
$$= g_2\bar{x}_2 + g_3\bar{x}_3 + \cdots + g_L\bar{x}_L - (g_2 + g_3 + \cdots + g_L)\bar{x}_1.$$

But $g_2 + g_3 + \cdots + g_2 = 1 - g_1$, so that

$$g_2(\bar{x}_2 - \bar{x}_1) + g_3(\bar{x}_3 - \bar{x}_1) + \cdots + g_L(\bar{x}_L - \bar{x}_1) = \sum_{j=1}^{L} g_j\bar{x}_j - \bar{x}_1 = y - \bar{x}_1.$$

Thus

$$\frac{\Delta y}{y} \leqslant \frac{\Delta g}{g} \frac{|y - \bar{x}_1|}{y}.$$

It is obvious that if the entire derivation is repeated but in so doing the error not in the coefficient g_1 but rather some other weight is eliminated, then a weighted value other than $\bar{x}_1$ will appear on the right-hand side of the inequality. Therefore the result obtained can be represented in the form

$$\frac{\Delta \bar{\bar{x}}}{\bar{\bar{x}}} \leqslant \frac{\Delta g}{g} \frac{|\bar{\bar{x}} - \bar{x}_j|}{\bar{\bar{x}}}.$$

The obtained inequality shows that the error introduced into the weighted mean as a result of the error of the weights is many times smaller than the latter error. The cofactor $|\bar{\bar{x}} - \bar{x}_j|/\bar{\bar{x}}$ can be assumed to be of the same order of magnitude as the relative error of the terms. Thus if this error is of the order of 0.01, then the error introduced into the weighted mean as a result of the error of the weights will be at least 100 times smaller than the latter.

8.4. Combining the results of measurements in which the random errors predominate

We shall study a variant that is possible in the case of multiple measurements with negligibly small systematic errors. Each result being combined in this case is usually the arithmetic mean of measurements, and the differences between them are explained by the random spread of the measurements in the groups. However, it must be verified that the true value of the measured quantity is the same for all groups. This problem is solved by the methods presented in Chap. 4. If it cannot be assumed that the same quantity is measured in all cases, then it is pointless to combine the measurements into groups.

If the unification of the groups is justified, then it is necessary to check the hypothesis that the variances of the measurements in the groups are equal. Methods for solving this problem are also presented in Chap. 4.

In the case when the variances of the groups can be assumed to be equal, the weights for each of the results are calculated from formula (8.7), the combined average is calculated from formula (8.1), and the variance of the combined average can be determined from formula (8.8).

When the variances of the groups cannot be taken to be equal to one another and the variances themselves and their ratios are unknown, the weights are sometimes found by substituting their estimates, instead of the variances, into formula (8.4). The variance of the weighted mean obtained is estimated by substituting into formula (8.5) estimates of the variances of the combined quantities, i.e., from the formula

$$S^2(\bar{\bar{x}}) = \frac{1}{\sum_{k=1}^{L} \frac{1}{S^2(\bar{x}_k)}}.$$

The estimates of the variances are random quantities and the weights obtained based on them are also random quantities, and although the weights need not be very accurate; in this case the weighted mean still is of a somewhat conditional character. As shown in Ref. 39, however, in the case when the observations are normally distributed, the weighted mean remains an unbiased estimate of the measured quantity. The error in estimating the standard deviation, obtained based on the formula presented for estimating the variance of the weighted mean, does not exceed 10% already with two groups of observations consisting of more than nine observations.

Example. The mass of some body is being measured. In one experiment the value $\tilde{m}_1 = 409.52$ g is obtained as the arithmetic mean of 15 measurements. The variance of the group of measurements is estimated to be $S_1^2 = 0.1\ g^2$. In a different experiment the value $\tilde{m}_2 = 409.44$ g was obtained with $n_2 = 10$ and $S_2^2 = 0.03\ g^2$. It is known that the systematic errors of the measurements are negligibly small, and the measurements can be assumed to be normally distributed. It is necessary to estimate the mass of the body using data from both experiments and to estimate the variance of the result.

We shall first determine whether the unification is justified, i.e., whether or not there is an inadmissible difference between the estimates of the measured quantity:

$$S^2(\bar{x}_1) = \frac{S_1^2}{n_1} = \frac{0.1}{15} = 0.0067, \quad S^2(\bar{x}_2) = \frac{0.03}{10} = 0.003,$$

$$S^2(\bar{x}_1 - \bar{x}_2) = S^2(\bar{x}_1) + S^2(\bar{x}_2) = 0.0097,$$

$$S(\bar{x}_1 - \bar{x}_2) = 0.098,$$

$$\bar{x}_1 - \bar{x}_2 = \tilde{m}_1 - \tilde{m}_2 = 0.08.$$

Since $\bar{x}_1 - \bar{x}_2 < S(\bar{x}_1 - \bar{x}_2)$, the unification is possible.

We shall check whether or not both groups of observations have the same variance (see Sec. 4.6):

$$F = S_1^2/S_2^2 = 0.1 : 0.03 = 3.3.$$

The degrees of freedom are $\nu_1 = 14$ and $\nu_2 = 9$. We shall assume that the significance level is 2%. In addition, $q = 0.01$ and $F_q = 5$ (see Table A.6). Since $F < F_q$, it can be assumed that the variances of the groups are equal.

We shall now find the weights of the arithmetic means. According to (8.7) we have $g_1 = 15/25 = 0.6$ and $g_2 = 10/25 = 0.4$. The weighted mean is $\bar{\bar{m}} = 0.6 \times 409.52 + 0.4 \times 409.44 = 409.49$ g. Now we find $S(\bar{\bar{m}})$. In accordance with formula (8.8) we have

$$S^2(\bar{\bar{m}}) = \frac{1}{25 \times 24}(14 \times 0.1 + 9 \times 0.03 + 15 \times 0.03^2 + 10 \times 0.05^2) = 28 \times 10^{-4}\ \text{g}^2,$$

$$S(\bar{\bar{m}}) = 5.3 \times 10^{-2}\ \text{g}.$$

If in addition to estimating the standard deviation it is also necessary to find the confidence error, then in order to use Student's distribution the effective number of degrees of freedom must be found using formula (6.8).

8.5. Combining the results of measurements containing both systematic and random errors

Let us assume that a quantity A is measured by several methods. Each method gives the result x_j ($j = 1,\dots,L$) with error ζ_j:

$$x_j = A + \zeta_j.$$

In order to combine in a well-founded manner the series of values of x and obtain a more accurate estimate of the measured quantity, one must have certain information about the errors ζ_j. We shall start from the condition that none of the measurements have absolutely constant systematic errors. However, this assumption must be checked. If it is not true, then the problem posed cannot be solved.

The error ζ_j is the sum of realizations of the conditionally constant systematic ϑ_j and random ψ_j errors: $\zeta_j = \vartheta_j + \psi_j$.

Having in mind a possible set of results with each method of measurement, the unknown systematic error ϑ_j of a concrete realization of the jth method of measurement can be regarded as the realization of a random quantity. Usually the limits θ_j of ϑ_j are estimated, and they are regarded as symmetric relative to the true value of the measured quantity: $|\vartheta_j| \leqslant \theta_j$ and $M[\vartheta_j] = 0$.

The random error ψ_j is assumed to be a centered quantity, i.e., $M[\psi_j] = 0$.

Thus when there are no absolutely constant errors, for example, methodological errors, we can write $M[x_j] = A$.

As follows from the theory of combining of the results of measurements, the weights are determined by the variances of these results. In our case we can write

$$D[x_j] = D[\vartheta_j] + D[\psi_j].$$

Therefore, given the variances $D[\vartheta_j]$ and $D[\psi_j]$, the problem can be solved exactly and uniquely. Unfortunately, the variances themselves are always unknown, and their estimates must be employed. To estimate the variances of conditionally constant systematic errors of each method of measurement, we shall use the assumption that the errors are uniformly distributed within the estimated limits.

Although the weights need not be found with high accuracy, it still casts some doubt on the fact that the weighted mean is a better estimate of the measured quantity than the combined results. This is why in metrology great care is taken in combining results of measurements.

Based on what was said above, when the results of measurements must be combined it is always necessary to check the agreement between the starting data and the obtained results. If some contradiction is discovered, for example, the combined average falls outside the permissible limits of some term, then the reason for this must be determined and the contradiction must be eliminated. Sometimes this is very difficult to do, and special experiments must be performed.

Great care must be exercised in combining the results of measurements because in this case information about the errors is employed to refine the result of the measurement and not to characterize its uncertainty, as is usually done.

It can happen, however, that the weighted mean is a natural estimate of the measured quantity. An example is the accurate measurement of the activity of a source of α particles. In order to increase the accuracy, the activity is measured at different distances from the source to the detector and with different diaphragms. The measured activity itself remains the same. However, the estimates of the activity obtained with different diaphragms are found to differ somewhat from one another. Their errors are also different. For this reason, when assigning weights for the obtained estimates of the measured quantity, in this example one must start from estimates of the variances of the total error of the measurement results being combined. This, undoubtedly, will lead to a more correct result than if all measurements were assumed to have equal weights or weights were assigned taking into account only the random errors.

So, in the case at hand, the weights of the measurements being combined should be calculated using the formula

$$g_j = \frac{\dfrac{1}{S^2(\vartheta_j)+S^2(\psi_j)}}{\sum\limits_{k=1}^{L} \dfrac{1}{S^2(\vartheta_k)+S^2(\psi_k)}}, \tag{8.9}$$

where $S^2(\vartheta_j)$ and $S^2(\vartheta_k)$ are estimates of the variances of the possible sets of systematic errors of the jth and kth measurement results, and $S^2(\psi_j)$ and $S^2(\psi_k)$ are estimates of the variances of the random errors of the same measurement results.

We shall now estimate the errors of the combined average. In solving this problem, since the errors of the weights are insignificant (see Sec. 8.3), we shall assume that the weights of the combined measurement results are known exactly.

In the case of multiple measurements one must have for each result x_j an estimate of the limits of the systematic error θ_j and an estimate of the standard deviation S_j of the random error. Then the corresponding indicators of accuracy of the combined average will be

$$\theta(\bar{\bar{x}}) = k \sqrt{\sum_{j=1}^{L} g_j^2 \theta_j^2},$$

$$S(\overline{\overline{x}})=\sqrt{\sum_{j=1}^{L} g_j^2 S_j^2}\,.$$

The confidence limits of the total error of the combined average can be found based on the estimates obtained for $\theta(\overline{\overline{x}})$ and $S(\overline{\overline{x}})$. The method for solving this problem was examined in detail in Chap. 5.

In the case of single measurements one usually knows only the estimates of the limits of the errors of the measurements being combined, i.e., Δ_j $(j=1,...,L)$. Based on available information about the form of the distribution of the possible sets of actual errors of each measurement result, it is necessary to transfer from the limits Δ_j of the errors to estimates of the variances of these errors. Once the variances have been obtained, it is not difficult to find the weights of the measurement results being combined. Next, the confidence limits of the error of the weighted mean can be calculated using the scheme developed for linear indirect measurements (see Chap. 6).

We shall discuss some particular cases of single measurements. We shall examine the measurement of one quantity with several instruments.

Let the random errors of the instruments be small compared with the limit of permissible errors, and let the permissible errors be the same and equal to Δ for all instruments.

Let the indications of the instruments be $x_1,...,x_n$ and the actual errors in the indications be $\vartheta_1,...,\vartheta_q$ $(|\vartheta_i| < \Delta)$. Then we can write

$$A=x_1-\vartheta_1,$$

$$\cdots$$

$$A=x_n-\vartheta_n. \tag{8.10}$$

The natural intuitive estimate of the true value of the measured quantity in the case when several instruments of equal accuracy are used to perform the measurements is the arithmetic mean of the instrumental indications:

$$\tilde{A}=\frac{\sum_{i=1}^{n} x_i}{n}\,.$$

It has been proved mathematically that in the class of linear estimates this is the best estimate.

We must estimate the error in the result obtained. Adding the left and right sides of Eqs. (8.10) and dividing them by n, we obtain

$$A=\frac{\sum_{i=1}^{n} x_i}{n}-\frac{\sum_{i=1}^{n} \vartheta_i}{n}\,.$$

We do not know the real errors of the instruments. We know only that $|\vartheta_i| \leqslant \Delta$ for all $i=1,...,n$.

To find the limits of the sum of the random quantities ϑ_i (and their errors over a set of instruments of a given type can be assumed to be random quantities), it is necessary to known their distribution functions. As pointed out above, these functions cannot be found from the experimental data. However, it can often be assumed that the errors of complicated instruments have symmetric distributions. The mathematical expectation of the distribution is close to the errors of working standards employed to calibrate these instruments. To a first approximation, we shall assume that $M[\vartheta_i] = 0$.

For example, if the errors of the instruments are distributed uniformly, then according to formula (5.3)

$$\theta_1 = k\sqrt{\sum_{i=1}^{n} \Delta^2} = k\Delta\sqrt{n}.$$

From here, the uncertainty of the estimate $\tilde{A}$ will be

$$\Delta_{\Sigma 1} = \frac{\theta_1}{n} = \frac{k\Delta}{\sqrt{n}}.$$

If the errors of the instruments are assumed to have a normal distribution and $\Delta = z_{(1+\alpha)/2}\sigma$, then

$$\theta_2 = z_{(1+\alpha)/2}\sqrt{\sum_{i=1}^{n} \sigma_i^2} = z_{(1+\alpha)/2}\sigma\sqrt{n} = \Delta\sqrt{n}.$$

Then

$$\Delta_{\Sigma 2} = \frac{\theta_2}{n} = \frac{\Delta}{\sqrt{n}}.$$

This estimate corresponds to the same probability α, which was used to establish the limit of permissible error Δ. Comparing $\Delta_{\Sigma 1}$ and $\Delta_{\Sigma 2}$ shows that they differ only by the factor k, which, depending on the confidence probability, can range from 1.1 ($\alpha = 0.95$) to 1.4 ($\alpha = 0.99$). As expected, the number of instruments plays the main role. Five to ten instruments are required in order to reduce the error by a factor of 2 or 3. But we have to stress here that a real improvement of the errors is limited with the errors of working standards employed to calibrate these instruments.

The problem can also be solved as follows. We choose the maximum and minimum indications of the instruments: x_{max} and x_{min}. We verify that

$$(x_{max} - x_{min}) \leqslant 2\Delta. \tag{8.11}$$

If inequality (8.11) is not satisfied, then one of the instruments used to perform the measurement has an inadmissably large error or the variation of some of the influence quantities is too large. The reason for this phenomenon must be determined and eliminated, i.e., inequality (8.11) must be satisfied.

Figure 8.1 illustrates the indications x_{max} and x_{min}, and the intervals corresponding to the limits of permissible errors $\pm\Delta$ are marked off.

The true value of the measured quantity must lie in the section of the tolerance field that belongs simultaneously to the instrument with indication x_{max} and the instrument with indication x_{min}. In the figure this section is hatched. Its boundaries determine more accurately the tolerance field.

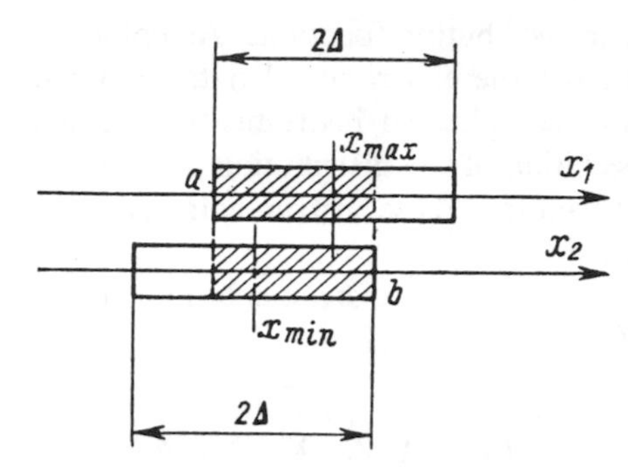

FIG. 8.1. The highest and lowest indications of the group of instruments and the intervals of their possible errors.

It is natural to take for the estimate of the measured quantity the center of the interval $x_{max} - x_{min}$, which is found from

$$\tilde{A}=\frac{x_{max}+x_{min}}{2}. \tag{8.12}$$

The coordinate of the point a, determining the left-hand boundary of the error in the result, will be $x_a = x_{max} - \Delta$. The coordinate of the point b, which determines the right-hand limit of error, is equal to $x_b = x_{min} + \Delta$.

Therefore the limit of error Δ_1 of the more accurate result is

$$\Delta_1 = |x_b - \tilde{A}| = |\tilde{A} - x_a|$$

or

$$\Delta_1 = \left| x_{min} + \Delta - \frac{x_{max}+x_{min}}{2} \right| = \left| \Delta - \frac{x_{max}-x_{min}}{2} \right|.$$

It is easy to see that in the limit $x_{max} - x_{min} = 2\Delta$ the error Δ_1 is formally equal to 0. It is clear, however, that the minimum value of Δ_1 cannot be less than the error of the working standard used to calibrate the instruments employed.

It is interesting to note that the estimate, based on Eq. (8.12), of the measured quantity mathematically gives the best approximation when the errors ϑ_i are distributed uniformly over the interval $[-\Delta, +\Delta]$.

Based on the foregoing arguments it can be shown that a well-known assumption of metrology is valid: when measuring instruments having different accuracy are used in parallel, the accuracy of the result is determined by the most accurate measuring instrument.

For example, assume that the voltage of some source was measured simultaneously with three voltmeters having different accuracy but the same upper limit of the measurement range 15 V. The measurements were performed under reference conditions. The following results were obtained.

(1) Class 0.5 voltmeter: $U_1 = 10.05$ V; the limit of permissible intrinsic error $\Delta_1 = 0.075$ V.

(2) Class 1.0 voltmeter: $U_2 = 9.9$ V; the limit of permissible intrinsic error $\Delta_2 = 0.15$ V.

(3) Class 2.5 voltmeter: $U_3 = 9.7$ V, the limit of permissible intrinsic error $\Delta_3 = 0.375$ V.

Since the measurements were performed under normal conditions, we shall assume that the limits of permissible intrinsic error of the instruments are equal to the limits of the errors of measurement.

Assume that the errors of the instruments of each type have a uniform distribution. Then

$$\sigma_i = \Delta_i / \sqrt{3}.$$

We shall find the weights of the results. Since the upper limit of the measurement range is the same for all instruments, the calculation can be performed based on the limits of reduced error of the instruments:

$$g_1' = \frac{1}{\Delta_1^2} = \frac{1}{0.25} = 4, \quad g_2' = \frac{1}{\Delta_2^2} = 1, \quad g_3' = \frac{1}{\Delta_3^2} = \frac{1}{6.25} = 0.16.$$

From here

$$g_1 = \frac{g_1'}{\sum_{i=1}^{3} g_i'} = \frac{4}{5.16} = 0.77,$$

$$g_2 = \frac{g_2'}{\sum_{i=1}^{3} g_i'} = \frac{0.20}{5.16} = 0.20, \quad g_3 = \frac{g_3'}{\sum_{i=1}^{3} g_i'} = \frac{0.16}{5.16} = 0.03.$$

Now we find the weighted mean

$$\tilde{U} = \sum_{i=1}^{3} g_i U_i = 0.77 \times 10.05 + 0.2 \times 9.9 + 0.03 \times 9.7 = 10.01 \text{ V}.$$

The limit of the error in the weighted mean can be found from the formula (6.9) for the error in the result of indirect measurements,

$$\begin{aligned} \Delta \tilde{U} &= k \sqrt{\sum_{i=1}^{3} g_i^2 \Delta_i^2} \\ &= k \sqrt{0.77^2 (7.5 \times 10^{-2})^2 + 0.2^2 (15 \times 10^{-2})^2 + 0.03^2 \times 0.375^2} \\ &= k \sqrt{(33 + 9 + 1.3) \times 10^{-4}} = 0.066k. \end{aligned}$$

Assuming, as usual, $\alpha = 0.95$, we take $k = 1.1$ and find $\Delta \tilde{U} = 0.07$ V.

In Fig. 8.2 the indications of all three instruments are plotted and the limits of permissible error of the instruments are marked. The value obtained for the weighted mean is also indicated there. This value remained in the tolerance field of the most accurate result, but was shifted somewhat in the direction of indications of the less accurate instruments; this is natural. The limits of error of the result decreased insignificantly compared with the error of the most accurate term.

If all distributions were assumed to be normal distributions, truncated at the same level by discarding instruments whose error exceeds

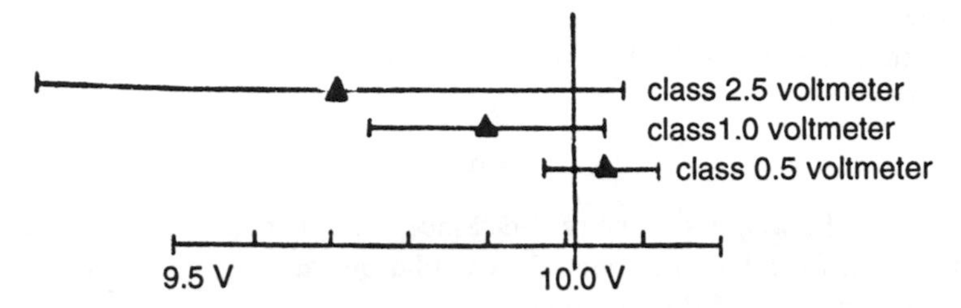

FIG. 8.2. Indications of the instruments and the intervals of their possible errors.

$z_{(1+\alpha)/2}\sigma$ ($z_{(1+\alpha)/2} = \text{const}$), then the weights would not change and the weighted mean would have the same value as we found above. Only the estimate of the error limits would change, since it must now be calculated from formula (6.11).

We obtain $(\Delta\widetilde{U})' = 0.066$ V. However, the difference between $\Delta\widetilde{U}$ and $(\Delta\widetilde{U})'$ is insignificant.

Let us see what would happen if different instruments had different distributions. For example, assume that the class 0.5 and 1.0 instruments have a uniform distribution and the class 2.5 instruments have a truncated normal distribution, and in addition $z_{(1+\alpha)/2} = 2.6$. Then

$$\sigma_1 = \frac{0.075}{\sqrt{3}} = 0.043, \quad g_1' = \frac{1}{\sigma_1^2} = 306,$$

$$\sigma_2 = \frac{0.15}{\sqrt{3}} = 0.087, \quad g_2' = \frac{1}{\sigma_2^2} = 78,$$

$$\sigma_3 = \frac{0.375}{2.6} = 0.144, \quad g_3' = \frac{1}{\sigma_3^2} = 42.$$

Therefore

$$g_1 = \frac{g_1'}{\sum_{i=1}^{3} g_i'} = \frac{306}{426} = 0.72,$$

$$g_2 = \frac{g_2'}{\sum_{i=1}^{3} g_i'} = \frac{78}{426} = 0.18,$$

$$g_3 = \frac{g_3'}{\sum_{i=1}^{3} g_i'} = \frac{42}{426} = 0.10.$$

From here

TABLE 8.1. The results of measurements of the activity of nuclides by different geometric factors.

Number of group i	Source–detector distance (mm)	Diaphragm radius (mm)	Estimate of measured quantity $x_i \times 10^5$	Estimates of the standard deviations: Random errors of the result (%)	Estimates of the standard deviations: Systematic errors of the result (%)
1	97.50	20.017	1.65197	0.08	0.52
2	97.50	12.502	1.65316	0.1	0.52
3	397.464	30.008	1.66785	0.16	0.22
4	198.00	20.017	1.66562	0.3	0.42
5	198.00	30.008	1.66014	0.08	0.42

$$\tilde{U}_1 = 0.72\times 10.05 + 0.18\times 9.9 + 0.1\times 9.7 = 9.99 \text{ V}.$$

The values obtained for $\tilde{U}$ and $\tilde{U}_1$ are very close. This indicates that a significant change in the form of the distribution functions in this case does not appreciably affect the result.

The foregoing example could also have been solved by a graphical-analytic method, similarly to the method used to solve the problem of combining the indications of equally accurate instruments. Now, however, the relation determining $\tilde{A}$ must contain the weights of the terms. In accordance with the foregoing considerations, these weights can be taken to be inversely proportional to the squares of the limits of permissible errors of the instruments.

8.6. Example: Measurement of the activity of nuclides in a source

We shall examine the measurement of the activity of nuclides by the method of absolute counting of α particles emitted by the source in a small solid angle. The measured activity is determined from the formula

$$A = GN_0\eta,$$

where G is the geometric factor of the apparatus, N_0 is the α-particle counting rate, and η is the α-particle detection efficiency.[14]

The geometric factor depends on the diameter of the source, the distance between the source and the detector, as well as the diameter of the diaphragm, and it is calculated from measurements of these quantities. In the course of a measurement G does not change, so that errors of G create a systematic error of measurement of the activity A. Measurements of the numbers of α particles, however, have random errors.

To reduce the error arising from the error of the geometric factor, the measurements were performed for different values of this factor (by changing the distance between the source and detector and the diameter of the diaphragm). All measurements were performed using the same source ^{239}Pu. Table 8.1 gives the five combinations of the geometric parameters studied. In each case 50 measurements were performed, and estimates of the measured quantity and the parameters of their errors, which are also presented in Table 8.1, were calculated. The rms deviations of the systematic errors of the results were calculated from the estimated limiting

TABLE 8.2. The estimates of generalized variances and weights of measurement results by different geometric factors.

Number of group	Estimate of generalized variance $S_\Sigma^2(\bar{x}_i)$	Weight g_i
1	0.28	0.12
2	0.28	0.12
3	0.07	0.46
4	0.27	0.12
5	0.18	0.18

values of all components under the assumption that they can be regarded as centered uniformly distributed random quantities.

The data in Table 8.1 show, first of all, that the systematic errors are much larger than the random errors, so that the number of measurements in the groups was sufficient. The observed difference between the obtained values of the activity of the nuclides in the groups can be explained by their different systematic errors.

In the example studied the same quantity was measured in all cases. For this reason, here the weighted mean is a well-founded estimate of the measured quantity. Based on the considerations presented in Sec. 8.5, we shall use formula (8.9) to calculate the weights. First, we shall calculate an estimate of the generalized variance

$$S_\Sigma^2(\bar{x}_i)=S^2(\psi_i)+S^2(\vartheta_i).$$

The results of the calculations are given in Table 8.2.

As an example we shall calculate g_1:

$$g_1=\frac{\frac{1}{0.28}}{\frac{1}{0.28}+\frac{1}{0.28}+\frac{1}{0.07}+\frac{1}{0.27}+\frac{1}{0.18}}=\frac{3.57}{30.7}=0.12.$$

Now we find the weighted mean: $\tilde{A}=\Sigma_{i=1}^{5}g_i\bar{x}_i=1.6625\times 10^5$.

It remains to estimate the error in the value obtained. We shall calculate an estimate of the generalized variance of the weighted mean with the help of the formula

$$S^2(\tilde{A})=\frac{1}{\sum_{i=1}^{5}[1/S_\Sigma^2(\bar{x}_i)]}=\frac{1}{30.7}=0.033.$$

From here $S_{\tilde{A}}=0.18\%$. Since the error of the weighted mean is determined by the systematic component, it is best presented in the form of limits (in this case, confidence limits). For the estimated value of the variance the limits are calculated for the normal distribution. In this case, this is all the more justified because, as we have seen, the error of the weighted mean consists primarily of five terms. Even if all the terms were uniformly distributed, the distribution of their composition would be virtually normal.

For the standard confidence probability $\alpha = 0.95$, $z_\alpha = 1.96$ and $\delta(0.95) = 1.96 \times 0.18 = 0.35\%$. In the form of absolute error we obtain $\Delta(0.95) = 0.006 \times 10^5$.

The result of the measurement can be given in the form

$$A_{0.95} = (1.662 \pm 0.006) \times 10^5.$$

One can see that in this example the simple arithmetic mean, equal to 1.660×10^5, of the estimates obtained for the measured quantity does not differ significantly from the weighted mean. This agreement, however, is purely accidental. In cases similar to the one examined above, the weighted mean, of course, is a better-founded estimate of the measured quantity than the simple arithmetic mean.

Chapter 9

Calculation of the errors of measuring instruments

9.1. The problems of calculating measuring instrument errors

Measuring instruments are extremely diverse, but because they are used for a common purpose, there exists a general theory of their errors. The central problem of this theory is to calculate the intrinsic error of measuring instruments, which is their most important metrological characteristic. The calculation of the additional errors, caused by controlled changes in the influence quantities, or influence coefficients and functions of the influence quantities, depends on the arrangement of a measuring instrument. For this reason, the calculation of these metrological characteristics falls within the purview of the theory of measuring instruments with a particular principle of operation.

In general form, a measuring instrument can be considered as a number of functionally related units that transform an input signal to an output signal. During the manufacturing process, a desirable shape of functional dependence between these signals (a transfer function) is first obtained by adjusting some of the units. Then, each instrument is graduated or calibrated. In essence, the purpose of these operations is to fix and to represent the obtained transfer function of the instrument by means of a scale or a graph or an equation.

No matter how accurate these operations are, the resulting instrument will have some errors for the following reasons:

(a) inaccuracy in fixing the transfer function, that is, inaccuracy in constructing the scale or the graph or the equation;

(b) imperfection of the reading device of the measuring instrument;

(c) variations of influence quantities (within limits of reference conditions);

(d) drifting of some properties of the measuring instrument units with time.

Each of the above contributes a component to the intrinsic error of a measuring instrument.

The errors of measuring instruments under normal operating conditions (i.e., when the influence quantities deviate from their reference values or when they exceed the limits of the range of reference values) and their calculation based on known properties of the measuring instruments and the conditions of measurement are regarded as an integral part of the problem of estimating measurement errors.

The problem of estimating the resulting error can be formulated for a separate instrument or a collection of instruments of a definite type. In the first case, the problem consists of estimating the error of a particular instrument from the known parameters of the components of this instrument. In so doing, one can find either the errors of the instrument on definite segments of the instrument scale or an estimate of the limits of error of the given instrument. These problems must be solved when designing unique measuring instruments and when performing an elementwise calibration.

In the second case, i.e., for a collection of measuring instruments, the problem is formulated differently. The limits of error of instruments be estimated based on the properties of the components of the instrument (direct problem). But most often the limits of instrument error are prescribed and it is required to find the percentage of instruments whose error will fall within these limits (inverse problem).[47]

Each of these problems, admittedly, can be formulated, with some modifications, for any type of measuring instrument—for standards, measuring transducers, or measuring instruments and systems—only if their errors are caused by deviations from the nominal values of the parameters of the components of the measuring instrument.

9.2. Methods for calculating instrument errors

We shall examine both methods successively, i.e., the direct and inverse problems, referring to collections of measuring instruments.

In the general form the output signal y of an instrument is related to the informative parameter of the input signal A, the parameters x_i of the components of the instrument $(i=1,...,n)$, noise, and other factors giving rise to errors z_j $(j=1,...,m)$, by the relation

$$y=f(A,x_i,z_j). \tag{9.1}$$

For each parameter we shall establish the nominal value, i.e., the value for which the measuring instruments would not have an error. The deviations of the real properties of the components from the nominal properties result in the instrument error. We shall call conventionally the deviation from the nominal values of the parameters of the components the errors of the components, and we shall assume that they are expressed in the form of relative errors

$$\varepsilon_i=\frac{x_{ir}-x_i}{x_i}, \tag{9.2}$$

where x_i is the nominal value and x_{ir} is the real value of the parameter of the component.

The effect of the error of each component on the instrument error is determined in the manner studied in Chap. 6 for determining the influence coefficients of the measurement error of the arguments on the error of an indirect measurement.

For relative errors, we can write

$$\frac{\Delta y}{y}=\frac{1}{y}\frac{\partial f}{\partial x_i}\Delta x_i=\frac{1}{y}\frac{\partial f}{\partial x_i}\varepsilon_i x_i.$$

From here, the influence coefficient of the error x_i will be, in relative form,

$$V_i = \frac{\Delta y/y}{\varepsilon_i} = \frac{\partial f}{\partial x_i}\frac{x_i}{y}. \tag{9.3}$$

We shall express the influence coefficients for sources of additive errors, since they cannot be represented as a deviation from some nominal values, in the standard form

$$W_j = \frac{\partial f}{\partial z_j}. \tag{9.4}$$

In what follows we shall refer, somewhat arbitrarily, to the factors responsible for the additive errors as noise.

The absolute error of an instrument at indication y is determined by the relation

$$\zeta = y\sum_{i=1}^{n} V_i\varepsilon_i + \sum_{j=1}^{m} W_j z_j. \tag{9.5}$$

We find Eq. (9.1) and the influence coefficients V_i and W_i based on the structural layout of the instrument. Having derived Eq. (9.5), we no longer need the structural layout of the instrument, and we need study only the components of the error. In Eq. (9.5) the error components are referred to the output of the instrument.

We shall now study in greater detail the direct problem, i.e., the problem of estimating instrument error. We have in mind estimation of errors at any point in the range of indication. If the point where the instrument error is maximum is known, then in many cases only the error for this point need be calculated. Often this point is the end point y_f of the instrument scale, since in this case the multiplicative components of the error are maximum.

The calculations are most conveniently performed for the relative and not the absolute errors. For $y = y_f$, the relative instrument error, as follows from formula (9.5), will be

$$\varepsilon_f = \sum_{i=1}^{n} V_i\varepsilon_i + \frac{1}{y_f}\sum_{j=1}^{m} W_j z_j. \tag{9.6}$$

We shall divide all errors appearing on the right-hand side of formula (9.5) or (9.6) into systematic and random. If some term in formula (9.5) has both systematic and random components, then we shall separate them and replace such a term in formula (9.5) with two terms. In so doing, the systematic components of the errors are assumed, as always, to have a fixed value for every sample of the instrument or for each component of the instrument.

The systematic components form the systematic instrument error and the random components form the random instrument error.

The random instrument error is manifested differently in each application of the instrument. For example, if we want to estimate the largest error of the instrument, then we must add to the estimate of the limits of its systematic error the maximum random error.

It should be noted that in the theoretical description of the random instrument error this error is regarded as a random quantity, and in addition it is most often assumed to have a normal distribution. Such a model admits the possibility of errors of any size, and it becomes unclear how to find the limiting random instrument error in the model.

It should be noted that different situations can arise when measurement errors and the errors of measuring instruments are estimated. In the first case, the random errors have already been realized, and for this reason the random and systematic components can be summed statistically. In the second case, we are estimating the largest error of the instrument that can be manifested in any future experiment, and for this reason the components must be added arithmetically.

The systematic errors of a collection of instrument components of the same type can be regarded as a set of realizations of a random quantity. This quantity is described statistically, for example, by a histogram. For components having a systematic error it is not difficult to construct the histogram of the distribution of systematic errors. For components with a random error such a description becomes more complicated, since it becomes two dimensional: the realization depends on both the sample of the component and the realization of the random error of this component in a given experiment. But all components of the same type can usually be assumed to have the same distribution of the random error, so that the differences of their errors are determined only by the change in some parameter of this distribution. This can be taken into account and one-dimensional distributions of the corresponding parameter and systematic error of the component can be studied instead of the two-dimensional distribution of the random quantity.

The summation of random quantities involves the construction of the composition of their distribution functions.

If the instrument consists of a large number of components and there are many terms in formula (9.5) or (9.6), then the composition of the error components will give, as is well known, a close-to-normal distribution. This makes it possible to simplify the solution, since it is not difficult to find the parameters of the resulting distribution:

$$M[\varepsilon_f]=\sum_{i=1}^{n} V_i M[\varepsilon_i]+\frac{1}{y_f}\sum_{j=1}^{m} W_j M[z_j],$$

$$D[\varepsilon_f]=\sum_{i=1}^{n} V_i^2 D[\varepsilon_i]+\frac{1}{y_f^2}\sum_{j=1}^{m} W_j^2 D[z_j]. \tag{9.7}$$

If some errors are correlated and $\rho_{\nu\lambda}$ is the corresponding correlation coefficient, then another term, equal to $2\rho_{\nu\lambda}V_\nu V_\lambda\sqrt{D_\nu D_\lambda}$, must be introduced in the expression for the variance.

Using the estimates of the mathematical expectations of the error components and their variances, the estimate of the mathematical expectation and variance of the resulting normal distribution can be calculated from the relations presented. This problem is solved first for systematic errors. Taking the percentage of instruments whose error must be less than the computed estimate as the confidence probability α, we find the corresponding limit of systematic error:

$$\theta_r = \widetilde{M}[\varepsilon_f] + z_{(1+\alpha)/2}\widetilde{\sigma}(\varepsilon_f),$$

$$\theta_l = \widetilde{M}[\varepsilon_f] - z_{(1+\alpha)/2}\widetilde{\sigma}(\varepsilon_f). \tag{9.8}$$

Next we estimate the practically limiting random error. Usually the number of terms here is very small, and this error is not calculated at all but rather it is estimated based on the experimental data. If we find an estimate of the standard deviation of the resulting random error, then for the practically limiting value we take $\Psi = t_q\widetilde{\sigma}(y)$, where $q = 1 - \alpha$, and we find t_q from a table of Student's distribution taking into account the degree of freedom from the experimental data.

Next, in accordance with what was said above, we find the practically limiting errors of the instruments:

$$\Delta_r = \theta_r + \Psi, \quad \Delta_l = \theta_l - \Psi. \tag{9.9}$$

Usually $M[\varepsilon_f] = 0$ and $|\theta_r| = |\theta_l|$, so that $|\Delta_r| = |\Delta_l|$.

When the number of terms is small, the problem must be solved by constructing a composition of the distributions of the terms.

It should be noted that in the general case the probability adopted for calculating the limits of systematic error may not be equal to the probability corresponding to the practically limiting random error. Both these probabilities should be indicated. However, this is difficult and is not normally done.

If the terms are given by their permissible limits and there are no data favoring some one distribution, then the corresponding errors are best assumed to be uniformly distributed. In this case the confidence limit of the systematic instrument error can be found from formula (6.9), transformed somewhat:

$$\theta = k\sqrt{\sum_{i=1}^{n} (V_i\delta_i)^2 + \frac{1}{y^2}\sum_{j=1}^{m} (W_j\Delta_j)^2}, \tag{9.10}$$

where the values of the coefficient k are presented in Table 3.1; δ_i $(i = 1,\ldots,n)$ are the limits of permissible systematic errors (of the instrument components) forming the multiplicative component of the instrument error; and Δ_j $(j = 1,\ldots,m)$ is the same, but for errors that make up the additive instrument error.

The practically limiting random error $\Psi(y)$ is found in the same manner as in other cases, and the total instrument error is calculated in the same way. Since systematic instrument errors are assumed to be random quantities, the confidence probability α used to calculate the limits of systematic error indicates the relative number of instruments whose systematic errors do not exceed these limits.

Often the errors of the instrument components are given by their permissible limits, including both systematic and random error components. In this case one can proceed in two ways. The error of each instrument component can be divided into separate components based on experiment, after which the problem is solved according to the scheme presented above. But the total errors of the instrument components can be assumed, without separating the random error, to be uniformly distributed within prescribed limits and they can be added statistically. In the absence of data for separating errors into components, the second method is preferable.

The calculations of the errors are repeated for a series of indications of the instrument. In the process, the confidence probabilities, one of which was used to

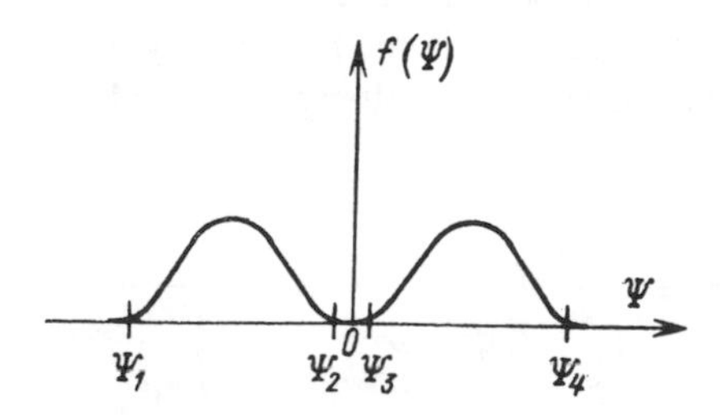

FIG. 9.1. Possible form of the probability density of the practically limiting random instrument error.

determine the limits of systematic error of a set of instruments and the other was used to determine the practically limiting random error of the instruments, should be kept constant. From the data obtained it is possible to construct a graph of the limiting instrument error as a function of instrument indication. The error can be expressed as an absolute or relative error, but absolute error is usually more convenient.

We now consider the second problem: the limit of permissible instrument error is prescribed and it is required to estimate the probability of encountering an instrument with an error less than this limit.

We shall outline the general scheme of the solution. The probability p_g of encountering an instrument whose error does not exceed the permissible limit is equal to

$$p_g = 1 - (p_r + p_l), \tag{9.11}$$

where p_r and p_l are the probabilities of encountering an instrument whose error exceeds the upper limit and drops below the lower limit, respectively.

We can write

$$p_r = P\{\zeta \geqslant (\Delta - \Psi)\}, \quad p_l = P\{\zeta \leqslant -(\Delta - \Psi)\},$$

where Δ is the limit of permissible instrument error and Ψ is the practically limiting random error.

To solve the problem it is necessary to know the distribution function of the systematic instrument errors over the entire collection of the instruments.

But the practically limiting random error cannot always be assumed to be the same for all instruments; it is usually different for different instruments. To obtain a more accurate solution of the problem it is necessary to find the distribution of the practically limiting random instrument error. An example of a possible density of such a distribution is presented in Fig. 9.1. The probabilities p_l and p_r must now be calculated from the formulas

$$p_l = K_l \int_{\Psi_1}^{\Psi_2} P\{\zeta < -(\Delta - \Psi)\} f(\Psi)\, d\Psi,$$

$$p_r = K_r \int_{\Psi_3}^{\Psi_4} P\{\zeta > (\Delta - \Psi)\} f(\Psi)\, d\Psi, \tag{9.12}$$

in which the values of the probability density $f(\Psi)$ play the role of weights and K_l and K_r are normalization factors. If the probability density is $f(\Psi)$ (it is usually symmetric relative to the ordinate axis), then $p_l = p_r$ and $K_l = K_r = 2$. In the general case, however,

$$K_l = \frac{1}{\int_{\Psi_1}^{\Psi_2} f(\Psi)d\Psi}, \quad K_r = \frac{1}{\int_{\Psi_3}^{\Psi_4} f(\Psi)d\Psi}.$$

The coefficients K_l and K_r were introduced in connection with the fact that by construction of the distribution functions the area under the entire curve of the probability density is equal to unity, and we require that the area under each branch of the curve be equal to unity (for $\Psi < 0$ and for $\Psi > 0$).

It is not difficult to derive formulas (9.12) if several values of Ψ and the percentage of cases when each of the selected values occurs are given. Having found p_l for each Ψ, it is natural to add them, weighting each one by a weight proportional to the percentage of times it is encountered. From here, extrapolating to a continuous distribution of Ψ, we arrive at formulas (9.12).

Thus we can find the probability for manufacturing a measuring instrument whose error is less than a fixed limit, if it has a unique output signal. Examples of such measuring instruments are single-valued measures.

Much more often, however, measuring instruments have a definite range of measurement. In this case, the probability of getting a good instrument is equal to the probability that the error of the selected instrument over the entire measurement range is less than the prescribed limit. How does one estimate this probability?

One would think that for a pointer-type instrument the probability p_{gi} can be calculated for each marker of the instrument scale and p_g can be found by multiplying the probabilities together. However, one cannot proceed in this manner, because the errors at different points of the scale are not independent of one another, primarily because of the multiplicative component. In addition, such calculations would be too laborious, since the instrument scale often has 100 to 150 markers.

We shall examine a different method for solving the problem. For definiteness, we shall consider a pointer-type instrument with one measurement range. The method described above makes it possible to find all components of the instrument error at any marker of the instrument scale. We shall calculate them for the final value of the scale y_f. The random component usually varies insignificantly along the instrument scale. For this reason, once the practically limiting random error for the final value of the scale has been estimated, we can assume that we have it for the entire scale of the instrument.

We shall assume at first that the modulus of the practically limiting random error, equal to Ψ, is the same for all instruments studied.

Before summing the systematic error components, we separate them into additive and multiplicative components, after which we add them separately. The addition is performed by statistical methods. As a result we obtain the probability density of the additive and multiplicative error components for $y = y_f$, i.e., for the final value of the scale.

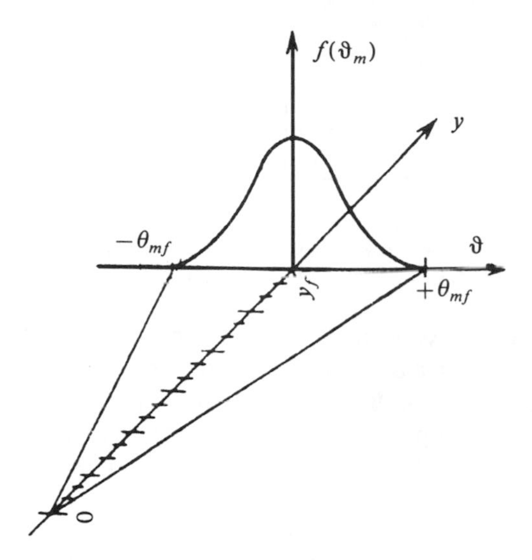

FIG. 9.2. Example of the representation of multiplicative instrument errors together with their probability density.

However, the instrument can have an inadmissibly large error at any marker of the scale, and this must be taken into account.

The systematic error ϑ_y of an arbitrarily chosen instrument at the point y of the instrument scale consists of the multiplicative component ϑ_{my} and the additive component ϑ_{ay}:

$$\vartheta_y = \vartheta_{my} + \vartheta_{ay}.$$

In addition,

$$|\vartheta_{my}| \leqslant \theta_{my},$$

where $\theta_{my} = \theta_{mf} y/y_f$ and θ_{mf} is the largest multiplicative component of the error (neglecting the sign) at $y = y_f$.

An example of the change in the multiplicative errors of an instrument along the instrument scale is shown in Fig. 9.2 together with the probability density of this error at $y = y_f$.

We recall that the multiplicative error increases from the beginning to the end of the scale in proportion to the indications of the instrument.

The additive components of the error, however, vary along the scale in a random manner, but so that $|\vartheta_{ay}| \leqslant \theta_{ay}$, where $\theta_{ay} = \varphi(y)$ is the largest additive component of the error (neglecting the sign) at the point y of the scale. For this reason the additive component must be regarded as a two-dimensional random quantity: It changes in each section of the scale (at each scale marker) as well as along the scale.

For our problem it is best to study not the entire collection of additive components of systematic instrument errors, but rather only the collection of the largest and smallest errors, chosen separately for each instrument. Statistically (over the

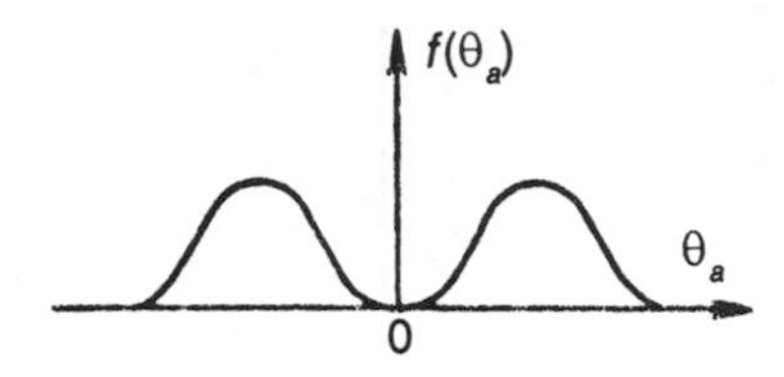

FIG. 9.3. Possible form of the probability density of the extremal additive instrument errors.

set of instruments) these extremal additive components of the systematic error θ_a are determined by two distribution functions, shown in Figs. 9.3 and 9.4.

The graph presented in Fig. 9.3 permits finding the probability of the extremal values of θ_a, and the graph presented in Fig. 9.4 permits finding the probability of encountering an extremal value (positive or negative) on one or another section of the scale of the instrument.

Given these dependences, we can find the probability that the error of the manufactured instrument will be less than a predetermined limit at any point of its scale. The solution is obtained by numerical methods. The scheme of the calculations is as follows.

(1) We transfer from the continuous distributions $f(\vartheta_m)$ and $f(\theta_a)$ to discrete distributions. For this the ranges of possible values of the components $[-\theta_{mf}, +\theta_{mf}]$ and $[-\theta_{af}, +\theta_{af}]$ are divided into a number of intervals so that each interval can be replaced by the average error on it θ_{mi} $(i=1,...,h)$ and θ_{aj} $(j=1,...,t)$.

We set the probability that each average will appear in the error interval equal to the area under the curve of the probability density of the corresponding error on this interval.

Thus we obtain a series of multiplicative and a series of additive errors and the corresponding probabilities:

$$\theta_{mi}, \quad p_{mi}, \quad i=1,...,h,$$

$$\theta_{aj}, \quad p_{aj}, \quad j=1,...,t.$$

We note that the multiplicative errors θ_{mi} correspond to $y = y_f$ and the additive errors θ_{ai} are not related with any section of the scale.

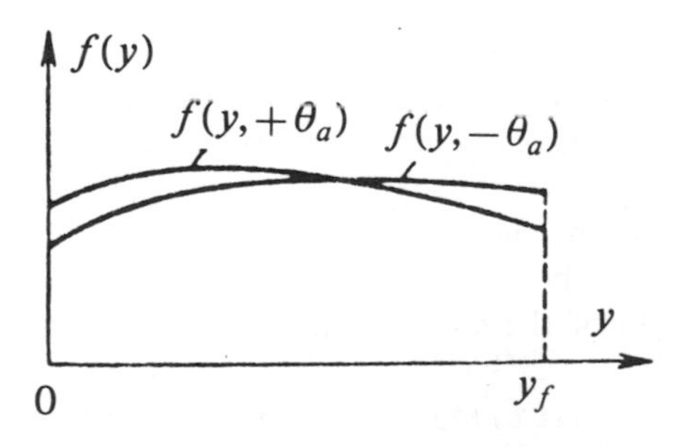

FIG. 9.4. Possible form of the probability density of the extremal positive and negative additive instrument errors along the scale.

(2) We find the section of the scales on which inadmissibly large errors can appear.

Inadmissably large errors, by definition, are errors that satisfy the inequalities

$$\zeta_l \leqslant -\Delta, \quad \zeta_r \geqslant +\Delta. \tag{9.13}$$

We take a pair of components θ_{aj} and θ_{mi} and find a point on the scale of the instrument y_{ij} such that for $y > y_{ij}$ one of the inequalities (9.13) can be satisfied. It is best to study at the same time only positive or only negative components.

Based on the foregoing arguments regarding the separation of the random errors the inequalities (9.13) will assume the form $\theta_{aj} + \vartheta_{myi} \geqslant \Delta - \Psi$ for positive θ_{aj} and θ_{mi} and the form $\theta_{aj} + \vartheta_{myi} \leqslant -\Delta + \Psi$ for negative components. Here

$$\vartheta_{myi} = \theta_{mi} y_i / y_f.$$

The solution of the inequalities gives

$$y_{ij} \geqslant \frac{\Delta - \Psi - \theta_{aj}}{\theta_{mi}} y_f \begin{cases} \theta_{aj} > 0, \\ \theta_{mi} > 0, \end{cases}$$

$$y_{ij} \geqslant \frac{\Delta - \Psi + \theta_{aj}}{|\theta_{mi}|} y_f \begin{cases} \theta_{aj} < 0, \\ \theta_{mi} < 0. \end{cases} \tag{9.14}$$

Thus the section of the scale where inadmissibly large errors can appear for each pair of components is $y_f - y_{ij}$.

(3) We shall calculate the probability that an instrument with an error less than the prescribed limit is manufactured.

The probability that the selected pair of components θ_{mi} and θ_{aj} appears simultaneously is equal to (since they are independent)

$$p = p_{mi} p_{aj}. \tag{9.15}$$

The probability that the selected additive component θ_{aj} will appear in the section of the scale $y_f - y_{ij}$ is determined based on one of the two functions $f(y)$ and is equal to

$$p = \int_{y_{ij}}^{y_f} f(y) dy.$$

From here we find the probability that an instrument with an inadmissibly large error for chosen θ_{mi} and θ_{aj} will be manufactured:

$$p_{ij} = p_{mi} p_{aj} \int_{y_{ij}}^{y_f} f(y) dy. \tag{9.16}$$

To each pair of components there is associated a unique probability p_{ij}. The calculation must be performed separately for positive and negative pairs of components, since in the computational scheme presented above the left- and right-hand branches of the distribution function of the resulting systematic instrument error are taken into account separately.

If we studied the positive components θ_{mi} and θ_{aj}, then the calculations for the negative components have to be repeated because of the unsymmetric function $f(\vartheta_m)$ and $f(\theta_a)$ and the inequality $f(y)_{+\theta_a} \neq f(y)_{-\theta_a}$ different probabilities will correspond to positive and negative errors. But if the functions are symmetric and $f(y)_{+\theta_a} = f(y)_{-\theta_a}$, then the calculations can be shortened by setting

$$p_{ij} = 2p_{mi}\,p_{aj} \int_{y_f - y_{ij}}^{y_f} f(y)\,dy, \tag{9.17}$$

where the indices i and j now enumerate the positive and negative errors with the same magnitude.

As a result of the calculation we find

$$p_l = \sum_{\theta<0} p_{ij}, \quad p_r = \sum_{\theta>0} p_{ij},$$

and then the probability p_g, which we seek, of manufacturing an instrument with an error less than the prescribed limit is

$$p_g = 1 - (p_l + p_r).$$

We obtained the answer for the case when the practically limiting random error Ψ is the same for all instruments. Often, however, as we have already noted, it is necessary to take into account the fact that different instruments can have different random errors, i.e., different Ψ.

To solve this problem formulas (9.12) must be used. It is first necessary to establish a number of discrete values of Ψ_s and the corresponding probabilities p_s. Next, using the scheme presented above, p_{gs} can be found for each Ψ_s. Averaging the obtained probabilities with the weights p_s, we obtain the solution

$$p_g = \sum_s p_s p_{gs}. \tag{9.18}$$

Thus it is possible to find the probability of manufacturing an instrument with an error less than the prescribed limit if the instrument has one measurement range. The solution of this problem for instruments with many measuring ranges is, in principle, the same as the solution presented above.

Instruments with voltage dividers, shunts, measuring transformers, and similar instruments with a variable transfer coefficient have several measurement ranges. Every instrument has one sample of such instruments of one type or another.

One would think that in order to estimate the probability of getting an instrument with an inadmissibly large error it is necessary to have the distribution function of the errors of all instrument components. Given these functions, all possible combinations of errors in one instrument must then be sorted through, and the particular combinations that give errors that are less (or greater) than a prescribed limit are then selected. This is possible, but complicated.

The problem simplifies significantly if the distribution of the highest and lowest errors of the instrument components (voltage dividers or shunts) that give the instruments several measurement ranges are studied separately.[16] We shall call these devices multivalued units.

Each multivalued unit (a particular sample of the unit) can be described by only two errors with the largest modulus: positive and negative errors. A set of units of one type will then correspond to the distribution function of the largest and smallest errors.

The largest (smallest) error of each instrument consists of the largest (smallest) errors of its components. Correspondingly, the distribution function of the largest (smallest) instrument errors can be constructed according to analogous functions of the components. For multivalued units this is the distribution function of the largest and smallest errors over the set of units. A single-valued unit has one error distributed over the set of units. After the distribution function of the largest errors of the instruments has been determined, we find the probability sought.

The foregoing solution contains one fundamental inconsistency. The essence of this inconsistency is that instrument errors must be calculated when the instruments are developed, and the calculation is based on data on the physically nonexistent units of these instruments. It is possible to get out of this difficulty by focusing on units of analogous instruments which are already being produced. The distribution functions of the parameters of such units can be estimated. Of course, a specialist can introduce into these data certain corrections so as to be able to extrapolate them to the parameters of the units being designed.

In conclusion it should be added that the accuracy of the calculations in which continuous distributions are replaced by discrete distributions depends on the number of discrete intervals, which can be made very large, if computers are employed. However the probability of manufacturing an instrument with an error less than the prescribed limit need not be found with high accuracy.

If, however, the starting data are represented in the form of histograms, then in the solution under study all information contained in them is employed and the computational uncertainty is determined primarily by the uncertainty of the histograms.

9.3. Calculation of the errors of electric balances (unique instrument)

Electric balances are an instrument in which the force of interaction of the moving and nonmoving coils, through which the same constant current flows, is balanced by the force of gravity of the weights. Standards of the unit of electric current strength (ampere) have been developed in the U.S. and the former USSR based on this principle. (Now the ampere is reproduced with higher accuracy using standards of the volt and the ohm. Therefore the electric balance lost its value.)

We shall calculate the errors of Soviet-made electric balances. We shall use the data presented in Ref. 31.

The current strength at the point of equilibrium of the electric balance is determined by the expression

$$I = \sqrt{mg/F}, \tag{9.19}$$

where m is the mass of the balancing weight, g is the acceleration of gravity, and F is the constant of the electric balance.

The constant of the electric balance is equal to the derivative of the mutual inductance of the two coils (mobile and immobile) with respect to the vertical displacement of the mobile coil and is calculated from their geometric dimensions.

The difference between the value of the current strength calculated using formula (9.19) and its true value, i.e., the error of the electric balance, is determined

TABLE 9.1. Limits of the components of the errors $\varepsilon(F)$.

Reason for error	Limits of error in the constant $\delta F \times 10^6$
Uncertainty in the measurement of the radial dimensions	
Immobile coil $\delta F(r_{\text{im}})$	± 3
First part of the moving coil $\delta F(r_{m1})$	± 3
Second part of the moving coil $\delta F(r_{m2})$	± 2
Uncertainty in the measurement of the axial dimensions	
Immobile coil $\delta F(\ell_{\text{im}})$	± 2
First part of the moving coil $\delta F(\ell_{m1})$	± 1.3
Second part of the moving coil $\delta F(\ell_{m2})$	± 0.7
Deviation of the coils from the cylindrical shape $\delta F(R)$	± 2

by the uncertainty in all the quantities entering into this formula as well as by the effect of the field of the wires carrying the current to the mobile coil. These sources of error create the systematic error of the electric balance.

The equilibration of the balance, however, is also accompanied by random errors, which are caused by friction in the supports of the cross arm of the balance, fluctuations of the ambient air temperature, changes in the external magnetic field, effect of air flows, and some other factors.

The systematic error of the electric balance must be estimated by a computational method; it cannot be determined experimentally (as long as one is not concerned with comparing the national standard of the unit of current strength with the standard of this unit in other countries). It is, however, virtually impossible to calculate the random error, but it can be estimated based on the experimental data. For our balances the relative standard deviation of the current strength is $S_{\text{rel}} = 2 \times 10^{-6}$.

The uncertainty of the quantities entering into formula (9.19) is characterized by the following data. For the mass of the balancing load the relative error falls within the limits $\pm$ 1.25 $\times$ 10^{-6} and for the acceleration of gravity the error falls within the limits $\pm$ 4 $\times$ 10^{-6}. (At the present time this error can be significantly smaller.)

The error in the constant of the electric balance is in turn caused by a number of factors. Table 9.1 gives the limits of the errors introduced into the constant of the electric balance by each of these factors.[31]

We shall find the influence coefficients of the relative errors of the measurements of the mass $\varepsilon(m)$, acceleration of gravity $\varepsilon(g)$, and the calculation of the constant of the electric balance $\varepsilon(F)$ in accordance with formula (9.3).

We represent expression (9.19) in the form of the product of the arguments:

$$I = m^{1/2} g^{1/2} F^{-1/2}.$$

As shown in Sec. 6.6, in this case the influence coefficients are equal to the powers of the corresponding arguments, i.e.,

$$V_m = \tfrac{1}{2}, \quad V_g = \tfrac{1}{2}, \quad V_F = -\tfrac{1}{2}.$$

Aside from the enumerated and estimated components of the error, it is also necessary to take into account the error mentioned above due to the influence of the field generated by the wires conducting the current to the mobile coil. Experiments show that this field creates an additional force on the mobile part that falls within $\pm 2 \times 10^{-6}$ times the nominal strength of the interaction of the coils.

Since the influence coefficient of the force of interaction (mq) is $V_{mq} = 1/2$, this error has the same influence coefficient $V_H = 1/2$.

According to formula (9.6) the total systematic error of the electric balance (in the relative form) will be

$$\varepsilon_\Sigma = V_H\varepsilon(H) + V_m\varepsilon(m) + V_g\varepsilon(g) + V_F\varepsilon(F),$$

where $|\varepsilon(H)| \leqslant 2 \times 10^{-6}$, $|\varepsilon(m)| \leqslant 1.25 \times 10^{-6}$, $|\varepsilon(g)| \leqslant 4 \times 10^{-6}$, and a series of components was given for the error $\varepsilon(F)$.

All components of the error ε_Σ are determined by their limits. For this reason, we shall use formula (9.10) and find the confidence systematic error of the electric balance. We shall take $\alpha = 0.95$ and $k = 1.1$. Then

$$\theta_{0.95} = 1.1\sqrt{(\tfrac{1}{2})^2(2^2 + 1.25^2 + 4^2 + 2\times 3^2 + 3\times 2^2 + 1.3^2 + 0.7^2)\times 10^{-12}}$$

$$= 1.1\times 10^{-6}\sqrt{13.5} = 4\times 10^{-6}.$$

The practically limiting random error can be estimated if we know S_{rel} and have some idea about the form of the distribution of the experimental data. If it can be assumed that the data correspond to a normal distribution and the confidence probability is also taken to be 0.95, then $\Psi = 2S_{\text{rel}} = 4 \times 10^{-6}$. Then the practically limiting component of the error of the electric balance in the single-balancing regime will be, according to formula (9.9),

$$\delta I_{0.95} = \theta + \Psi = 8\times 10^{-6}.$$

When measuring with the help of electric balances the emf of standard cells several balancings can be performed and the error in the result can be reduced by averaging the data obtained.

We underscore the fact that the result obtained pertains to a specific sample of electric balances, since it was obtained using data on the parameters of the components of this instrument.

9.4. Calculation of the error of ac voltmeters (mass-produced instrument)

We shall study the inverse problem. The limit of permissible instrument error as the fiducial error is given, and it is required to calculate the percentage of instruments satisfying this requirement when they are manufactured using a technology with a prescribed degree of development.

We shall study the voltmeter in a ferrodynamic system. Figure 9.5 shows the block diagram of this instrument. This diagram was constructed in accordance with the theory of instruments of this system. Figure 9.6 shows a graphical representation of the instrument scale. Block 1 converts the measured voltage U_x into the current of strength

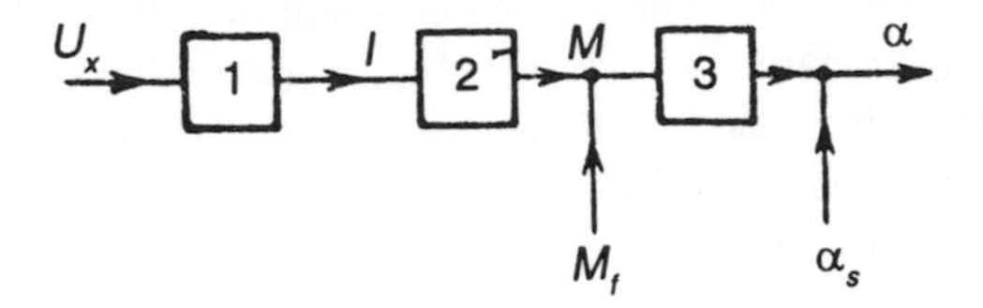

FIG. 9.5. Block diagram of voltmeter.

$$I = U_x/R,$$

where R is the resistance of the input circuit of the voltmeter.

The current I is converted with the help of block 2 into a torque

$$M_t = KI^2,$$

where K is the electrodynamic constant of the instrument.

The block 3 generates a countertorque

$$M_c = W\alpha,$$

where W is the stiffness of the spring and α is the angle of rotation of the moving part.

When the moving part is in a position of equilibrium $M_t = M_c$, and from here

$$\alpha = \frac{K}{WR^2} U_x^2. \tag{9.20}$$

When the instrument is manufactured, the particular combination of the parameters K, W, and R that is realized in the instrument is fixed by regulating the instrument and calibrating its scale. For this reason, instrument errors will arise only as a result of changes in the stiffness of the spring and the input resistance relative to their values at the moment of regulation. The constant K, however, is virtually unchanged and does not give rise to any errors.

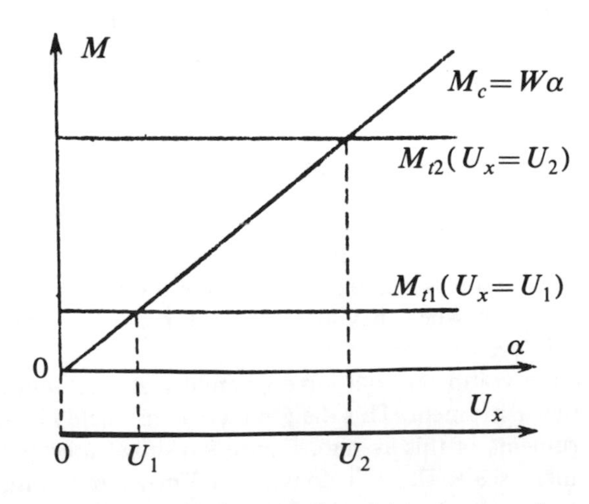

FIG. 9.6. Graphical construction of the voltmeter scale.

As regards the parameters W and R, formula (9.20) is exact and permits finding the instrument error introduced by changes in these parameters. Structurally it is identical to formula (6.23). For this reason, the values of the influence coefficients for the relative changes in the stiffness W and resistance R can be written down immediately:

$$V_W = -1 \quad \text{and} \quad V_R = -2.$$

In addition to the instability of the parameters of the blocks, errors can also appear owing to the friction in the supports of the moving part and the uncertainty of the scale. The sources of these errors are indicated in Fig. 9.5. Since these errors are additive, it is best to express them as absolute errors. We shall express them in units of the angle of rotation of the moving part.

Friction introduces a random error (M_f is the friction moment). It is customarily described as a dead zone, i.e., by the difference of the indications of the instrument that is obtained by approaching continuously from the right and left a particular marker on the scale. The largest dead zone is determined (see, for example, the standards for electric measuring instruments). The largest random error due to friction is equal to one-half the dead zone. Therefore the limits Ψ_t of this error are also known. It is assumed that the random errors of each instrument are uniformly distributed within these limits. The limits themselves, however, can be different for different instruments.

The error in the scale of the instrument α_s for each scale marker of a particular instrument is a systematic error. But this error varies from one marker to another. It also varies from one instrument to another. For each particular instrument it is possible to find the largest error of the scale. It can be assumed that this error is encountered with equal probability on any scale marker. The set of instruments is characterized by the distribution of these largest scale errors.

So, the components of the instrument errors are as follows:

(a) the error from the variation of spring stiffness $\vartheta_1 = -\varepsilon_W$,

(b) the error due to the variation of the input resistance $\vartheta_2 = -2\varepsilon_R$,

(c) the error due to the friction $\psi = \alpha_f$, and

(d) the error due to scale inaccuracies $\vartheta_3 = \alpha_s$.

The errors ϑ_1 and ϑ_2 are multiplicative and are expressed as a percentage; the errors ϑ_3 and ψ are additive and are expressed in units of the angle of rotation of the moving part of the voltmeter (in degrees), i.e., they are referred to the output. For this reason, the instrument error scaled to the output is given by the relation

$$\zeta_\alpha = \frac{\alpha}{100}(\vartheta_1 + \vartheta_2) + \vartheta_3 + \psi, \tag{9.21}$$

where α is the angle of rotation of the moving part of the instrument, which corresponds to its indication U_x for which the error is calculated.

Let us assume that the limits of intrinsic error are given and they are $\pm 1\%$ (as the fiducial errors). Next, we assume that the data given in Table 9.2 are known for each component of the error. These data characterize the degree of development of the manufacturing technology.

We shall also assume that in 30% of the instruments the practically largest dead zone does not exceed 0.8° and in 70% of the instruments it does not exceed 0.4°. Thus the random error falls within the limits $\Psi_1 = \pm 0.4°$ for 30% of the instru-

TABLE 9.2. Starting data on the sources of systematic instrument errors.

Source of error or error	Interval of probability distribution		Frequency of occurrence of the interval
	Left limit	Right limit	
Relative change in spring stiffness ε_W	−0.3%	−0.2%	0.2
	−0.2%	−0.1%	0.5
	−0.1%	0	0.3
Relative change in resistance ε_R	−0.3%	−0.1%	0.2
	−0.1%	+0.1%	0.2
	+0.1%	+0.3%	0.6
Absolute error of the instrument scale α_s	−0.6°	−0.2°	0.5
	−0.2°	+0.2°	0
	+0.2°	+0.6°	0.5

ments and within the limits $\Psi_2 = \pm 0.2°$ for 70% of the instruments. Given these data, we must find the probability that the fiducial error of the voltmeters falls within the limits $\Delta = \pm 1\%$.

The fiducial error must be put into the form of an absolute error. We shall express it in degrees of rotation of the moving part. This can be done with the help of a graph similar to that presented in Fig. 9.6. Assume that in our case the limit of permissible error in degrees $\Delta = 1°$ (neglecting the sign).

Focusing on formula (9.21), we shall first find the composition of the multiplicative errors ϑ_1 and ϑ_2. Using the data presented in Table 9.2 and the influence coefficients found, it is not difficult to describe the histograms of the distributions of these errors. These descriptions are given in Table 9.3.

It is convenient to solve the problem by the method of sorting, described in Sec. 3.6. For this the histograms must be replaced by discrete distributions. To each

TABLE 9.3. Description of histograms of distributions of multiplicative components of the instrument error.

Error	Interval of error distribution (%)		Frequency of occurrence of interval
	Left limit	Right limit	
ϑ_1	+0.2	+0.3	0.2
	+0.1	+0.2	0.5
	0	+0.1	0.3
ϑ_2	+0.2	+0.6	0.2
	−0.2	+0.2	0.2
	−0.6	−0.2	0.6

TABLE 9.4. Discrete representation of the distribution of the multiplicative instrument error.

Number	$\eta = \eta_1 + \eta_2$	$p = p_1 p_2$
1	$+0.25 + 0.4 = +0.65$	0.04
2	$+0.25 + 0.0 = +0.25$	0.04
3	$+0.25 - 0.4 = -0.15$	0.12
4	$+0.15 + 0.4 = +0.55$	0.10
5	$+0.15 + 0.0 = +0.15$	0.10
6	$+0.15 - 0.4 = -0.25$	0.30
7	$+0.05 + 0.4 = +0.45$	0.06
8	$+0.05 + 0.0 = +0.05$	0.06
9	$+0.05 - 0.4 = -0.35$	0.18

interval there is assigned an error equal to the center of the interval. The probability of the appearance of this error is assumed to be equal to the frequency of this interval.

Let the error ϑ_1 be represented by the discrete random quantity η_1 and the error ϑ_2 by the discrete random quantity η_2. We obtain the following:

$$\begin{array}{cccc} \eta_1 & +0.25 & +0.15 & +0.05 \\ p_1 & 0.2 & 0.5 & 0.3 \end{array},$$

$$\begin{array}{cccc} \eta_2 & +0.4 & 0 & -0.4 \\ p_2 & 0.2 & 0.2 & 0.6 \end{array}.$$

The random quantity $\eta = \eta_1 + \eta_2$ corresponds to the error $\vartheta_1 + \vartheta_2$. Its realizations are presented in Table 9.4.

The limiting values of the total error $\eta_{min} = -0.6\%$ and $\eta_{max} = +0.9\%$ (see Table 9.3); these errors correspond to probabilities of 0 and 1, respectively. The probability distribution is constructed based on the obtained data. The numerical values are summarized in Table 9.5.

TABLE 9.5. Table of the computed values of the probability distribution of multiplicative instrument error.

η	-0.6	-0.35	-0.25	-0.15	$+0.05$
p	0	0.18	0.30	0.12	0.06
Σp	0	0.18	0.48	0.60	0.66

η	$+0.15$	$+0.25$	0.45	0.55	0.65	0.9
p	0.10	0.04	0.06	0.10	0.04	0
Σp	0.76	0.80	0.86	0.96	1.0	1.0

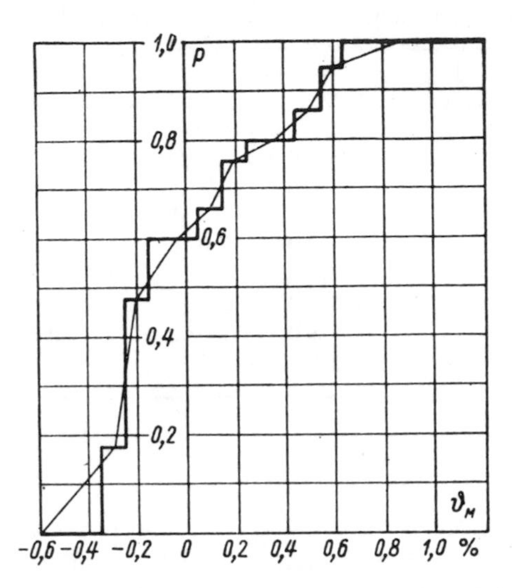

FIG. 9.7. Step and linear approximations of the distribution function of the multiplicative errors of voltmeters.

Based on these data we construct a step curve as a first approximation to the distribution function sought for the multiplicative error of the instruments, after which the function is smoothed by the method of linear approximation. The distribution function so obtained is presented in Fig. 9.7.

We shall now express the multiplicative error in the form of absolute errors—as fractions of the angle of rotation of the mobile part. We shall find the largest error, i.e., the error corresponding to maximum deflection. We assume that $\alpha_{\max} = 100°$. Then the numerical values of the error $\vartheta_m = (\vartheta_1 + \vartheta_2)(\alpha_{\max}/100)$ will be equal to the values given in Table 9.5. Using these data and the graph in Fig. 9.7, we construct a histogram of the multiplicative error of the instrument with maximum angle of rotation of the mobile part. The data for this histogram are as follows:

Interval number i	1	2	3	4	5
Limits of interval for the error ϑ_m (degrees)					
left	+0.6	+0.3	+0	−0.3	−0.6
right	+0.9	+0.6	+0.3	0	−0.3
Average value of θ_{mi}	+0.75	+0.45	+0.15	−0.15	−0.45
Probability of falling within the interval p_{mi}	0.05	0.15	0.20	0.42	0.18

We obtain, analogously, from the data in Table 9.2 the average values of the intervals of the distribution of the largest scale errors of the instruments and the corresponding probabilities:

Interval number j	1	2	3
Average value θ_{aj} of the error in the interval	+0.4	0	−0.4
Probability of falling within the interval P_{aj}	0.5	0	0.5

In accordance with formulas (9.14), we find the sections of the scale where the instrument can be rejected. In our case $\Delta_1 = \Delta - \Psi_1 = 0.6°$ and $\Delta_2 = \Delta - \Psi_2 = 0.8°$ (for 30% and 70% of the instruments, respectively).

For $\theta_{aj} > 0, \theta_{mi} > 0$, and $\Delta_1 = 0.6°$ we obtain:

$$y_{11}=\frac{0.6-0.4}{0.75}y_f=0.27y_f,\quad y_{21}=\frac{0.6-0.4}{0.45}y_f=0.45y_f,$$

$$y_{12}=\frac{0.6}{0.75}y_f=0.8y_f.$$

The remaining combinations give $y_{ij} > y_f$, which means that it is impossible to obtain an inadmissibly large error.

We shall assume that the distribution function of the additive errors of the instruments along the scale is uniform and identical for positive and negative errors: $f(y) = 1/y_f$. Then for each scale section studied we obtain

$$p'_{ij}=\int_{y_{ij}}^{y_f}\frac{1}{y_f}dy=\left(1-\frac{y_{ij}}{y_f}\right).$$

Therefore

$$p'_{11}=(1-0.27)=0.73,\quad p'_{21}=(1-0.45)=0.55,\quad p'_{12}=(1-0.80)=0.20.$$

For $\theta_{aj} < 0, \theta_{mi} < 0$, and Δ_1 we obtain

$$y_{53}=\frac{0.6-0.4}{0.45}y_f=0.45y_f,\quad p'_{53}=0.55.$$

From here we find the probability that an instrument is rejected for each combination of instrument components:

$$p_{11}=0.73\times0.05\times0.5=0.018,$$
$$p_{21}=0.55\times0.15\times0.5=0.041,$$
$$p_{12}=0,$$
$$p_{53}=0.55\times0.18\times0.5=0.050.$$

Therefore

$$p_l=p_{53}=0.050,\quad p_r=p_{11}+p_{21}+p_{12}=0.059.$$

The probability of manufacturing a high-quality instrument is

$$p_{g1}=1-(0.050+0.059)=0.89.$$

Analogous calculations for $\Delta_2 = 0.8$ give

$$y_{11}=\frac{0.8-0.4}{0.75}y_f=0.53y_f, \quad p'_{11}=0.47,$$

$$y_{21}=\frac{0.8-0.4}{0.45}y_f=0.89y_f, \quad p'_{21}=0.11,$$

$$y_{53}=\frac{0.8-0.4}{0.45}y_f=0.89y_f, \quad p'_{53}=0.11;$$

$$p_{11}=0.47\times0.05\times0.5=0.012,$$

$$p_{21}=0.11\times0.15\times0.5=0.008,$$

$$p_{53}=0.11\times0.18\times0.5=0.010.$$

Now $p_l = 0.010$ and $p_r = 0.020$, and $p_{g2} = 0.97$.

The weighted-mean probability of manufacturing an instrument whose error is less than the prescribed limit is equal to

$$p_g=0.3p_{g1}+0.7p_{g2}=0.3\times0.89+0.7\times0.97=0.95.$$

Therefore for the properties of the instrument components and the scale fabrication quality presented above, approximately 95% of the instruments will have a fiducial error not exceeding 1%.

This calculation was performed for reference conditions and determines the limits of intrinsic instrument error.

For a prescribed limit of instrument error the obtained percentage of rejections can serve as a basis for increasing the quality requirement for one or another of the instrument components, improving the technology used to fabricate the components, etc. The limits of permissible errors of all components can be calculated uniquely if the weights are assigned for their errors. These weights are apparently difficult to determine objectively, and sometimes it is impossible to do so. For this reason, the main method is to estimate the percentage of rejects and select specifications of the instrument components so that the percentage of rejects is acceptable.

9.5. Calculation of the error of digital thermometers (mass-produced instrument)

Digital thermometers are usually constructed according to a scheme in which the digital–analog integrator converts the emf of the thermocouple into a corresponding voltage, after which this voltage is converted into a proportional time interval and thus into the indication of the instrument. This is explained by the graphs presented in Fig. 9.8.

The graph in Fig. 9.8(a) refers to an analog–digital integrator. The integration time is maintained strictly constant, and the slope of the straight lines is proportional to the emf at the input of the integrator (in accordance with the principle of

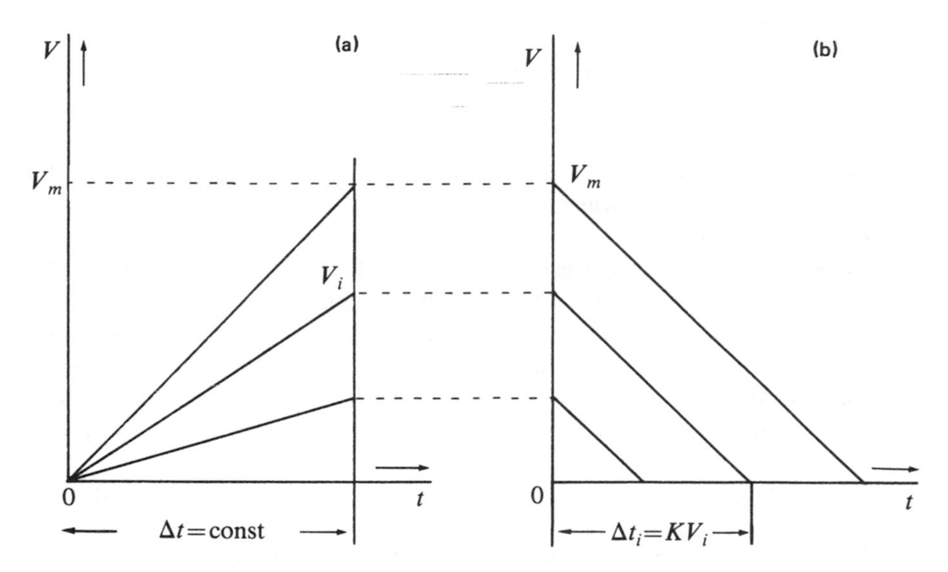

FIG. 9.8. Graphs explaining the principle of operation of a digital integrating instrument.

operation of the integrator). For this reason, the voltage at the output of the integrator U_i is proportional to the emf of the thermocouple.

The graph in Fig. 9.8(b) shows how the voltage U_i is converted into a proportional time interval Δt_i; this is equivalent to conversion into a number—the indication of the instrument.

Since the temperature dependence of the thermocouple emf is known, the indications of the instrument give the measured temperature.

Two special procedures are realized in the process of the conversions. One is linearization of the temperature dependence of the thermocouple emf. The other is compensation of the effect of a deviation of the temperature at the so-called cold ends of the thermocouple from the reference level. The latter is assumed to be 0 °C, the melting point of ice.

The specifications provided by manufacturers of thermocouple thermometers still do not give the user clear indications of the temperature measurement accuracy that can be achieved with the instrument. One would think that the well-known Fluke Company, which introduced the concept of total accuracy of digital thermometers—total instrument accuracy for digital thermometers[12]—would have filled in this omission. The catalog of the firm shows, however, that it gives the thermometer error without taking into account the thermocouple error, i.e., it does not complete the solution of the problem.

We shall study the calculation of the error of digital thermometers based on data on the accuracy of their components and the calibration accuracy. We shall take a thermometer with a thermocouple of type J and a measurement range of 0–750 °C. The instrument can be used in the range $+25 \pm 10$ °C, and after calibration it can be used for one year. We shall consider the direct problem.

Focusing on Ref. 12, we shall assume that the following data are known (the numbers for the calculation here are arbitrary):

1. Linearity: The limits of instrument error due to deviation from linearity of the characteristic $\theta_L = \pm 0.1$ °C.

2. Reference junction.

2.1. The limits of instrument error due to the effect of a deviation of the temperature by 10 °C from the normal temperature (+25 °C) on this circuit $\theta_{TJ} = \pm 0.2$ °C.

2.2. The limits of instrument error due to the change in the parameters of this circuit over a period of one year (instability) $\theta_{SJ} = \pm 0.2$ °C.

3. Reference voltage.

3.1. The limits of instrument error due to the effect of a deviation of the temperature by 10 °C from the normal temperature on this circuit $\theta_{TV} = \pm 0.5$ °C.

3.2. The limits of instrument error due to a change in the parameters of this circuit over a period of one year (instability) $\theta_{SV} = \pm 0.5$ °C.

4. Correspondence to NBS (NIST) data. The limits of instrument error due to inaccurate linearization of the standard characteristic of the thermocouple $\theta_{sc} = \pm 0.15$ °C.

5. The limits of instrument error due to the discreteness of the indications $\Psi_D = \pm 0.5$ °C.

6. The limits of instrument calibration error $\theta_C = \pm 0.15$ °C.

The absolutely constant error, i.e., the error that is the same for all instruments of a given type, will be the error due to inconsistency with the NBS (NIST) data.

The random error will be the error introduced by the discreteness of the instrument indications. All other errors must be regarded as conditionally constant.

We shall regard conditionally constant errors as uniformly distributed random quantities, as has already been assumed above. Their characteristic feature, in our case, is that some of them depend on one another. Taking this into account, the total conditionally constant error of the thermometers must be calculated using the following formula (for probability $\alpha = 0.95$):

$$\theta_1 = 1.1\sqrt{\theta_L^2 + (\theta_{TJ} + \theta_{TV})^2 + (\theta_{SJ} + \theta_{SV})^2 + \theta_C^2} = 1.1\sqrt{1.01} = 1.1\ ^\circ\text{C}.$$

The limits of absolutely constant error must be summed with the limits θ_1 arithmetically. We obtain

$$\theta_2 = \theta_{sc} + \theta_1 = 0.15 + 1.1 = 1.25\ ^\circ\text{C}.$$

The limits of random error must also be taken into account by arithmetic summation, since we are estimating the instrument that which can be obtained for any use of the instrument in the future:

$$\Delta = \theta_2 + \Psi = 1.25 + 0.5 = 1.75\ ^\circ\text{C}.$$

Although Δ is expressed in units of the measured quantity, this is still not the total error of the instrument as a thermometer, since it does not include the thermocouple error, and without the thermocouple the instrument cannot operate as a thermometer.

According to ANSI Standard MC 96.1 the limits of error of thermocouples of type J in the temperature range 0–750 °C are ±2.2 °C or ±0.75%, whichever is greater. It is easy to calculate that up to 300 °C the limits of error will be ±2.2 °C, after which they must be calculated using the relation $\pm$ 0.75 $\times$ 100 $\times$ T_x, where T_x is the measured temperature. At T_x = 750 °C this error will fall within the limits $\pm$5.6 °C.

The thermocouple errors must be regarded as conditionally constant, and for this reason they must be taken into account when calculating θ_1. Taking this component into account, we obtain $\theta_1' = 1.1\sqrt{1.01+5.6^2} = 1.1\sqrt{32.4} = 6.3$ °C.

After this we find the modulus of the limit of the total instrument error:

$$\Delta' = \theta_1' + \theta_{sc} + \Psi = 6.3 + 0.15 + 0.5 = 6.95 \approx 7\ °C.$$

If this limit is represented in the form of fiducial error, we obtain $\gamma = \pm 7\ °C/750\ °C \times 100 \approx \pm 1\%$.

In this case this value of the fiducial error will be identical to the limit of relative error in the range 300–750 °C. Under this temperature range, the limit of relative error starts to increase. Thus the limit of instrument error found in the range 300–750 °C is also an estimate of the limit of minimum error in measuring the temperature that one can count on when using a thermometer consisting of a thermocouple of type J and the digital indicator under study. In other words, this estimate of the error characterizes the maximum temperature measurement accuracy expected for this thermometer.

Normalization of the inaccuracy of the indicator of the thermometer is not only insufficient for the user, but can even mislead the user, because its error is expressed in units of measurement of temperature, while it characterizes only the indicator of the thermometer. The difference is very significant: $\Delta = 1.75$ °C while $\Delta' = 7$ °C.

Chapter 10

Problems in the theory of calibration

10.1. Types of calibration

In order for measurements to be unified, measuring instruments must be unified. This means that all measuring instruments must carry legal units, and the errors of the measuring instruments must not exceed predetermined limits. To achieve these ends, all measuring instruments in use are periodically checked. In the process, reference standards and working standards are used either to verify that the errors of the measuring instruments being checked do not exceed their limits or the measuring instruments are recalibrated.

The general term for the above procedures is *calibration*. But one should distinguish between a real calibration and a simplified calibration.

Real calibration results in the determination of a relation between the indications of a measuring instrument and the corresponding true values of a measurand. This relation can be expressed in the form of a table, or as a graph, or as a function. It can also be expressed in the form of the table of corrections to the indications of the measuring instrument.

The simplified calibration (also called verification) simply reveals whether the errors of a measuring instrument exceed their specified limits.

Essentially, verification is a specific case of quality control, much like quality control in manufacturing. And since it is quality control, verification results do have some rejects.

In addition, a check of an entire set of elements is distinguished from a check of a single element. In a complete check the error of the checked measuring instrument is determined as a whole, while in the case of an elementwise check the errors of the elements of the measuring instrument being checked are determined. A complete check is always preferable; such a check gives the most reliable solution of the problem. In some cases, however, a complete check is impossible to perform and one must resort to an elementwise check.

In an elementwise check, the error of the measuring instrument being checked is calculated by means of the same methods that were examined in Sec. 9.2 for solving the direct problem of calculating the errors of a measuring instrument from the errors of its components. The data required for the calculation are obtained by measuring the parameters of the components of the measuring instrument being checked. Usually, however, this problem is solved completely only once, and in so doing the standards for the errors of the components are determined. In the future, when a check is performed only the parameters of the components and the serviceability of the measuring instrument are checked. If the parameters of the components satisfy the standards established for them, then the error of the measuring instrument checked in this manner falls within the established limits.

Elementwise calibration is often employed to check the measuring systems when the entire system cannot be delivered to a standard laboratory and the laboratory does not have necessary working standards that could be transported to the system's site.

The standardization of the metrological properties of the units of a system does not present any difficulties, and the units must be checked by standard methods.

As a rule, the operation of measuring systems cannot be interrupted and interruption for checking is inadmissible. For this reason, in most cases systems are assembled with a redundant set of units, so that units that are removed for checking could be replaced with units that are known to be serviceable. During the regular check of the system the units are once again interchanged.

When a system is checked, however, in addition to checking the units it is also necessary to check the serviceability of the system as a whole. The methods for solving this problem depend on the arrangement of the system, and it is hardly possible to make general recommendations here. For example, the following procedure can be used for a system with a temperature measuring channel.

After the serviceability of all units of the system has been checked, we note some indication of the instrument at the output of the system. Assume that the indication is $+470$ °C. Then we find from the nominal calibration characteristic of the primary measuring transducer the output signal that should be observed for the given value of the measured quantity. Thus, if a platinum–rhodium–platinum thermocouple was used as the measuring transducer, then when a temperature of $+470$ °C is measured, the emf at the output of the thermocouple must be equal to 3.916 mV. Next, disconnecting the wires from the thermocouple and connecting them to a voltage exactly equal to the nominal output signal of the thermocouple, we once again note the indication of the system. If it remains the same or has changed within the limits of permissible error of the thermocouple and voltmeter, then the system is serviceable.

Of course, this method of checking will miss the case when the error of the thermocouple is greater than the permissible error and the same is true for the voltmeter and these errors mutually cancel. However, this can happen only very rarely. Moreover, such a combination of errors is in reality permissible for the system.

At the same time, this method of checking permits evaluating at the same time the state of the thermocouple. The error of the thermocouple usually makes the largest contribution to the error of the measuring system. For this reason, the difference of the indications—observed at the moment of the check and obtained after the thermocouple is disconnected and its output signal is replaced by a nominal signal—must be less than the limit of permissible error of the thermocouples; it is permissible for this limit to be exceeded by an insignificant amount, which is determined by the accuracy of the system.

The foregoing method of checking the metrological state of measuring systems based on the use of redundant units is also promising in application to many other complicated modern measuring devices, which it is technically difficult or impossible to transport for checking to metrological organizations, as well as to devices whose operation cannot be interrupted.

10.2. Estimation of the errors of measuring instruments in verification

The error ζ of a measuring instrument is defined by the formula

$$\zeta = A_c - A,$$

where A_c is the indication of the instrument being checked, the nominal value of the standard, etc., and A is the true value of the measured quantity, the quantity reproduced by the standard being checked, etc.

The true value A is always unknown. If instead of the true value the corresponding indication of a working standard A_r (the real value of the measured quantity) is used, then instead of ζ we obtain

$$\zeta' = A_c - A_r. \tag{10.1}$$

In order to estimate the error ζ by ζ' the difference $\zeta' - \zeta$ must be small. The error of the working standard is

$$\gamma = A_r - A.$$

For this reason

$$\zeta' - \zeta = \gamma. \tag{10.2}$$

Most often, it is known only that the error of the working standard does not exceed the limit Δ_s established for it. Then

$$|\zeta' - \zeta| \leqslant \Delta_s.$$

In the relative form, the error of the error ζ, as follows from the expression

$$\varepsilon = \frac{|\zeta' - \zeta|}{\zeta},$$

depends on the error ζ itself and increases as ζ decreases.

It is natural to estimate this relative error as

$$\tilde{\varepsilon} = \Delta_s / \zeta'.$$

When a working standard is chosen, the limit Δ of permissible error of the measuring instrument being checked usually serves as the starting point. In this case the ratio

$$k = \Delta_s / \Delta$$

comes into play.

The relative error of the error can be expressed in terms of k:

$$\tilde{\varepsilon} = k \frac{\Delta}{\zeta'}. \tag{10.3}$$

For example, for $k=0.1$ the error $\zeta \approx 0.3\Delta$ is estimated with a relative error reaching 30%.

The errors need not be estimated very accurately, but the error in estimating errors does not exceed 30%. When the errors exceed this limit and taking into account the instability of the measuring instruments, the estimates obtained rapidly become meaningless.

In practice the value $k=0.3$ is often used. Then, as follows from the relation (10.3), an error of only $\zeta \approx \Delta$ can be estimated with an error not exceeding 30%.

It is interesting to extend the foregoing arguments to the case when the measuring instruments have significant random errors.

Random errors cause the indications of instruments to be non-single-valued and they make it difficult both to check and use instruments. If, for example, when checking a pointer-type instrument one need only check whether or not its errors do not exceed the limit established for them, then the measurements must be repeated several times and the largest errors must be found. In many fields of measurement the input to an instrument can be varied continuously. In such cases, in order to determine the largest error at each scale marker checked, two measurements are often sufficient: one by approaching the scale marker from below and the other by approaching from above .

We shall examine a check in which the same quantity is measured simultaneously with a working standard and the instrument being checked. Let y denote the indications of the working standard and x those of the instrument being checked. The difference of the indications of the two devices is

$$z=x-y. \tag{10.4}$$

In the general case

$$x_i=A+\vartheta_x+\psi_{xi}, \quad y_i=A+\vartheta_y+\psi_{yi}, \tag{10.5}$$

where ϑ_x and ϑ_y are the systematic errors and ψ_{xi} and ψ_{yi} are the random errors of the instruments in the ith check.

For the random instrument errors we have

$$M[\psi_{xi}]=0, \quad M[\psi_{yi}]=0.$$

Assume that in order to find the corrections the indications of the instruments are averaged. Using the relations (10.4) and (10.5), we obtain

$$\frac{\sum_{i=1}^{n} z_i}{n}=\vartheta_x+\frac{\sum_{i=1}^{n} \psi_{xi}}{n}-\left(\vartheta_y+\frac{\sum_{i=1}^{n} \psi_{yi}}{n}\right).$$

For a sufficiently large number of observations the effect of the random errors of the instrument being checked becomes insignificant. Hence

$$\frac{\sum_{i=1}^{n} \psi_{xi}}{n} \ll \frac{\sum_{i=1}^{n} z_i}{n}.$$

Assuming that the error of the working standard is also small, we obtain the answer

$$\tilde{C} = -\bar{z}.$$

The obtained estimate was found with an error not less than the systematic error of the working standard. This error can be estimated by the method described in Sec. 5.6. The required number of observations can be found with the help of the criterion presented in the same section. For this, $S(\bar{z})$ must be compared with the limit Δ_s of permissible error of the working standard:

$$\frac{\Delta_s}{S(\bar{z})} \geqslant 7.$$

From here

$$n \approx 7 \frac{\sqrt{\sum_{i=1}^{n} (z_i - \bar{z})^2}}{\Delta_s}.$$

If for the working standard the limit of systematic error θ_s is known

$$|\vartheta_y| \leqslant \theta_s,$$

then in the relations presented θ_s must be substituted for Δ_s.

The checking method studied above is convenient for analysis, but in practice it is avoided, since it is difficult to read accurately fractions of a graduation on the scale of the instrument being checked. The results presented for this method of checking are, however, general. In particular, they will also be valid for the main method of verification, in which the indicator of the instrument being checked is set every time on the scale marker being checked, and the corresponding real value of the measured quantity is found based on the indications of a working standard.

If the check is made with a reduced number of measurements, thanks to the smooth approach from both sides of the same scale marker of the instrument being checked, then the estimate of the correction is found based on the arithmetic mean of the two estimates obtained for the error.

Even though the correction was estimated by averaging the indications of instruments, it can then be introduced into each separate indication. After the correction has been introduced, it can be assumed that for the random errors $M[\psi] = 0$.

When the random errors are significant, it is sometimes desirable to estimate the variance or the standard deviation of this error. If the check is made by measuring an unchanged and known quantity, then Eq. (10.4) is valid and it is obvious that

$$S(x_i) = \sqrt{\frac{\sum_{i=1}^{n} (x_i - \bar{x})^2}{n-1}}$$

and formally there are no difficulties in solving the problem; however, it is difficult to obtain readings x_i that are accurate enough.

If, however, in each observation the indicator of the instrument being checked is set on the same scale marker, then from experiment we will not obtain the data required to solve the problem. In this case, according to Eq. (10.1),

$$\zeta_i' = A_c - y_i', \tag{10.6}$$

and $A_c = \text{const}$. Therefore

$$D[\zeta_i'] = D[y_i']$$

and the estimate obtained for the variance in accordance with this relation depends on both the random error of the instrument being checked and the random error of the working standard. To solve this problem, it is necessary to have an estimate of the standard deviation of the working standard $S(y)$.

Let the same quantity having the true value A be provided for both instruments. Then $A = x - \zeta = y - \gamma$, and therefore

$$\zeta - \gamma = x - y.$$

From here, based on relations (10.2) and (10.4), we obtain

$$\zeta_i' = z_i,$$

where $z_i = x_i - y_i$.

Therefore, for the indications obtained in accordance with formula (10.6) we have

$$D[\zeta_i'] = D[y_i'] = D[x_i] + D[y_i].$$

Correspondingly,

$$S^2(y') = S^2(x) + S^2(y).$$

Knowing $S^2(y)$ and having $S^2(y')$ based on the experimental data, we find

$$S^2(x) = S^2(y') - S^2(y). \tag{10.7}$$

Relation (10.7) indicates that is desirable to know the standard deviation for working standard. However, measuring instruments for which the standard deviations and, for example, the limits of error of the corrections are known, are difficult to use for checking measuring instruments that are to be made for single measurements. In such cases, for the working standard it would be helpful to know, in addition to the characteristics mentioned above, the limits of total error also. For this it is not necessary to determine separately both the components and limits of total error in each check. Obviously, the relation between the total error and its components for each type of measuring instruments is the same. For this reason, if such a relation is established in the course of the investigations, performed, for example, while certifying working standards, then in the future, when measuring instruments are routinely checked, it could be sufficient to determine only part of the errors under study.

10.3. Rejects of verification and ways to reduce their number

Because of the errors of working standards some fraction of serviceable instruments, i.e., instruments whose errors do not exceed the limits established for them, is rejected in a verification—rejection of the first kind—and some fraction of instruments that are in reality unserviceable are accepted—rejection of the second kind. This situation is typical for monitoring production quality and, just as with quality control, here a probabilistic analysis of the procedure is interesting.

Suppose that the same quantity is measured simultaneously by a working standard and the instrument being checked. As pointed out above, for analysis such a scheme is simpler than other schemes, but this is not reflected in the generality of the obtained results. In accordance with the conditions of the experiment we have

$$A=x-\zeta=y-\gamma,$$

where x and y are the indications of the checked and working standard and ζ and γ are the errors of the checked and working standard. From here

$$z=x-y=\zeta-\gamma. \tag{10.8}$$

We are required to show that $|\zeta|\leqslant\Delta$, where Δ is the limit of permissible error of the checked instrument. From the experimental data we can find z; we shall assume that if $|z|\leqslant\Delta$, then the checked instrument is serviceable, and if $|z|>\Delta$, then it is not serviceable.

To perform probabilistic analysis in this way it is necessary to know the probability distribution for the errors of the checked and standard instruments. Let us suppose we know them.

The probability of a rejection of the first kind is

$$p_1=P\{|\zeta-\gamma|>\Delta|_{|\zeta|\leqslant\Delta}\},$$

and the probability of a rejection of the second kind is

$$p_2=P\{|\zeta-\gamma|\leqslant\Delta|_{|\zeta|>\Delta}\}.$$

A rejection of the first kind is obtained for $|\zeta|\leqslant\Delta$, i.e., $\zeta\leqslant\Delta$ and $\zeta\geqslant-\Delta$, when $|\zeta-\gamma|>\Delta$, i.e.,

$$\zeta-\gamma>\Delta, \quad \zeta-\gamma<-\Delta$$

or

$$\gamma<\zeta-\Delta, \quad \gamma>\zeta+\Delta.$$

If the probability distribution of the errors of the checked and working standard are $f(\zeta)$ and $\varphi(\gamma)$, respectively, then

$$p_1=\int_{-\Delta}^{\Delta} f(\zeta)\left(\int_{-\infty}^{\zeta-\Delta}\varphi(\gamma)d\gamma+\int_{\zeta+\Delta}^{+\infty}\varphi(\gamma)d\gamma\right)d\zeta.$$

A rejection of the second kind is possible when $|\zeta|>\Delta$, i.e., when $\zeta>+\Delta$ and $\zeta<-\Delta$. In this case, $|\zeta-\gamma|\leqslant\Delta$, i.e.,

$$\zeta-\gamma\leqslant\Delta, \quad \zeta-\gamma\geqslant-\Delta.$$

From here $\zeta-\Delta\leqslant\gamma\leqslant\zeta+\Delta$. Therefore

$$p_2 = \int_{-\infty}^{-\Delta} f(\zeta)\left(\int_{\zeta-\Delta}^{\zeta+\Delta} \varphi(\gamma)d\gamma\right)d\zeta + \int_{\Delta}^{+\infty} f(\zeta)\left(\int_{\zeta-\Delta}^{\zeta+\Delta} \varphi(\gamma)d\gamma\right)d\zeta.$$

Thus if the probability densities and their parameters are known, then the corresponding values of p_1 and p_2 can be calculated, and their dependence on the relations between the limits of the permissible errors of the standard and checked instruments can be traced.

If, in addition, cost considerations are added, then, one would think, the problem of choosing this relation can be solved uniquely. In reality, when the accuracy of working standards is increased, the cost of the check increases also. A rejection also has a certain cost. Therefore, by varying the limits of error of working standards it is possible to find the minimum losses and this variant is regarded as optimal.

The mathematical relations for solving the problem can be easily derived. Unfortunately, however, in the general case it is impossible to estimate the losses due to the use of instruments whose errors exceed the established limits. In general, it is difficult to express in terms of money the often significant economic effect of increasing measurement accuracy. For this reason, it is only in exceptional cases that economic criteria can be used to justify the choice of the relation between the limits of permissible error of the working standard and checked instruments.

In addition, as has already been pointed out above, the fundamental problem is to determine the probability distribution of the errors of the instruments. The results, presented in Chap. 2, of statistical analysis of data from a check of a series of instruments showed that the sample data are unstable. Therefore the distribution function of the instrument errors cannot be found from these data. However, there are no other data; they simply cannot be obtained anywhere.

Moreover, the fact that the sampling data are unstable could mean that the distribution functions of the errors of the instruments themselves change in time. There are definite reasons for this supposition.

Suppose that the errors of a set of measuring instruments of some type at the moment they are manufactured have a truncated normal distribution with zero mean. For measures (measuring resistors, shunts, weights, etc.) a too large error of the same sign results in certain rejection. This is taken into account when manufacturing measures, and as a result the distribution of the intrinsic errors of measures is usually unsymmetric. For example, if when a weight is manufactured its mass is found to be even slightly less than the nominal mass, then the weight is discarded. Figure 10.1 shows both variants of the distributions.

Instrument errors change in the course of use. Usually the errors only increase. In those cases when, like for weights, the direction of the change of the errors is known beforehand and this is taken into account by the rules of manufacturing, the errors can at first be reduced, but then they will still increase. Correspondingly, changes in the instrument errors deform the distribution functions of the errors. This process, however, does not occur only spontaneously. At the time of routine checks measuring instruments whose errors exceed the established limits are discarded. Figure 10.1 shows the approximate general picture of the changes occurring in the probability distribution in time. The process ultimately terminates when the measuring instruments under study no longer exist: either their errors exceed the established limits or they are no longer serviceable for other reasons.

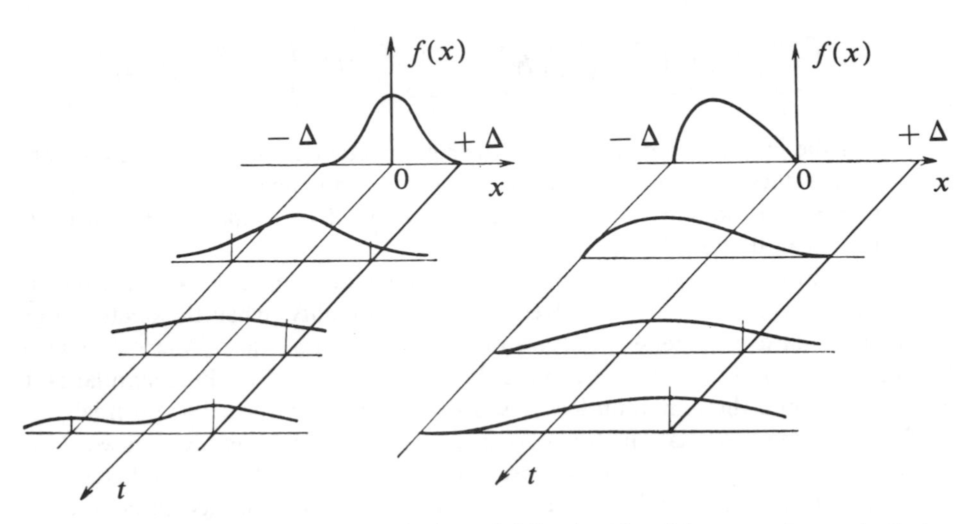

FIG. 10.1. Examples of possible changes in the probability densities of the errors of measuring instruments in time.

The actual picture is still more complicated, since the stock of measuring instruments of each type can also change periodically as a result of the appearance of new measuring instruments.

It should be noted that the properties of measuring instruments such as influence functions and influence factors of different influence quantities, as a rule, do not change with time, and for this reason there is a much better foundation for describing them with the help of a probabilistic model.

The foregoing considerations show that the probabilities of rejection in a check must be calculated very carefully.

It is nonetheless of interest to analyze purely abstract situations and to examine a series of models in order to cast light on the general laws. Such an analysis has been performed by a number of authors.

E. F. Dolinskiĭ obtained the following results under the assumption that the errors of standard and checked instruments have normal distributions.[23]

(i) Rejection in a check depends primarily on the relation between the limit of permissible error of the checked instruments and the standard deviation of the errors of these instruments; as the standard deviation decreases the number of instruments rejected in a check decreases.

(ii) The relation between the errors of the checked and standard instruments (between the permissible limits of their errors or between the standard deviations of these errors) affects the number of rejections in a check much less than do the properties of the distribution of the errors of the checked instruments.

In Ref. 29 the dependences of rejections of the first and second kinds were constructed for normal, uniform, and arcsine distributions of the errors of the checked and standard instruments. It is shown that rejections of the second kind are significantly reduced in number if when checking the instruments the interval of the permissible instrument errors is made smaller than the officially established interval.

Digressing from the statistical instability of the distribution of errors of checked instruments, it should be noted that this approach toward describing the quality of a check has a fundamental drawback. Assume that we have the distribution function of the errors of the checked instrument, i.e., we know the error of the entire collection of instruments. For each specific batch of instruments, however, the number rejected will depend on how many of the instruments in the batch are unserviceable. Therefore the probability of rejection is not a good indicator of the checking effectiveness, since an indicator of the checking effectiveness should not depend on whether the number of bad instruments being checked is large or small.

If batches of instruments were checked, then one could talk about distribution functions for each batch and correspondingly about rejections for each batch and the average number of rejections. But instruments are checked separately or in small batches (several instruments at a time), so that in this approach one cannot talk about distribution functions.

This contradiction can be resolved by resorting to some conditionally chosen distributions. To obtain an estimate of the highest probability of a rejection of the second kind one can, for example, take the distribution of errors of bad instruments, i.e., instruments whose error exceeds the permissible limits. Since in practice bad instruments are not the only instruments that are checked, it is clear that in reality the probability of a rejection of the second kind will always be less than the value obtained by this method.

Analogously, to estimate the upper limit of the probability of a rejection of the first kind one can take some distribution of errors that do not exceed the limits of permissible errors.

When the problem is solved in this manner there arises the problem of choosing the form of the worst distributions. This question cannot be solved objectively, and very many variants can be proposed. Examples of such test distributions are as follows: for "bad" instruments—the symmetric distribution, constructed from positive and negative branches of the normal distribution, separated by 2Δ, with standard deviation $\sigma = \Delta/\sqrt{3}$; for "good" instruments—the uniform distribution with the limits $\pm\Delta$. The distribution of the errors of working standards, out of caution, should be taken as uniform with permissible limits of $\pm\ \Delta_s$.

However, this supposition cannot be justified, and for this reason one cannot insist upon it.

Thus based on the widely used checking method examined above it is impossible to find a sufficiently convincing method for choosing in a well-founded manner the relation between the errors of the standard and the checked instruments. For this reason, in practice this question is solved by a volitional method by standardizing the critical relation between the limits of permissible errors. Thus in electric measuring techniques it is assumed that the error of working standards must not be less than 5 times less than the limit of permissible error of the checked instruments. In standards based on electronic instruments, the accuracy requirements for working standards are not as stringent: this ratio is usually equal to 3. Other ratios (for example, 1:10) are rarely encountered.

The ratios 1:10 and 1:5 usually are not objectionable, but it is often technically difficult to realize them. The ratio of 1:3, however, is always criticized as being inadequate.

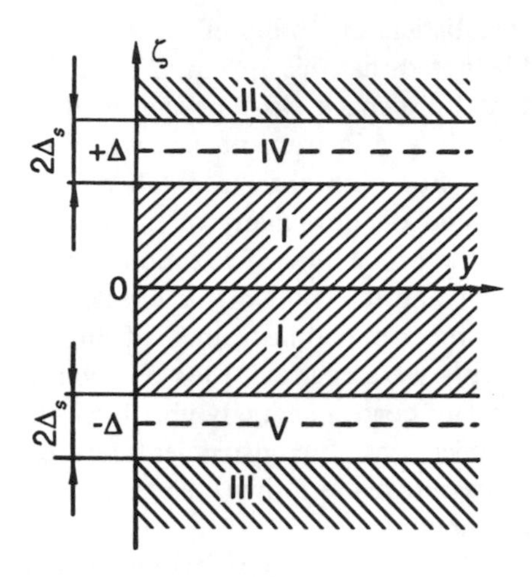

FIG. 10.2. Zones of definite serviceability (I), definite rejection (II and III), and uncertainty (IV and V) when verification of measuring instruments with the limit Δ of permissible error based on a working standard whose limit of permissible error is Δ_s.

Is it possible to choose a different rule for singling out unserviceable instruments, so as to avoid the difficulties connected with justifying on the basis of probability the choice of the ratio between the errors of standard and checked instruments? This problem can, in principle, be solved as follows.

It follows from the definition that a serviceable instrument is an instrument for which $|x-A| \leqslant \Delta$ and an instrument is unserviceable if $|x-A| > \Delta$.

Analogous inequalities are also valid for a working standard: $|y - A| \leqslant \Delta_s$, if the instrument is serviceable and $|y - A| > \Delta_s$ if it is not serviceable.

For $x > A$, for a serviceable instrument $x - A \leqslant \Delta$. But $y - \Delta_s \leqslant A \leqslant y + \Delta_s$. For this reason, replacing A by $y - \Delta_s$, we obtain for a serviceable instrument

$$x-y \leqslant \Delta - \Delta_s. \tag{10.9}$$

Analogously, for $x < A$, for a serviceable instrument

$$x-y \geqslant -(\Delta - \Delta_s). \tag{10.10}$$

Repeating the calculations for an unserviceable instrument, it is not difficult to derive the corresponding inequalities:

$$x-y > \Delta + \Delta_s, \tag{10.11}$$

$$x-y < -(\Delta + \Delta_s). \tag{10.12}$$

Figure 10.2 graphically depicts the foregoing relations. Let the scale of the checked instrument be the abscissa axis. On the ordinate axis we mark the points $+\Delta$ and $-\Delta$, and around each of these points we mark the points displaced from them by $+\Delta_s$ and $-\Delta_s$. If Δ and Δ_s remain the same for the entire scale of the instrument, then we draw from the marked points on the ordinate axis straight lines parallel to the abscissa axis.

Region I corresponds to inequalities (10.9) and (10.10). The instrument for which the differences $x-y$ fall within this region are definitely serviceable irrespective of the ratio of the errors of the standard and checked instruments.

Inequalities (10.11) and (10.12) correspond to the regions II and III. The instruments for which the differences $x-y$ fall within the region II or III are definitely unserviceable.

Some checked instruments can have errors such that

$$\Delta-\Delta_s<|x-y|<\Delta+\Delta_s.$$

These errors correspond to the regions IV and V in Fig. 10.2. Such instruments essentially cannot be either rejected or judged to be serviceable, since in reality they include both serviceable and unserviceable instruments. If they are assumed to be serviceable, then the user will get some unserviceable instruments. This rejection can harm the user. If, however, all such doubtful instruments are rejected, then in reality some serviceable instruments will be rejected. For instruments that are doubtful when they are manufactured or when they are checked after servicing it is best that they be judged unserviceable. This tactic is helpful for anyone using instruments and forces the manufacturers to use more accurate standard instruments. But this is not always possible to do in regular checks.

In those cases when the percentage of doubtful instruments is significant and the instruments are expensive and are difficult to fix, it is best to check them again. Here several variants are possible. One variant is to recheck the doubtful instruments with the help of more accurate working standards.

In those cases when more accurate instruments cannot be used for one reason or another, the check can also be made with the help of other samples of working standards that are rated at the same accuracy as those used in the initial check. Since different working standards have somewhat different errors, the results of comparing the checked instruments with them will be somewhat different. As a result of this, some of the doubtful instruments will be judged absolutely serviceable and some will be confidently rejected.

The best method is to increase the accuracy of the working standard. However, there then arises the question as to how much the accuracy of the standard instruments should be increased.

If there are no technical limitations, then the accuracy of the working standard can be increased until the instrument being checked can be judged as being either serviceable or unserviceable. If, however, the limit of permissible error of the standard instrument becomes 5–10 times less than the limit of permissible error of the checked instrument, then the accuracy of the working standard should not be increased further: the errors of instruments are usually not stable enough to be estimated with high accuracy.

For a 5–10-fold difference in the errors, the error of working standard can usually always be neglected. This practice can be justified by the fact that in this case the probability of rejection is always low (because the zone of uncertainty is narrow, the percentage of instruments incorrectly judged to be serviceable or rejected is always low) as well by the fact that only those instruments whose errors do not differ much from the limit established for them can be incorrectly judged as serviceable.

Rejection of instruments in checks is eliminated completely if instead of verification the instruments are recalibrated. The accuracy of the new calibration characteristic can be almost equal to the accuracy of the working standard. This makes this method extremely attractive.

The drawback of this method is that the new calibration characteristic is most often constructed with the help of a table of corrections to the old calibration characteristic. This is not convenient for using the instrument. More importantly, in this method the stability of the instrument is concealed. The possessor of the instrument must accumulate calibration results and analyze them. Analysis makes it possible to judge the stability of an instrument, how often the instrument should be recalibrated, and thereby the desirability of continued use of the instrument, if it must be calibrated very often.

10.4. Calculation of a necessary number of standards

Calibration, testing, and verification are metrological operations, with whose help the dimensions of decreed units of physical quantities are transferred to all measuring instruments. The units themselves, however, are reproduced with the help of reference standards.

Reference standards are not created for all units. The circumstances under which reference standards need to be created deserve discussion.

First, we note that reference standards are always necessary for the units of the basic quantities. The question of whether or not it is desirable to create reference standards pertains only to the units of derived quantities. Derived physical quantities include quantities measured only by indirect methods, for example, the area. It is clear that reference standards are not required for the units of such quantities.

But reference standards are also not always required for the units of quantities measured by direct methods. A reference standard is not necessary if the instrument used to measure a given quantity can be checked with adequate accuracy and efficiency with the help of working standards for other quantities. For example, to check tachometers, it is sufficient to have a device for rotating the shaft of the tachometer and a stroboscopic timer; a standard is not required in this case.

When a reference standard is created, the reproduction of the unit is centralized. On one hand, this complicates the measures that must be taken to ensure unity of measuring instruments, since some of the standard measuring instruments (at a minimum, the working reference standards) must be compared with the reference standard. On the other hand, it is found that usually complicated indirect measurements, with whose help the starting primary setups are certified when the reproduction of units is not centralized, are possible only for certifying one measuring instrument—the primary reference standard.

The question of whether or not a reference standard should be created is answered by comparing these contradictory factors. Thus the solution of this question is based on technical and economic considerations; this is why it is difficult to solve the problem.

It should also be noted that in many cases indirect measurements, required for reproducing the units of a derived quantity, do not provide the necessary accuracy. The creation of a reference standard, i.e., centralization of reproduction of the unit, makes it possible in this case to achieve greater unity of measuring instrument than in the absence of a reference standard, since when the size of the unit reproduced with the help of the reference standard is transferred the systematic error of the standard can be neglected. This circumstance is often exploited, although it is possible that the most accurate measuring instruments in one country can have a significant systematic error compared with the analogous measuring instruments in another country. Comparing reference standards from different countries makes it possible to avoid misunderstandings that can arise because of this.

The sizes of the units reproduced with the help of primary reference standards are transferred to the working standards with the help of a system of standards. The metrological coordination of standards, their relation with the working instruments, and the principles of the methods of comparison employed in the C.I.S. are customarily represented with the help of so-called checking or calibration schemes.

Physical standards are divided into ranks. The number of a rank indicates the number of steps included in transferring the size of a unit from the primary reference standard to a given working standard.

One of the most difficult questions arising in the construction of checking schemes is the question of how many ranks of standards should be provided. As the number of ranks increases, the error with which the size of a unit is transferred by the working measuring instrument increases. For this reason, in order to obtain high accuracy the number of ranks of standards should be reduced to a minimum.

The higher the accuracy of standards, the more expensive they are. In addition, more accurate measurements usually are more difficult to perform. Increasing the number of ranks makes it possible to have less accurate standards together with more accurate standards and makes the entire system of transferring the size of a unit more economical. For this reason, in the fields of measurement where there is a large margin in the accuracy of reference standards, the number of ranks of standards can be equal to the number of gradations of accuracy of the working measuring instruments.

Enlarging the stock of standards and increasing the number of ranks, i.e., the number of gradations of accuracy, make the work of calibration laboratories more difficult and involve a certain cost. At the same time the operations of calibration of measuring instruments themselves usually become more efficient. One would think that it is possible to find an economically optimal number of ranks of the checking scheme. This, however, requires information about the dependence of the cost of the equipment and labor on the accuracy. This information is usually not available. For this reason, in practice the optimal checking schemes cannot be determined.

Checking schemes are usually constructed when reference standards and working standards are partially already available and it is only necessary to arrange them in a hierarchical order. In this case, it can be assumed that the number of working measuring instruments, the frequency with which they must be calibrated, and the permissible number of annual comparisons of the most accurate measuring instruments with the reference standard are known. In addition, the time required to calibrate one sample of each type of measuring instrument and working standard or the limiting number of calibration permitted by the reference standards within a prescribed period of time can be estimated. This information makes it possible to

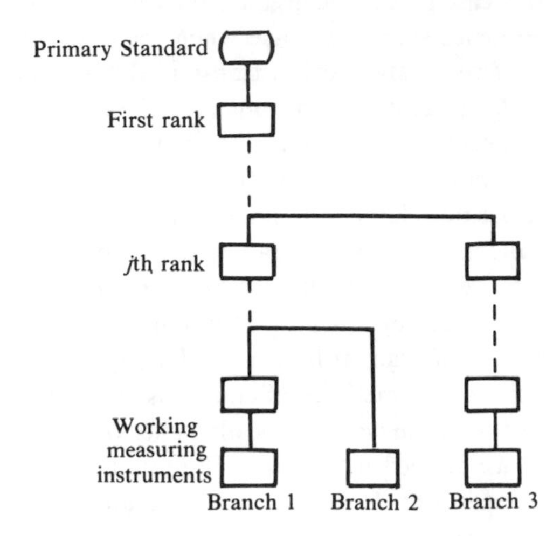

FIG. 10.3. Typical structure of checking schemes.

find the minimum necessary number of ranks of the checking scheme. The problem can be solved by the method of successive approximations.

In the general case checking schemes can be assumed to have the structure shown in Fig. 10.3. We shall first study the case when the checking scheme has only one vertical (i.e., it does not have the branches 2 and 3 shown in Fig. 10.3).

If the jth rank has N_j standards, then the maximum number of standards in the rank $(j+1)$ will be

$$N_{j+1}=N_j\frac{\eta_j T_{j+1}}{t_{j+1}}, \tag{10.13}$$

where η_j is the utilization factor of the standards of rank j, T_{j+1} is a time equal to the time interval between calibrations of the measuring instrument of rank $j+1$, and t_{j+1} is the time necessary to calibrate one measuring instrument in the rank $(j+1)$.

When calculating the coefficients η_j the utilization time of the measuring instrument must be compared with the calendar time, and the losses of working time must be taken into account. For example, if some apparatus is used eight hours per day and one hour is required for preparation and termination, and in addition preventative maintenance, servicing, and checking reduce the working time by 10%, then

$$\eta=\frac{8-1}{24}\times 0.9=0.2625.$$

Transferring, in accordance with the checking scheme, from the primary reference standard $(j=0, N_0=1)$ to the working measuring instrument, we determine the maximum number of standards of each rank and then the number of working measuring instruments N_m guaranteed by calibration:

$$N_m = N_0 N_1 \cdots N_{m-1} = \prod_{j=0}^{m-1} \eta_j \frac{T_{j+1}}{t_{j+1}}, \tag{10.14}$$

where m is the total number of steps in transferring the size of a unit from the reference standard to the working measuring instrument, inclusively.

To solve the problem we first choose some number of ranks $j_0 = m - 1$. In principle, it is possible to start with the minimum number of ranks $j_0 = 1$. For given j_0 we find $N_m^{(0)}$. If $N_m^{(0)}$ is less than the number of working measuring instruments that need to be calibrated, then either the number of ranks or η_j—the utilization factor of the standards—and the efficiency of the calibration operations must be increased, i.e., the time t_j $(j = 0,1)$ must be reduced. If the problem cannot be solved in this manner, then the number of ranks must be increased.

We calculate $N_m^{(1)}$ for the new number of ranks. As soon as the value of $N_m^{(i)}$ obtained is greater than the number of working measuring instruments that must be calibrated, the number of ranks can be regarded as sufficient.

For a more general checking scheme (Fig. 10.3) the possibilities of each branch of the checking scheme must be checked. It is best to start the calculation with the branch adjoining a standard of high rank, and to perform the calculation in the reverse order as compared with the method examined above, i.e., it is best to start from a fixed number of working measuring instruments in each class that must be checked along a given branch. We shall use formula (10.13) to calculate the number of working standards N_j for the lth branch that is required to calibrate N_{j+1} measuring instruments. For this it is useful to write formula (10.13) in the form

$$N_{jl} = N_{j+1,l} \frac{t_{j+1,l}}{\eta_{jl} T_{j+1,l}}.$$

The calculation proceeds up to standards servicing several branches of the checking scheme. In the process, the number of these standards that is required for the lth branch is determined. Then, for the rest of the checking scheme there remain $N_j - N_{jl}$ standards of the given rank. The number of measuring instruments that must be calibrated according to the vertical of the scheme (branch 1 in Fig. 10.3) is found using formula (10.14) taking into account the branching losses:

$$N_m^{(1)} = N_0 (N_1 - N_{12}) \cdots (N_j - N_{jl}) \cdots N_{m-1}.$$

The number $N_m^{(1)}$ obtained must be greater than the prescribed number of measuring instruments that must be calibrated along this branch. This condition is necessary for all branches. If it is not satisfied, then the number of ranks must be increased. As the number of ranks increases, the efficiency of the checking network, represented by the checking scheme, increases rapidly. The checking schemes employed have the maximum number (five) of ranks of standards, even for the most developed fields of measurement.

The relations presented above pertained to the simplest case, when at each step of transfer of the size of the unit the period of time between calibrations and the calibration time itself were the same for all measuring instruments. In reality, these time intervals can be different for different measuring instruments. Taking all this into account makes the calculations more complicated, but does not change their essential features.

It is necessary to transfer from different time intervals between calibrations to one conditionally standard T_{cs} time interval and to find the number of measuring instruments of each type N_k^{cs} that must be checked within this period. This is done with the help of the obvious formula

$$N_k^{cs} = N_k \frac{T_{cs}}{T_k}.$$

Next it is necessary to find the average time t_j^{av} required to check one measuring instrument for each step of the checking scheme:

$$t_j^{av} = \frac{\sum_{k=1}^{n} t_k N_k^{cs}}{\sum_{k=1}^{n} N_k^{cs}}. \tag{10.15}$$

Here n is the number of different types of measuring instruments at the jth step of the checking scheme.

We shall give a numerical example. Suppose it is required to organize a calibration of instruments of the types A and B and the following data are given.

(1) Instruments of type A: $N_A = 3 \times 10^4$; the time interval between calibrations $T_{A1} = 1$ yr for $N_{A1} = 2.5 \times 10^4$ and $T_{A2} = 0.5$ yr for $N_{A2} = 5 \times 10^3$; the calibration time $t_A = 5$ h.

(2) Instruments of type B: $N_B = 10^5$; $T_B = 1$ yr; the calibration time $t_B = 2$ h.

(3) Primary reference standard: Four comparisons per year are permitted; the frequency of the calibration of the most accurate measuring instruments, which can be working standards of rank 1, is 2 yr, i.e., $T_1 = 2$ yr; for them, $\eta_1 = 0.25$. For measuring instruments that can be working standards of rank 2, $T_2 = 2$ yr, $t_2 = 40$ h, and $\eta_2 = 0.25$.

The possible number of first-rank standards is

$$N_1 = N_0 f T_1 = 8,$$

since $N_0 = 1$, $f = 4$ is the maximum number of comparisons with a reference standard per year, and $T_1 = 2$.

It is obvious that eight standards are not enough to check 130 000 working instruments. We shall check to see if three ranks of standards are sufficient for this.

Since the time interval between checks is different for different instruments, we introduce the conditionally standard time interval between checks $T_{cs} = 1$ yr and find the number of instruments that must be checked within this time period. Conversion is necessary only for instruments of the type A with $T_{A2} = 0.5$ yr:

$$N_{A2}^{cs} = N_{A2} \frac{T_{cs}}{T_{A2}} = 5 \times 10^3 \times \frac{1}{0.5} = 10 \times 10^3.$$

Therefore

$$\sum_{k=A,B} N_k^{cs} = N_{AB} = N_{A1} + N_{A2}^{cs} + N_B = 135 \times 10^3$$

instruments must be calibrated within the time T_{cs}.

Different amounts of time are required to check instruments of the types A and B. We shall find the average checking time t_w^{av} of these working instruments. In accordance with formula (10.15)

$$t_w^{av} = \frac{(N_{A1} + N_{A2}^{cs})t_A + N_B t_B}{N_{AB}} = \frac{35 \times 10^3 \times 5 + 100 \times 10^3 \times 2}{135 \times 10^3} = 2.78 \text{ h.}$$

Now, using formula (10.13), we shall find the required number of second-rank standards:

$$N_2^{(1)} = \frac{N_{AB} t_w^{av}}{\eta_2 T_{cs}} = \frac{135 \times 10^3 \times 2.78}{0.25 \times 6 \times 10^3} = 250.$$

Here it was assumed that $T_{cs} = 250 \times 24 = 6 \times 10^3$ h.

It remains to verify that all working standards of the second rank can be checked. For this, we calculate from formula (10.13) the maximum possible number of standards of this rank:

$$N_2 = N_1 \frac{\eta_1 T_2}{t_2} = 8 \times \frac{0.25 \times 2 \times 6 \times 10^3}{40} = 600.$$

Since $N_2 > N_2^{(1)}$, in this case two ranks of working standards are sufficient.

With the help of calculations similar to those presented in the foregoing example it is possible to choose in a well-founded manner the structure of a checking scheme and to estimate the required number of working standards of each rank.

In calculating the checking scheme we did not take into account explicitly the accuracy of the measuring instruments. However, the contemplated scheme must be metrologically realizable. This means that the discrepancy between the accuracy of the primary reference standard and the accuracy of the working measuring instruments must make it possible to insert between them the required number of ranks of working standards. The problem facing instrument makers and metrologists is to provide the combination of accuracy and efficiency required to implement the checking scheme by designing working standards and reference standards.

Checking schemes usually have extra calibration possibilities. This makes it possible to distribute reference and working standards so as to limit their transport, to maximize the efficiency of calibration laboratories, and to take into account other practical considerations.

Chapter 11

Methods for calculating the correlation coefficient and accounting of dependencies between components of measurement uncertainty

11.1. Introduction

To calculate the uncertainty of an indirect measurement result, it is necessary to take into consideration possible dependencies between random errors of measurement results of the arguments. A common way to do that is by estimating and taking into account the correlation coefficient between the above errors.

While the mathematical foundations for this problem are well-developed, it still causes much difficulty and confusion in practice. We agree with R. H. Dieck that "probably one of the most misunderstood and misused statistics is the correlation coefficient."* Thus, the first goal of this chapter is to give formulas for *practical* calculations of the correlation coefficient as applied to measurements and to show the relation of these practical methods with the formal mathematical foundations on which they are based.

Our second goal is to illustrate the advantages of an alternative method of estimation of uncertainty of indirect measurements. This method, which we call a *method of reduction*, is described in section 6.2, with an example given in section 6.6 of this book. This method reduces the problem of obtaining the estimation of the indirect measurement result and its uncertainty to the problem of estimation of measurement results and uncertainties of a set of direct measurements. The methods for solving the latter are well-understood and simple.

In cases where using correlation coefficient is possible, the method of reduction gives the same results in a much simpler way. In addition, our method is free of some of the limitations of the correlation coefficient method. In fact, we argue the method of reduction presented here can be considered a general method for processing experimental data in indirect measurements.

We will limit the discussion here to the methods of obtaining the estimations of results of indirect measurements and their standard deviation. In practice, the next necessary steps are the estimation of systematic and random components of the

*R. H. Dieck. *Measurement Uncertainty. Methods and Applications.* ISA, 1992.

uncertainty of the measurement result, and then calculation of the overall combined uncertainty. The detailed treatment of these problems can be found in sections 6.4 and 6.6 in this book.

11.2. Mathematical basics

Mathematical formulas that define the essence of the correlation coefficient can be found in many books on the Theory of Probability and Mathematical Statistics. I refer to *Statistics: Probability, Inference and Decision* by Winker and Hays.[60]

Consider two random quantities X and Y with mathematical expectations equal to zero ($M[X]=0$ and $M[Y]=0$) and finite variances. Denote their variances as $D[X]=\sigma_X^2$ and $D[Y]=\sigma_Y^2$.

The variance of a random quantity $Z=X+Y$ can be calculated using the equation

$$D[Z]=M[(X+Y)^2]=M[X^2]+M[Y^2]+2M[XY].$$

The last term $M[XY]$ is named *second mixed moment* or *covariance*.

The covariance divided by the square root of the product of variances $\sigma_X^2\sigma_Y^2$ gives the correlation coefficient ρ_{XY}:

$$\rho_{XY}=\frac{M[XY]}{\sigma_X\sigma_Y}. \tag{11.1}$$

The value of the correlation coefficient always lies within $[-1,+1]$, and if $|\rho_{XY}|=1$ then there is linear functional dependency between X and Y. When $|\rho_{XY}|=0$, X and Y are not correlated. But it does not mean they are independent. Otherwise, when $0<|\rho_{XY}|<1$, the nature of the dependency between X and Y cannot be determined unambiguously: it can be stochastic as well as functional non-linear dependency. Therefore, in the last case, if the knowledge about the nature of the dependency between X and Y is required, it can only be obtained based on physical properties of the problem rather than inferred mathematically.

From above formulas we obtain

$$\sigma_Z^2=\sigma_X^2+\sigma_Y^2+2\rho_{XY}\sigma_X\sigma_Y. \tag{11.2}$$

In practice, we have to work not with the exact values of parameters of random quantities but with their estimations. So, instead of variances σ_Z^2, σ_X^2, σ_Y^2 and the correlation coefficient ρ_{XY}, we have to use their estimations S_Z^2, S_X^2, S_Y^2 (we will also use interchangeably $S^2(A)$ to denote an estimation of the variance of random quantity A), and r_{XY}. If n is the number of measured pairs (x_i, y_i) of random quantities X and Y $(i=1,\ldots,n)$, and $\bar{x}$ and $\bar{y}$ are averages over n observed values of X and Y, then

$$S_X^2=\frac{\Sigma_{i=1}^n(x_i-\bar{x})^2}{n-1}; \quad S_Y^2=\frac{\Sigma_{i=1}^n(y_i-\bar{y})^2}{n-1}. \tag{11.3}$$

The estimation of $M[XY]$ will be

$$m_{XY}=\frac{\Sigma_{i=1}^n(x_i-\bar{x})(y_i-\bar{y})}{n}.$$

Then, $r_{XY}=m_{XY}/S_X S_Y$.

Thus, the formulas for calculations are:

$$r_{XY}=\frac{\Sigma_{i=1}^{n}(x_i-\bar{x})(y_i-\bar{y})}{nS_X S_Y}; \tag{11.4}$$

$$S_Z^2=S_X^2+S_Y^2+2r_{XY}S_X S_Y. \tag{11.5}$$

The estimation of the variance of an average value $\bar{x}=1/n\ \Sigma_{i=1}^{n}x_i$ is known to be

$$S_{\bar{x}}^2=\frac{S_X^2}{n}. \tag{11.6}$$

Then, by dividing Eq. 11.5 by n, we obtain the estimation of the variance of the mean value of Z:

$$S_{\bar{z}}^2=S_{\bar{x}}^2+S_{\bar{y}}^2+2r_{XY}S_{\bar{x}}S_{\bar{y}}. \tag{11.7}$$

The correlation coefficient estimation here is the same as in Eq. 11.4. One can also use $S_{\bar{x}}$ and $S_{\bar{y}}$ for the calculation of the correlation coefficient estimation utilizing the fact that $S_X S_Y=nS_{\bar{x}}S_{\bar{y}}$. Then, the Eq. 11.4 will change to the following:

$$r_{XY}=\frac{\Sigma_{i=1}^{n}(x_i-\bar{x})(y_i-\bar{y})}{n^2 S_{\bar{x}}S_{\bar{y}}}. \tag{11.8}$$

It is necessary to stress that the number of realizations of X and Y (e.g., the number of measurements of X and Y) must be the same. Moreover, each pair of these realizations must be obtained under the same conditions, for example, at the same time, at the same temperature, using measuring instruments with the same dynamic characteristics (in the case of dynamic measurements), etc.

11.3. Indirect measurements and Taylor's series

Consider an indirect measurement with a measurand A_t and measuring arguments $\{A_j\}$ $(j=1,...,m)$ and assume that the dependency between A_t and $\{A_j\}$ is known:

$$A_t=f(A_1,...,A_m). \tag{11.9}$$

Having estimations $\tilde{A}_j$ of A_j and using Eq. 11.9, we obtain estimation $\tilde{A}_t$ of A_t:

$$\tilde{A}_t=f(\tilde{A}_1,...,\tilde{A}_m).$$

Inaccuracy of this estimation is usually calculated using Taylor's series. Assuming for simplicity the case of $m=2$, we obtain:

$$\zeta_t=f(\tilde{A}_1,\tilde{A}_2)-f(A_1,A_2)=\frac{\partial f}{\partial A_1}\zeta_1+\frac{\partial f}{\partial A_2}\zeta_2+R_2,$$

where ζ_t, ζ_1, and ζ_2 are the errors of $\tilde{A}_t$, $\tilde{A}_1$, and $\tilde{A}_2$ correspondingly, and R_2 is a residual term that can be usually neglected provided the errors ζ_1 and ζ_2 are small. Then, the variance of error ζ_t is:

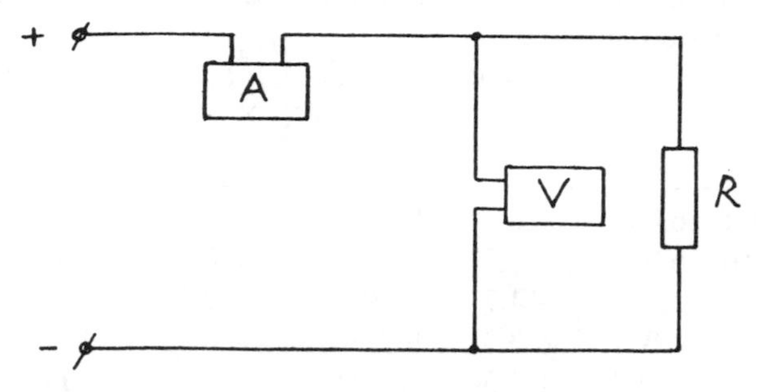

FIG. 11.1. The connections for the indirect measurement of an electrical resistance.

$$D[\zeta_t]=M\left[\left(\frac{\partial f}{\partial A_1}\zeta_1+\frac{\partial f}{\partial A_2}\zeta_2\right)^2\right].$$

With W_1 and W_2 to denote the influence coefficients $\partial f/\partial A_1$ and $\partial f/\partial A_2$, we obtain:

$$D[\zeta_t]=W_1^2M[\zeta_1^2]+W_2^2M[\zeta_2^2]+2W_1W_2M[\zeta_1\zeta_2].$$

Note that since all measuring arguments A_j are constant, the only random component in the estimates $\tilde{A}_j$ are ζ_j. Therefore, the estimations of variances of $\tilde{A}_j$, $S^2(\tilde{A}_j)$, are equal to those of ζ_j, $S^2(\zeta_j)$. Then, the estimation of the variance of $\tilde{A}_t$ is:

$$S^2(\tilde{A}_t)=W_1^2S^2(\tilde{A}_1)+W_2^2S^2(\tilde{A}_2)+2r_{1,2}W_1W_2S(\tilde{A}_1)S(\tilde{A}_2), \quad (11.10)$$

where $r_{1,2}$ is determined by Eq. 11.4 or 11.8.

If there are not two but three arguments ($m=3$), the Eq. 11.10 becomes more complicated:

$$S^2(\tilde{A}_t)=W_1^2S^2(\tilde{A}_1)+W_2^2S^2(\tilde{A}_2)+W_3^2S^2(\tilde{A}_3)+2r_{1,2}W_1W_2S(\tilde{A}_1)S(\tilde{A}_2)$$
$$+2r_{1,3}W_1W_3S(\tilde{A}_1)S(\tilde{A}_3)+2r_{2,3}W_2W_3S(\tilde{A}_2)S(\tilde{A}_3). \quad (11.11)$$

So, with the number of arguments increasing, the complexity of calculations increases rapidly.

11.4. Example: An indirect measurement of the electrical resistance of a resister

Consider the measurement of electrical resistance using an ammeter and a voltmeter. The connections of the instruments and the resister are shown in Fig. 11.1. Assume that the measurement was performed under reference conditions for the instruments, and the input resistance of the voltmeter is so high that its influence on the accuracy of the measurement can be neglected.

The results of measurements of the strength of current and voltage are given in Table 11.1. In accordance with the note given at the end of section 11.2, all results from Table 11.1 were obtained in pairs, which means here that the results with the same number were obtained at the same time.

TABLE 11.1 Data processing for indirect measurement of electrical resistance using the correlation analysis.

Num.	I_i A	U_i V	$(I_i-\overline{I})$ $\times 10^{-5} A$	$(I_i-\overline{I})^2$ $\times 10^{-10} A^2$	$(U_i-\overline{U})$ $\times 10^{-3} V$	$(U_i-\overline{U})^2$ $\times 10^{-6} V^2$	$(I_i-\overline{I})(U_i-\overline{U})$ $\times 10^{-8} AV$
1	2	3	4	5	6	7	8
1	0.05996	6.003	− 3.7	13.69	+ 2.82	7.95	− 10.4
2	0.06001	6.001	+ 1.3	1.69	+ 0.82	0.67	+ 1.1
3	0.05998	5.998	− 1.7	2.89	− 2.18	4.75	+ 3.7
4	0.06003	6.001	+ 3.3	10.89	+ 0.82	0.67	+ 2.7
5	0.06001	5.997	+ 1.3	1.69	− 3.18	10.11	− 4.1
6	0.05998	5.999	− 1.7	2.89	− 1.18	1.39	+ 2.0
7	0.06003	6.004	+ 3.3	10.89	+ 3.82	14.59	+ 12.6
8	0.05995	5.997	− 4.7	22.09	− 3.18	10.11	+ 14.9
9	0.06002	6.001	+ 2.3	5.29	+ 0.82	0.67	+ 1.9
10	0.06001	6.003	+ 1.3	1.69	+ 2.82	7.95	+ 3.7
11	0.05999	5.998	− 0.7	0.49	− 2.18	4.75	+ 1.5
Sum	0.65997	66.002		74.19		63.61	+ 29.6

The Eq. 11.9 now gets the form

$$R=\frac{U}{I}. \tag{11.12}$$

The influence coefficients are

$$W_1=\frac{\partial R}{\partial U}=\frac{1}{I}, \quad W_2=\frac{\partial R}{\partial I}=-\frac{U}{I^2}.$$

Using the values of $n\overline{U}$ and $n\overline{I}$ given in Table 11.1 (columns 2 and 3, the last row), we obtain from Eq. 11.12 the estimation for R:

$$\tilde{R}=\frac{\overline{U}}{\overline{I}}=\frac{n\overline{U}}{n\overline{I}}=\frac{66.002}{0.65997}=100.0075\approx 100.01\Omega.$$

Now we must calculate the variance and the standard deviation of this result.

First, we will estimate the variances of $\overline{I}$, $\overline{U}$, their standard deviations, and the correlation coefficient. Using Eqs. 11.3 and 11.6, we obtain:

$$S^2(\overline{I})=\frac{\Sigma_{i=1}^{n}(I_i-\overline{I})^2}{n(n-1)}=\frac{74.19\times 10^{-10}}{11\times 10}=0.674\times 10^{-10}A^2;$$

$$S^2(\overline{U})=\frac{\Sigma_{i=1}^{n}(U_i-\overline{U})^2}{n(n-1)}=\frac{63.61\times 10^{-6}}{11\times 10}=0.578\times 10^{-6}V^2.$$

The estimations of standard deviations are

$$S(\overline{I})=0.82\times 10^{-5}A; \quad S(\overline{U})=0.76\times 10^{-3}V.$$

Using Eq. 11.8, we obtain the correlation coefficient

$$r_{I,U}=\frac{\Sigma_{i=1}^{n}(I_i-\bar{I})(U_i-\bar{U})}{n^2S(\bar{I})S(\bar{U})}=\frac{29.6\times10^{-8}}{121\times0.82\times10^{-5}\times0.76\times10^{-3}}=0.39.$$

It is interesting to note that this value is statistically insignificant. Indeed, applying a standard method of,[21,60] we can check the hypothesis H_0: $\rho_{I,V}=0$ against H_1: $\rho_{I,V}\neq0$. The degree of freedom here is $\nu=11-2=9$, and we will take the significance level to be $q=0.05$ as usual. This gives the critical values $t_q=2.26$ and $r_q=t_q/\sqrt{t_q^2+\nu}=0.60$. Because $0.39<0.60$, we must accept H_0 and conclude that the obtained value $r_{I,V}=0.39$ is not significant. This means that, when the number of measurements n increases, the estimation $r_{I,V}$ of the correlation coefficient will in general decrease. However, it does not mean that the value of $r_{I,V}$ obtained for a *specific sample* can be neglected. To the contrary, it must be taken into consideration when calculating the estimation of variance for that sample.

In our example, inserting the obtained values into Eq. 11.10, we can calculate the desired estimation $S(\tilde{R})$:

$$S^2(\tilde{R})=\left(\frac{\bar{U}}{\bar{I}^2}\right)^2\times S^2(\bar{I})+\frac{1}{\bar{I}^2}\times S^2(\bar{U})-r_{I,U}\frac{\bar{U}}{\bar{I}^2}\times\frac{1}{\bar{I}}\times S(\bar{I})S(\bar{U})$$

$$=\left(\frac{6}{36\times10^{-4}}\right)^2\times0.674\times10^{-10}+\frac{1}{36\times10^{-4}}\times0.578\times10^{-6}$$

$$-2\times0.39\times\frac{6}{36\times10^{-4}}\times\frac{1}{6\times10^{-2}}\times0.82\times10^{-5}\times0.76\times10^{-3}$$

$$=1.87\times10^{-4}+1.61\times10^{-4}-1.35\times10^{-4}=2.13\times10^{-4}\Omega^2, \qquad (11.13)$$

and

$$S(\tilde{R})=\sqrt{S^2(\tilde{R})}=1.46\times10^{-2}\Omega.$$

11.5. Method of reduction: A general way of accounting for the dependencies between errors of measuring arguments

We now turn to an alternative method of accounting for the dependencies between argument errors in indirect measurements, which we call the *method of reduction*. This method does not rely on the estimation of correlation coefficients. As we saw in section 11.3, the method based on the correlation coefficients involves complex calculations, with the complexity increasing rapidly when the number of arguments increases. Moreover, as one of the examples in this section will demonstrate, the correlation coefficient-based method is not applicable in certain situations, which currently have to be dealt with using a variety of different methods.* Our method of reduction is free of these limitations of the correlation coefficient-based method and thus provides a general and uniform way of obtaining the results of indirect measurements and estimating their uncertainty.

---

*In fact, these limitations of the correlation coefficient-based method are often overlooked, which leads to much confusion and wrong results.

TABLE 11.2. Data processing for indirect measurement of electrical resistance using the method of reduction.

Num.	I_i A	U_i V	R_i Ω	$(R_i-\overline{R})$ Ω	$(R_i-\overline{R})^2$ $\times 10^{-2}\ \Omega^2$
1	2	3	4	5	6
1	0.05996	6.003	100.117	+0.109	1.188
2	0.06001	6.001	100.000	−0.002	0
3	0.05998	5.998	100.000	−0.002	0
4	0.06003	6.001	99.967	−0.041	0.168
5	0.06001	5.997	99.933	−0.075	0.562
6	0.05998	5.999	100.017	+0.009	0.008
7	0.06003	6.004	100.017	+0.009	0.008
8	0.05995	5.997	100.033	+0.025	0.0625
9	0.06002	6.001	99.983	−0.025	0.0625
10	0.06001	6.003	100.033	+0.025	0.0625
11	0.05999	5.998	99.983	−0.025	0.0625
Sum			1100.083		2.184

The new method is very simple. Let $A_{1i},...,A_{mi}$ be measurement results of the arguments obtained by a coordinated measurement *i*. By inserting these measurement results into Eq. 11.9, we get one value of the measurand A_{ti}. For *n* coordinated measurements of the arguments, we will similarly obtain *n* values of the measurand A_t. This set $\{A_{ti}, i=\overline{1,n}\}$ can be used to calculate the uncertainty of the measurement result as if it was obtained by direct measurements of A_t, utilizing simple and well-understood methods for direct measurements.

Let us first apply this method to the example from section 11.4. The initial data are repeated in columns 2 and 3 of Table 11.2. The calculated values of R_i, ($i = 1,...,11$) are given in column 4. Treating these values as if they were obtained by direct measurements, we obtain immediately the estimation of R as

$$\overline{R}=\frac{1}{n}\sum_{i=1}^{n} R_i=100.75\approx 100.01\Omega$$

and the estimation of its variance and standard deviation as

$$S^2(\overline{R})=\frac{1}{n(n-1)}\sum_{i=1}^{n}(R_i-\overline{R})^2=\frac{2.184\times 10^{-2}}{11\times 10}=1.99\times 10^{-4}\Omega^2;$$

$$S(\overline{R})=1.41\times 10^{-2}\Omega.$$

The coincidence of $\widetilde{R}$ from section 11.4 and $\overline{R}$ is natural in this case. Note also that the difference between $S(\widetilde{R})$ and $S(\overline{R})$ is negligibly small.

Now let us slightly change the example from section 11.4. Again, the electrical resistance of a register is measured indirectly by measuring the voltage and the strength of current using connections shown in Fig. 11.1. But now, while *n* pairs of values $\{U_i,I_i\}$, $i=\overline{1,n}$ were being obtained, the voltage was changing from U_0 to U_f.

Since the electrical resistance does not depend on the value of voltage, it is still possible to measure the resistance in this case. However, the correlation coefficient will now reflect the functional dependency between U_i and I_i rather than the relationship between random errors of measurements of these quantities. Therefore, the correlation analysis cannot be applied here, and a different approach should be used. In particular, in our case, we can exploit the fact that

$$U_i = RI_i, \tag{11.14}$$

and so the dependency between U_i and I_i is linear. Therefore, the estimation of R can be obtained using the regression method.[60]

The equation of the linear regression is

$$Y = \beta_1 + \beta_2 X. \tag{11.15}$$

The comparison of Eq. 11.14 and 11.15 gives

$$Y = U,\ X = I,\ \beta_1 = 0,\ \beta_2 = R.$$

Applying the regression method, we obtain

$$\tilde{R} = \frac{n\Sigma_{i=1}^{n} U_i I_i - \Sigma_{i=1}^{n} U_i \Sigma_{i=1}^{n} I_i}{n\Sigma_{i=1}^{n} I_i^2 - (\Sigma_{i=1}^{n} I_i)^2};$$

$$S(\tilde{R}) = \frac{S_Y(1 - r_{XY})}{S_x\sqrt{n-2}} = \frac{S_U(1 - r_{I,U})}{S_I\sqrt{n-2}}.$$

In addition to having to apply a different method in this case, a disadvantage of the above method is that the correlation coefficient $r_{I,U}$ here is very close to 1. It is not equal exactly to 1 only because of the errors in measurements of U_i and I_i. So, the accuracy of calculation of $r_{I,U}$ needs special attention.

On the other hand, we can apply here the method of reduction to obtain R and S_R in exactly the same way as in the case of a constant voltage. Indeed, n pairs of $\{U_i, I_i\}$ give a set of $\{R_i\}$, $i = \overline{1,n}$, and this produces

$$\bar{R} = \frac{1}{n}\sum_{i=1}^{n} R_i;\ S(\bar{R}) = \sqrt{\frac{\Sigma_{i=1}^{n}(R_i - \bar{R})^2}{n(n-1)}}.$$

It is obvious that the method of reduction is much simpler than the regression method. In addition, it appears to be more accurate due to the above mentioned disadvantage of the regression method. However, further research is required to justify this claim.

Appendix

TABLE A.1. Values of the probability density for the normal distribution $f(z) = 1/\sqrt{2\pi}\ e^{-z^2/2}$.

z	0	1	2	3	4	5	6	7	8	9
0.0	0.3989	0.3989	0.3989	0.3988	0.3986	0.3984	0.3982	0.3980	0.3977	0.3973
0.1	0.3970	0.3965	0.3961	0.3956	0.3951	0.3945	0.3939	0.3932	0.3925	0.3918
0.2	0.3910	0.3902	0.3894	0.3885	0.3876	0.3867	0.3857	0.3847	0.3836	0.3825
0.3	0.3814	0.3802	0.3790	0.3778	0.3765	0.3752	0.3739	0.3726	0.3712	0.3697
0.4	0.3683	0.3668	0.3653	0.3637	0.3621	0.3605	0.3589	0.3572	0.3555	0.3538
0.5	0.3521	0.3503	0.3485	0.3467	0.3448	0.3429	0.3410	0.3391	0.3372	0.3352
0.6	0.3332	0.3312	0.3292	0.3271	0.3251	0.3230	0.3209	0.3187	0.3166	0.3144
0.7	0.3123	0.3101	0.3079	0.3056	0.3034	0.3011	0.2989	0.2966	0.2943	0.2920
0.8	0.2897	0.2874	0.2850	0.2827	0.2803	0.2780	0.2756	0.2732	0.2709	0.2685
0.9	0.2661	0.2637	0.2613	0.2589	0.2565	0.2541	0.2516	0.2492	0.2468	0.2444
1.0	0.2420	0.2396	0.2371	0.2347	0.2323	0.2299	0.2275	0.2251	0.2227	0.2203
1.1	0.2179	0.2155	0.2131	0.2107	0.2083	0.2059	0.2036	0.2012	0.1989	0.1965
1.2	0.1942	0.1919	0.1895	0.1872	0.1849	0.1826	0.1804	0.1781	0.1758	0.1736
1.3	0.1714	0.1691	0.1669	0.1647	0.1626	0.1604	0.1582	0.1561	0.1539	0.1518
1.4	0.1497	0.1476	0.1456	0.1435	0.1415	0.1394	0.1374	0.1354	0.1334	0.1315
1.5	0.1295	0.1276	0.1257	0.1238	0.1219	0.1200	0.1182	0.1163	0.1145	0.1127
1.6	0.1109	0.1092	0.1074	0.1057	0.1040	0.1023	0.1006	0.0989	0.0973	0.0957
1.7	0.0940	0.0925	0.0909	0.0893	0.0878	0.0863	0.0848	0.0833	0.0818	0.0804
1.8	0.0790	0.0775	0.0761	0.0748	0.0734	0.0721	0.0707	0.0694	0.0681	0.0669
1.9	0.0656	0.0644	0.0632	0.0620	0.0608	0.0596	0.0584	0.0573	0.0562	0.0551
2.0	0.0540	0.0529	0.0519	0.0508	0.0498	0.0488	0.0478	0.0468	0.0459	0.0449
2.1	0.0440	0.0431	0.0422	0.0413	0.0404	0.0396	0.0387	0.0379	0.0371	0.0363
2.2	0.0355	0.0347	0.0339	0.0332	0.0325	0.0317	0.0310	0.0303	0.0297	0.0290
2.3	0.0283	0.0277	0.0270	0.0264	0.0258	0.0252	0.0246	0.0241	0.0235	0.0229
2.4	0.0224	0.0219	0.0213	0.0208	0.0203	0.0198	0.0194	0.0189	0.0184	0.0180
2.5	0.0175	0.0171	0.0167	0.0163	0.0158	0.0154	0.0151	0.0147	0.0143	0.0139
2.6	0.0136	0.0132	0.0129	0.0126	0.0122	0.0119	0.0116	0.0113	0.0110	0.0107
2.7	0.0104	0.0101	0.0099	0.0096	0.0093	0.0091	0.0088	0.0086	0.0084	0.0081
2.8	0.0079	0.0077	0.0075	0.0073	0.0071	0.0069	0.0067	0.0065	0.0063	0.0061
2.9	0.0060	0.0058	0.0056	0.0055	0.0053	0.0051	0.0050	0.0048	0.0047	0.0046
3.0	0.0044	0.0043	0.0042	0.0040	0.0039	0.0038	0.0037	0.0036	0.0035	0.0034
3.1	0.0033	0.0032	0.0031	0.0030	0.0029	0.0028	0.0027	0.0026	0.0025	0.0025
3.2	0.0024	0.0023	0.0022	0.0022	0.0021	0.0020	0.0020	0.0019	0.0018	0.0018
3.3	0.0017	0.0017	0.0016	0.0016	0.0015	0.0015	0.0014	0.0014	0.0013	0.0013
3.4	0.0012	0.0012	0.0012	0.0011	0.0011	0.0010	0.0010	0.0010	0.0009	0.0009
3.5	0.0009	0.0008	0.0008	0.0008	0.0008	0.0007	0.0007	0.0007	0.0007	0.0006
3.6	0.0006	0.0006	0.0006	0.0005	0.0005	0.0005	0.0005	0.0005	0.0005	0.0004
3.7	0.0004	0.0004	0.0004	0.0004	0.0004	0.0004	0.0003	0.0003	0.0003	0.0003
3.8	0.0003	0.0003	0.0003	0.0003	0.0003	0.0002	0.0002	0.0002	0.0002	0.0002
3.9	0.0002	0.0002	0.0002	0.0002	0.0002	0.0002	0.0002	0.0002	0.0001	0.0001

TABLE A.2. Values of the normalized Laplace function $\Phi(z) = 1/\sqrt{2\pi}\int_0^z e^{-y^2/2}dy$.

z	0	1	2	3	4	5	6	7	8	9
0.0	0.000 00	0.003 99	0.007 98	0.011 97	0.015 95	0.019 94	0.023 92	0.027 90	0.031 88	0.035 86
0.1	0.039 83	0.043 80	0.047 76	0.051 72	0.055 67	0.059 62	0.063 56	0.067 49	0.071 42	0.075 35
0.2	0.079 26	0.083 17	0.087 06	0.090 95	0.094 83	0.098 71	0.102 57	0.106 42	0.110 26	0.114 09
0.3	0.117 91	0.121 72	0.125 52	0.129 30	0.133 07	0.136 83	0.140 58	0.144 31	0.148 03	0.151 73
0.4	0.155 42	0.159 10	0.162 76	0.166 40	0.170 03	0.173 64	0.177 24	0.180 82	0.184 39	0.187 93
0.5	0.191 46	0.194 97	0.198 47	0.201 94	0.205 40	0.208 84	0.212 26	0.215 66	0.219 04	0.222 40
0.6	0.225 75	0.229 07	0.232 37	0.235 65	0.238 91	0.242 15	0.245 37	0.248 57	0.251 75	0.254 90
0.7	0.258 04	0.261 15	0.264 24	0.267 30	0.270 35	0.273 37	0.276 37	0.279 35	0.282 30	0.285 24
0.8	0.288 14	0.291 03	0.293 89	0.296 73	0.299 55	0.302 34	0.305 11	0.307 85	0.310 57	0.313 27
0.9	0.315 94	0.318 59	0.321 21	0.323 81	0.326 39	0.328 94	0.331 47	0.333 98	0.336 46	0.338 91
1.0	0.341 34	0.343 75	0.346 14	0.348 50	0.350 83	0.353 14	0.355 43	0.357 69	0.359 93	0.362 14
1.1	0.364 33	0.366 50	0.368 64	0.370 76	0.372 86	0.374 93	0.376 98	0.379 00	0.381 00	0.382 98
1.2	0.384 93	0.386 86	0.388 77	0.390 65	0.392 51	0.394 35	0.396 17	0.397 96	0.399 73	0.401 47
1.3	0.403 20	0.404 90	0.406 58	0.408 24	0.409 88	0.411 49	0.413 09	0.414 66	0.416 21	0.417 74
1.4	0.419 24	0.420 73	0.422 20	0.423 64	0.425 07	0.426 47	0.427 86	0.429 22	0.430 56	0.431 89
1.5	0.433 19	0.434 48	0.435 74	0.436 99	0.438 22	0.439 43	0.440 62	0.441 79	0.442 95	0.444 08
1.6	0.445 20	0.446 30	0.447 38	0.448 45	0.449 50	0.450 53	0.451 54	0.452 54	0.453 52	0.454 49
1.7	0.455 43	0.456 37	0.457 28	0.458 18	0.459 07	0.459 94	0.460 80	0.461 64	0.462 46	0.463 27
1.8	0.464 07	0.464 85	0.465 62	0.466 38	0.467 12	0.467 84	0.468 56	0.469 26	0.469 95	0.470 62
1.9	0.471 28	0.471 93	0.472 57	0.473 20	0.473 81	0.474 41	0.475 00	0.475 58	0.476 15	0.476 70
2.0	0.477 25	0.477 78	0.478 31	0.478 82	0.479 32	0.479 82	0.480 30	0.480 77	0.481 24	0.481 69
2.1	0.482 14	0.482 57	0.483 00	0.483 41	0.483 82	0.484 22	0.484 61	0.485 00	0.485 37	0.485 74
2.2	0.486 10	0.486 45	0.486 79	0.487 13	0.487 45	0.487 78	0.488 09	0.488 40	0.488 70	0.488 99
2.3	0.489 28	0.489 56	0.489 83	0.490 10	0.490 36	0.490 61	0.490 86	0.491 11	0.491 34	0.491 58
2.4	0.491 80	0.492 02	0.492 24	0.492 45	0.492 66	0.492 86	0.493 05	0.493 24	0.493 43	0.493 61
2.5	0.493 79	0.493 96	0.494 13	0.494 30	0.494 46	0.494 61	0.494 77	0.494 92	0.495 06	0.495 20
2.6	0.495 34	0.495 47	0.495 60	0.495 73	0.495 85	0.495 98	0.496 09	0.496 21	0.496 32	0.496 43
2.7	0.496 53	0.496 64	0.496 74	0.496 83	0.496 93	0.497 02	0.497 11	0.497 20	0.497 28	0.497 36
2.8	0.497 44	0.497 52	0.497 60	0.497 67	0.497 74	0.497 81	0.497 88	0.497 95	0.498 01	0.498 07
2.9	0.498 13	0.498 19	0.498 25	0.498 31	0.498 36	0.498 41	0.498 46	0.498 51	0.498 56	0.498 61

Note. The values of $\Phi(z)$ for $z = 3.0$–4.5 are as follows:

z	$\Phi(z)$	z	$\Phi(z)$	z	$\Phi(z)$
3.0	0.498 65	3.4	0.499 66	3.8	0.499 93
3.1	0.499 03	3.5	0.499 77	3.9	0.499 95
3.2	0.499 31	3.6	0.499 84	4.0	0.499 968
3.3	0.499 52	3.7	0.499 89	4.5	0.499 997

TABLE A.3. Critical values of the distribution of $T_n = (x_n - \bar{x})/S$ or $T_1 = (\bar{x} - x_1)/S$ (with unilateral check).

Number of observations, n	Upper 0.5% significance level	Upper 1% significance level	Upper 5% significance level
3	1.155	1.155	1.153
4	1.496	1.492	1.463
5	1.764	1.749	1.672
6	1.973	1.944	1.822
7	2.139	2.097	1.938
8	2.274	2.221	2.032
9	2.387	2.323	2.110
10	2.482	2.410	2.176
11	2.564	2.485	2.234
12	2.636	2.550	2.285
13	2.699	2.607	2.331
14	2.755	2.659	2.371
15	2.806	2.705	2.409
16	2.852	2.747	2.443
17	2.894	2.785	2.475
18	2.932	2.821	2.504
19	2.968	2.854	2.532
20	3.001	2.884	2.557
21	3.031	2.912	2.580
22	3.060	2.939	2.603
23	3.087	2.963	2.624
24	3.112	2.987	2.644
25	3.135	3.009	2.663
26	3.157	3.029	2.681
27	3.178	3.049	2.698
28	3.199	3.068	2.714
29	3.218	3.085	2.730
30	3.236	3.103	2.745

TABLE A.4. Values of the q percent points of Student's distribution.

Number of degrees of freedom $\nu = n - 1$	Significance level $q = (1 - \alpha) \times 100$ (%)		
	10	5	1
1	6.31	12.71	63.66
2	2.92	4.30	9.92
3	2.35	3.18	5.84
4	2.13	2.78	4.60
5	2.02	2.57	4.03
6	1.94	2.45	3.71
7	1.90	2.36	3.50
8	1.86	2.31	3.36
9	1.83	2.26	3.25
10	1.81	2.23	3.17
12	1.78	2.18	3.06
14	1.76	2.14	2.98
16	1.75	2.12	2.92
18	1.73	2.10	2.88
20	1.72	2.09	2.84
22	1.72	2.07	2.82
24	1.71	2.06	2.80
26	1.71	2.06	2.78
28	1.70	2.05	2.76
30	1.70	2.04	2.75
∞	1.64	1.96	2.58

TABLE A.5. Values of the q percent points of the χ^2 distribution $P\{\chi^2 > \chi_q^2\}$.

Number of degrees of freedom $\nu = n - 1$	Significance level q (%)									
	99	95	90	80	70	30	20	10	5	1
1	0.000 16	0.003 93	0.0158	0.0642	0.148	1.074	1.642	2.706	3.841	6.635
2	0.0201	0.103	0.211	0.446	0.713	2.408	3.219	4.605	5.991	9.210
3	0.115	0.352	0.584	1.005	1.424	3.665	4.642	6.251	7.815	11.345
4	0.297	0.711	1.064	1.649	2.195	4.878	5.989	7.779	9.488	13.277
5	0.554	1.145	1.610	2.343	3.000	6.064	7.289	9.236	11.070	15.086
6	0.872	1.635	2.204	3.070	3.828	7.231	8.558	10.645	12.592	16.812
7	1.239	2.167	2.833	3.822	4.671	8.383	9.803	12.017	14.067	18.475
8	1.646	2.733	3.490	4.594	5.527	9.524	11.030	13.362	15.507	20.090
9	2.088	3.325	4.168	5.380	6.393	10.656	12.242	14.684	16.919	21.666
10	2.558	3.940	4.865	6.179	7.267	11.781	13.442	15.987	18.307	23.209
11	3.053	4.575	5.578	6.989	8.148	12.899	14.631	17.275	19.675	24.725
12	3.571	5.226	6.304	7.807	9.034	14.011	15.812	18.549	21.026	26.217
13	4.107	5.892	7.042	8.634	9.926	15.119	16.985	19.812	22.362	27.688
14	4.660	6.571	7.790	9.467	10.821	16.222	18.151	21.064	23.685	29.141
15	5.229	7.261	8.547	10.307	11.721	17.322	19.311	22.307	24.996	30.578
16	5.812	7.962	9.312	11.152	12.624	18.418	20.465	23.542	26.296	32.000
17	6.408	8.672	10.085	12.002	13.531	19.511	21.615	24.769	27.587	33.409
18	7.015	9.390	10.865	12.857	14.440	20.601	22.760	25.989	28.869	34.805
19	7.633	10.117	11.651	13.716	15.352	21.689	23.900	27.204	30.144	36.191
20	8.260	10.851	12.443	14.578	16.266	22.775	25.038	28.412	31.410	37.566
21	8.897	11.591	13.240	15.445	17.182	23.858	26.171	29.615	32.671	38.932
22	9.542	12.338	14.041	16.314	18.101	24.939	27.301	30.813	33.924	40.289
23	10.196	13.091	14.848	17.187	19.021	26.018	28.429	32.007	35.172	41.638
24	10.856	13.848	15.659	18.062	19.943	27.096	29.553	33.196	36.415	42.980
25	11.524	14.611	16.473	18.940	20.867	28.172	30.675	34.382	37.652	44.314
26	12.198	15.379	17.292	19.820	21.792	29.246	31.795	35.563	38.885	45.642
27	12.879	16.151	18.114	20.703	22.719	30.319	32.912	36.741	40.113	46.963
28	13.565	16.928	18.939	21.588	23.647	31.391	34.027	37.916	41.337	48.278
29	14.256	17.708	19.768	22.475	24.577	32.461	35.139	39.087	42.557	49.588
30	14.953	18.493	20.599	23.364	25.508	33.530	36.250	40.256	43.773	50.892

TABLE A.6. Values of the upper 1% of points of the distribution $F_{0.01} = S_1^2/S_2^2$.

	Number of degrees of freedom ν_1											
ν_2	1	2	3	4	5	6	8	12	16	24	50	∞
2	98.49	99.00	99.17	99.25	99.30	99.33	99.36	99.42	99.44	99.46	99.48	99.50
3	34.12	30.81	29.46	28.71	28.24	27.91	27.49	27.05	26.83	26.60	26.35	26.12
4	21.20	18.00	16.69	15.98	15.52	15.21	14.80	14.37	14.15	13.93	13.69	13.46
5	16.26	13.27	12.06	11.39	10.97	10.67	10.29	9.89	9.68	9.47	9.24	9.02
6	13.74	10.92	9.78	9.15	8.75	8.47	8.10	7.72	7.52	7.31	7.09	6.88
7	12.25	9.55	8.45	7.85	7.46	7.19	6.84	6.47	6.27	6.07	5.85	5.65
8	11.26	8.65	7.59	7.01	6.63	6.37	6.03	5.67	5.48	5.28	5.06	4.86
9	10.56	8.02	6.99	6.42	6.06	5.80	5.47	5.11	4.92	4.73	4.51	4.31
10	10.04	7.56	6.55	5.99	5.64	5.39	5.06	4.71	4.52	4.33	4.12	3.91
11	9.65	7.20	6.22	5.67	5.32	5.07	4.74	4.40	4.21	4.02	3.80	3.60
12	9.33	6.93	5.95	5.41	5.06	4.82	4.50	4.16	3.98	3.78	3.56	3.36
13	9.07	6.70	5.74	5.20	4.86	4.62	4.30	3.96	3.78	3.59	3.37	3.16
14	8.86	6.51	5.56	5.03	4.69	4.46	4.14	3.80	3.62	3.43	3.21	3.00
15	8.68	6.36	5.42	4.89	4.56	4.32	4.00	3.67	3.48	3.29	3.07	2.87
16	8.53	6.23	5.29	4.77	4.44	4.20	3.89	3.55	3.37	3.18	2.96	2.75
17	8.40	6.11	5.18	4.67	4.34	4.10	3.79	3.45	3.27	3.08	2.86	2.65
18	8.28	6.01	5.09	4.58	4.25	4.01	3.71	3.37	3.20	3.00	2.79	2.57
19	8.18	5.93	5.01	4.50	4.17	3.94	3.63	3.30	3.12	2.92	2.70	2.49
20	8.10	5.85	4.94	4.43	4.10	3.87	3.56	3.23	3.05	2.86	2.63	2.42
21	8.02	5.78	4.87	4.37	4.04	3.81	3.51	3.17	2.99	2.80	2.58	2.36
22	7.94	5.72	4.82	4.31	3.99	3.76	3.45	3.12	2.94	2.75	2.53	2.31
23	7.88	5.66	4.76	4.26	3.94	3.71	3.41	3.07	2.89	2.70	2.48	2.26
24	7.82	5.61	4.72	4.22	3.90	3.67	3.36	3.03	2.85	2.66	2.44	2.21
25	7.77	5.57	4.68	4.18	3.86	3.63	3.32	2.99	2.81	2.62	2.40	2.17
26	7.72	5.53	4.64	4.14	3.82	3.59	3.29	2.96	2.78	2.58	2.36	2.13
27	7.68	5.49	4.60	4.11	3.78	3.56	3.26	2.93	2.74	2.55	2.33	2.10
28	7.64	5.45	4.57	4.07	3.75	3.53	3.23	2.90	2.71	2.52	2.30	2.06
29	7.60	5.42	4.54	4.04	3.73	3.50	3.20	2.87	2.68	2.49	2.27	2.03
30	7.56	5.39	4.51	4.02	3.70	3.47	3.17	2.84	2.66	2.47	2.24	2.01
35	7.42	5.27	4.40	3.91	3.59	3.37	3.07	2.74	2.56	2.37	2.13	1.90
40	7.31	5.18	4.31	3.83	3.51	3.29	2.99	2.66	2.48	2.29	2.05	1.80
45	7.23	5.11	4.25	3.77	3.45	3.23	2.94	2.61	2.43	2.23	1.99	1.75
50	7.17	5.06	4.20	3.72	3.41	3.19	2.89	2.56	2.38	2.18	1.94	1.68
60	7.08	4.98	4.13	3.65	3.34	3.12	2.82	2.50	2.32	2.12	1.87	1.60
70	7.01	4.92	4.07	3.60	3.29	3.07	2.78	2.45	2.28	2.07	1.82	1.53
80	6.96	4.88	4.04	3.56	3.26	3.04	2.74	2.42	2.24	2.03	1.78	1.49
90	6.92	4.85	4.01	3.53	3.23	3.01	2.72	2.39	2.21	2.00	1.75	1.45
100	6.90	4.82	3.98	3.51	3.21	2.99	2.69	2.37	2.19	1.98	1.73	1.43
125	6.84	4.78	3.94	3.47	3.17	2.95	2.66	2.33	2.15	1.94	1.69	1.37
∞	6.64	4.60	3.78	3.32	3.02	2.80	2.51	2.18	1.99	1.79	1.52	1.00

TABLE A.7. Values of the upper 5% of points of the distribution $F_{0.05} = S_1^2/S_2^2$.

	Number of degrees of freedom ν_1											
ν_2	1	2	3	4	5	6	8	12	16	24	50	∞
2	18.51	19.00	19.16	19.25	19.30	19.33	19.37	19.41	19.43	19.45	19.47	19.50
3	10.13	9.55	9.28	9.12	9.01	8.94	8.84	8.74	8.69	8.64	8.58	8.53
4	7.71	6.94	6.59	6.39	6.26	6.16	6.04	5.91	5.84	5.77	5.70	5.63
5	6.61	5.79	5.41	5.19	5.05	4.95	4.82	4.68	4.60	4.53	4.44	4.36
6	5.99	5.14	4.76	4.53	4.39	4.28	4.15	4.00	3.92	3.84	3.75	3.67
7	5.59	4.74	4.35	4.12	3.97	3.87	3.73	3.57	3.49	3.41	3.32	3.23
8	5.32	4.46	4.07	3.84	3.69	3.58	3.44	3.28	3.20	3.12	3.03	2.93
9	5.12	4.26	3.86	3.63	3.48	3.37	3.23	3.07	2.98	2.90	2.80	2.71
10	4.96	4.10	3.71	3.48	3.33	3.22	3.07	2.91	2.82	2.74	2.64	2.54
11	4.84	3.98	3.59	3.36	3.20	3.09	2.95	2.79	2.70	2.61	2.50	2.40
12	4.75	3.88	3.49	3.26	3.11	3.00	2.85	2.69	2.60	2.50	2.40	2.30
13	4.67	3.80	3.41	3.18	3.02	2.92	2.77	2.60	2.51	2.42	2.32	2.21
14	4.60	3.74	3.34	3.11	2.96	2.85	2.70	2.53	2.44	2.35	2.24	2.13
15	4.54	3.68	3.29	3.06	2.90	2.79	2.64	2.48	2.39	2.29	2.18	2.07
16	4.49	3.63	3.24	3.01	2.85	2.74	2.59	2.42	2.33	2.24	2.13	2.01
17	4.45	3.59	3.20	2.96	2.81	2.70	2.55	2.38	2.29	2.19	2.08	1.96
18	4.41	3.55	3.16	2.93	2.77	2.66	2.51	2.34	2.25	2.15	2.04	1.92
19	4.38	3.52	3.13	2.90	2.74	2.63	2.48	2.31	2.21	2.11	2.00	1.88
20	4.35	3.49	3.10	2.87	2.71	2.60	2.45	2.28	2.18	2.08	1.96	1.64
21	4.32	3.47	3.07	2.84	2.68	2.57	2.42	2.25	2.15	2.05	1.93	1.81
22	4.30	3.44	3.05	2.82	2.66	2.55	2.40	2.23	2.13	2.03	1.91	1.78
23	4.28	3.42	3.03	2.80	2.64	2.53	2.38	2.20	2.11	2.00	1.88	1.76
24	4.26	3.40	3.01	2.78	2.62	2.51	2.36	2.18	2.09	1.98	1.86	1.73
25	4.24	3.38	2.99	2.76	2.60	2.49	2.34	2.16	2.07	1.96	1.84	1.71
26	4.22	3.37	2.98	2.74	2.59	2.47	2.32	2.15	2.05	1.95	1.82	1.69
27	4.21	3.35	2.96	2.73	2.57	2.46	2.30	2.13	2.03	1.93	1.80	1.67
28	4.20	3.34	2.95	2.71	2.56	2.44	2.29	2.12	2.02	1.91	1.78	1.65
29	4.18	3.33	2.93	2.70	2.54	2.43	2.28	2.10	2.00	1.90	1.77	1.64
30	4.17	3.32	2.92	2.69	2.53	2.42	2.27	2.09	1.99	1.89	1.76	1.62
35	4.12	3.26	2.87	2.64	2.48	2.37	2.22	2.04	1.94	1.83	1.70	1.57
40	4.08	3.23	2.84	2.61	2.45	2.34	2.18	2.00	1.90	1.79	1.66	1.51
45	4.06	3.21	2.81	2.58	2.42	2.31	2.15	1.97	1.87	1.76	1.63	1.48
50	4.03	3.18	2.79	2.56	2.40	2.29	2.13	1.95	1.85	1.74	1.60	1.44
60	4.00	3.15	2.76	2.52	2.37	2.25	2.10	1.92	1.81	1.70	1.56	1.39
70	3.98	3.13	2.74	2.50	2.35	2.23	2.07	1.89	1.79	1.67	1.53	1.35
80	3.96	3.11	2.72	2.49	2.33	2.21	2.06	1.88	1.77	1.65	1.51	1.32
90	3.95	3.10	2.71	2.47	2.32	2.20	2.04	1.86	1.76	1.64	1.49	1.30
100	3.94	3.09	2.70	2.46	2.30	2.19	2.03	1.85	1.75	1.63	1.48	1.28
125	3.92	3.07	2.68	2.44	2.29	2.17	2.01	1.83	1.72	1.60	1.45	1.25
∞	3.84	2.99	2.60	2.37	2.21	2.09	1.94	1.75	1.64	1.52	1.35	1.00

Glossary

Absolute error (of a measuring instrument): The difference between a value of a measurand obtained by a measuring instrument and the true value of this measurand. *Note*: The absolute error of a material measure is the difference between the nominal value of this measure and the true value of a quantity that was reproduced by this measure.

Absolutely constant elementary error: An elementary error that remains the same value in repeated measurements performed under the same conditions using an arbitrarily chosen measuring instrument of a given type. The value of an absolutely constant error is unknown but its limits can be estimated.

Accuracy class: A class of measuring instruments that meets certain metrological requirements that are intended to keep errors within specified limits (Ref. 1, no. 5.22).

Accuracy of measurement: A qualitative expression of the closeness of the result of a measurement to the true value of the measurand.

Accuracy of a measuring instrument: The ability of a measuring instrument to produce measurements whose results are close to the true value of a measurand.

Additional error of measuring instruments: The difference between an error of a measuring instrument when the value of one influence quantity exceeds its reference value and the error of the measuring instrument under reference condition. *Note*: Additional errors of indicating electrical instruments are generally called *variations.*

Calibration: The set of operations that establish, under specified conditions, the relationship between values indicated by a measuring instrument or measuring system or values represented by material measure and the corresponding known values of a measurand (Ref. 1, no. 6.13). Results of a calibration may be presented in the form of a calibration curve, or a statement that the errors of an instrument exceed or do not exceed certain limits.

Combined measurement: Simultaneous measurements of several quantities of the same kind based on results of direct measurements of different combinations of them (Ref. 3, no. 4.4). *Example*: Combined measurement is performed when the masses of separate weights from one set are found using the known value of one weight.

Conditionally constant elementary error (of a measurement): An elementary error that varies in repeated measurements performed under the same conditions or with different specimens of measuring instruments of the same type having certain limits. These limits can be calculated or estimated.

Dead zone: The range through which a stimulus can be varied without producing a change in the response of a measuring instrument (Ref. 1, no. 5.14).

Discrimination threshold: The smallest change in a stimulus that produces a perceptible change in the response of a measuring instrument (Ref. 1, no. 5.12).

Drift: A slow variation with time at an output of a measuring instrument that is independent of a stimulus.

Elementary error (of a measurement): A component of error or uncertainty of a measurement associated with a single source of inaccuracy of the measurement.

Error (of a measurement): The deviation of the result of a measurement from the true value of the measurand expressed in absolute or relative form.

Fiducial error: A ratio of limit of absolute error of a measuring instrument and a value specified for this instrument. The specified value is called the fiducial value. It may be, for example, the span or the upper limit of the nominal range of the measuring instrument. Fiducial error is expressed as a percentage.

Inaccuracy (of a measurement): A qualitative characteristic of the degree of deviation of a measurement result from the true value of the measurand. Quantitatively, inaccuracy can be characterized either as a measurement error or as a measurement uncertainty.

Indicating instrument: A measuring instrument that displays the value of a measurand.

Influence coefficient: A factor that is multiplied by a value of the variation of the influence quantity relative to its reference condition limits gives an additional error.

Influence function: A metrological characteristic of the measuring instrument expressing the relationship between the measuring instrument errors and an influence quantity.

Informative parameter (of an input signal): A parameter of an input process to be measured.

Intrinsic error: The error of a measuring instrument under reference conditions (Ref. 1, no. 5.27).

Limits of permissible error (of a measuring instrument): Extreme values of error of a given measuring instrument permitted by standard or specification (Ref. 1, no. 5.23).

Material measure: A measuring instrument that reproduces a physical quantity with known value.

Measurand: A value of physical quantity to be measured.

Measurement: The set of experimental operations that are performed using special technical products (measuring instruments) for the purpose of finding the value of a physical quantity (Ref. 3, no. 4.1).

Measuring instrument: A technical object developed for the purpose of measurements (Ref. 3, no. 5.1). The measuring instrument has standardized metrological characteristics.

Measuring standard: A measuring instrument intended to reproduce and/or conserve a unit of a physical quantity in order to transmit its value to other measuring instruments (Ref. 3, no. 10.1).

Measuring system: A complete set of measuring instruments and supplementary equipment assembled for obtaining measurement results in required form and for inputting data to a control system.

Measuring transducer: A measuring instrument that converts the measurement signals into a form suitable for transmission or processing. The signals at the output of a measuring transducer can not be directly observed.

Metrological characteristic of a measuring instrument: A characteristic of a measuring instrument that is necessary to judge the suitability of the instrument for performing measurements in a known range or that is necessary to estimate the inaccuracy of measurement results.

Metrology: The field of knowledge concerned with measurement (Ref. 1, no. 2.02]. Metrology is an applied science. It includes knowledge of measurements of any kind of physical quantities and with any level of accuracy.

Normal operating conditions: Conditions of use of measuring instruments giving the ranges of the influence quantities within which a measuring instrument is designed to operate and for which the metrological characteristics of this instrument lie within specified limits.

Primary standard: A measurement standard that has the highest accuracy in a country.

Random error (of a measurement): A component of the inaccuracy of a measurement that, in the course of a number of measurements of the same measurand under the same conditions varies in an unpredictable way.

Reference conditions: A complete set of values of influence quantities standardized for specific types of measuring instruments in order to ensure a maximum accuracy of the performing measurements and to calibrate these instruments.

Reference standard: A standard, generally of the highest metrological quality available at a given location, from which measurements at that location are derived (Ref. 1, no. 6.08). Reference standards are secondary standards, as a rule.

Relative error: Absolute error divided by a true value of the measurand (Ref. 1, no. 3.11). The measurement result substitutes the true value, in practice.

Repeatability of a measurement: The closeness of agreement among a number of consecutive measurements for the same measurand performed under the same operating conditions with the same measuring instruments, over a short period of time.

Reproducibility of a measurement: The closeness of agreement among repeated measurements for the same measurand performed in different locations, under different operating conditions, or over a long period of time.

Resolution: The smallest interval between two adjacent values of the output signal of a measuring instrument that can be distinguished.

Response time: The time interval between the instant when a measuring instrument gets a stimulus and the instant when the response reaches and remains within specified limits of its final steady value.

Result of measurement: The value of a measurand obtained by measurement (Ref. 3, no. 8.18). Measurement results are expressed as a number of proper units. For example, 1.25 m is the value of length of a body; 1.25 is here the numerical value.

Secondary standard: A measurement standard that obtains the value of a unit from the primary standard (Ref. 1, no. 6.05).

Sensitivity: The change in the response of a measuring instrument divided by the corresponding change in the stimulus (Ref. 1, no. 5.10).

Simultaneous measurement: The set of measurements of several quantities of different kinds performed under special plan. (Ref. 3, no. 4.5). *Example*: A measurement in which an electric resistance of a resistor and its temperature coefficients are found based on direct measurements of the resistance and temperature performed at different temperatures.

Span: The modulus of the difference between the two limits of a nominal range of a measuring instrument (Ref. 1, no. 5.02). Span is expressed in a unit of the measured quantity. *Example*: A range is − 15 to + 15 V; the span is 30 V.

Systematic error (of measurement): A component of the inaccuracy of measurement that, in the course of a number of measurements of the same measurand, remains constant or varies in a predictable way.

True value: The value of a measurand that being known would ideally reflect both qualitatively and quantitatively the corresponding property of a object (Ref. 3, no. 2.5).

Uncertainty of measurement: An interval within which a true value of a measurand lies with given probability. Uncertainty is defined with its limits and corresponding confidence probability. The limits are read out from a result of measurement. Uncertainty can be expressed in absolute or in relative form.

Verification: A kind of calibration that reveals whether or not the errors of a measuring instrument lie within their specified limits.

Working standard: A measurement standard that is used to calibrate measuring instruments.

References

Standards and Recommendations

1. International Vocabulary of Basic and General Terms in Metrology, ISO (1984).
2. Measurement Uncertainty, American National Standard ANSI/ASME PTC 19.1 (1985).
3. Metrology, Terms and Definitions, State Standard of the Soviet Union, GOST 16263-70, Moscow (1970).
4. Direct Measurements with Multiple Observations, Methods of Processing the Results of Observations: Basic Principles, State Standard of the Soviet Union, GOST 8.207-76, Moscow (1976).
5. National Measurement Accreditation Service, B 3003, English National Physical Laboratory, Teddington (1986).
6. Process Instrumentation Terminology, American National Standard ANSI/ISA S51.1 (1979).
7. Publication 51, PT 1-84 (1984), Direct Acting Indicating Analogue Electrical Measuring Instruments and Their Accessories, Bureau Central de la Comission Electrotechnique Internationale, Geneva (1984), 4th ed.
8. Recommandation Internationale 34, Classes de precision des Instruments de Mesurage, Organisation International de Metrologie Legale, Paris (1974).
9. Standard Practice for Use of the International System of Units (SI), American Society for Testing and Materials E 380-91 (1991).
10. Standard Practice for Dealing With Outlying Observations, American Society for Testing and Materials, E 178-80 (Reapproved 1989).
11. Standard Practice for the Evaluation of Single-Pan Mechanical Balances, American Society for Testing and Materials, E 319-85 (1985).
12. Technical Data Fluke, Application Information B0005, Total Instrument Accuracy for Digital Thermometers and Data Loggers (formerly AB-29).

Books and Articles

13. R. B. Abernethy *et al.*, Measurement Uncertainty Handbook, Instrument Society of America (1980).
14. V. Ya. Alekseev, F. M. Karavaev *et al.*, "Estimate of the measured quantity from results of measurements with different systematic errors," Metrologiya (Supplement to the journal Izmeritel'naya Tekhnika), No. 1, 18–24 (1978).
15. H. Bäckström, "Uber die dezimalgleichung beim ablesen von scalen," Zeitschrift fur Instrumentenkunde (1930–32).
16. V. V. Berezina and I. N. Rybakov, "Distribution of the resulting error, with multivalued components, of measuring instruments," Izmeritel'naya Tekhnika (Measurement Techniques), No. 12, 25–27 (1975).
17. G. D. Burdun and B. N. Markov, *Principles of Metrology* [in Russian], 2nd ed. (Izd. standartov, Moscow, 1975).
18. I. E. Burns, P. J. Campion, and A. Williams, "Error and uncertainty," Metrologia **9**, 101–104 (1973).
19. E. R. Cohen, "Uncertainty and error in physical measurements," Enrico Fermi Summer School, June 27-July 7, Lerici (1989). Italian Physical Society. Proceedings of the International School of

Physics, "Enrico Fermi". Course CX edited by L. Crovini and T. J. Quinn. Villa Marigola, Lerici, June 27-July 7, 1989. Metrology at the Frontiers of Physics and Technology (in press). North-Holland. Amsterdam-Oxford-New York-Tokyo.

20. E. R. Cohen and B. N. Taylor, "The 1986 adjustment of the fundamental physical constants," Rev. Mod. Phys. **59**, No. 4 (1987).
21. H. Cramer, *Mathematical Methods of Statistics* (Princeton University Press, Princeton, NJ, 1946).
22. C. Croarkin, *Measurement Assurance Programs,* Part II: *Development and Implementation,* NBS Special Publication 676-11 (U.S. GPO, Washington, DC, 1985).
23. E. F. Dolinskii, *Analysis of the Results of Measurements* [in Russian], 2nd ed. (Izd. standartov, Moscow, 1973).
24. L. I. Dovbeta *et al.*, "Estimation of errors in direct measurements with single observations," in *Methods for Analyzing the Results of Measurements,* Proceedings of the Metrological Institutes of the USSR, D. I. Mendeleev All-Union Scientific-Research Institute of Metrology [in Russian], No. 242 (302), (Scientific-Industrial Corporation "D. I. Mendeleev All-Union Scientific-Research Institute of Metrology," Leningrad, 1979), pp. 49–59.
25. L. I. Dovbeta, "On the conditions of measurements with single observations," *ibid.*, pp. 60–63.
26. J. W. M. DuMond, "Accurate measurements of fundamental physical constants," *Science and Humanity. Yearbook* [in Russian] (Znanie, Moscow, 1964).
27. B. D. Ellis, *Basic Concepts of Measurement* (Cambridge University Press, Cambridge, 1966).
28. C. Eisenhart, "Realistic evaluation of the precision and accuracy of instrument calibration systems," J. Res. Natl. Bur. Stand. C **67**, 161–187 (1963).
29. V. V. Frumkin and A. V. Kotlyar, "Relation between the errors of working and standard instruments for checking," Metrologiya (Supplement to the journal Izmeritel'naya Tekhnika), No. 3, 3–11 (1973).
30. P. Giacomo, "News from BIPM," Metrologia **17**, 69 (1981).
31. S. V. Gorbatsevich, "Analysis of errors in measurement of the emf of reference standard normal elements on electric balances," in *Investigations in the Field of Electric Measurements,* Proceedings of the All-Union Scientific-Research Institute of Metrology, No. 38 (98) [in Russian], (Standartgiz, Leningrad, 1959), pp. 5–20.
32. V. A. Granovskii, *Dynamic Measurements, Principles of Metrological Support* [in Russian] (Energoatomizdat, Leningrad, 1984).
33. I. A. Grachev and S. G. Rabinovich, "Approximate method for constructing the distribution function of the composite of several distributions," Izmeritel'naya Tekhnika, No. 1, 8–11 (1968).
34. A. Hald, *Statistical Theory With Engineering Applications* (Wiley, New York, 1952).
35. D. M. Himmelblau, *Process Analysis by Statistical Methods* (Wiley, New York, 1970).
36. D. Kamke and K. Kramer, *Physikalishe grundlagen der masseinheiten. Mit Einem anhang uber fehlerrechnung* (Teubner, Stuttgart, 1977).
37. A. M. Kagan and Yu. V. Linnik, "On some problems of mathematical statistics in metrology," in *Fifty Years of Metric Reform in the USSR,* Proceedings of the Metrological Institutes of the USSR [in Russian], D. I. Mendeleev All-Union Scientific-Research Institute of Metrology, No. 123 (183) (Leningrad, 1970), pp. 39–47.
38. F. M. Karavaev, *Measurements of the Activity of Nuclides* [in Russian] (Izd. standartov, Moscow, 1972).
39. Zh. F. Kudryashova, "On estimation of the measured quantity based on two groups of measurements," Avtometriya, No. 1, 13–18 (1972).
40. Zh. F. Kudryashova and T. N. Siraya, "Construction of the confidence intervals for the product of two random quantities," in *Methods for Analyzing the Results of Measurements,* Proceedings of the Metrological Institutes of the USSR [in Russian], D. I. Mendeleev All-Union Scientific-Research Institute of Metrology, No. 172 (234), Energiya, (Leningrad, 1975), pp. 63–69.
41. B. R. Levin, *Theoretical Principles of Statistical Radioelectronics* [in Russian] (Sov. Radio, Moscow, 1966).
42. Yu. V. Linnik, *The Method of Least Squares and the Principles of the Mathematical-Statistical Theory of the Analysis of Observations* [in Russian], 2nd ed. (Fizmatgiz, Moscow, 1962).
43. M. F. Malikov, *Principles of Metrology* [in Russian] (Committee on Measures and Measuring Devices at the Council of Ministers of the USSR, Moscow, 1949).

44. B. S. Massey, *Measures in Science and Engineering, Their Expression, Relation and Interpretation* (Wiley, New York, 1986).
45. A. I. Mekhannikov, "Methods for immediate processing of the results of observations with the help of ordering statistics," Metrologiya (Supplement to Izmeritel'naya Tekhnika), No. 4, 47–58 (1972).
46. B. W. Petley, *Physical Constants and the Frontier of Measurement* (National Physical Laboratory) (Adam Hilger, Bristol and Philadelphia, 1988).
47. S. G. Rabinovich, "Toward the calculation of the errors of measuring instruments," Izmeritel'naya Tekhnika, No. 2, 15–18 (1968).
48. S. G. Rabinovich, "Method for calculating the measurement errors," Metrologiya (Supplement to Izmeritel'naya Tekhnika), No. 1, 3–12 (1970).
49. S. G. Rabinovich, *Galvanometric Self-Balancing Instruments* [in Russian] (Izd. standartov, Moscow, 1972).
50. S. G. Rabinovich and T. L. Yakovleva, "Analysis of the temporal stability of the distributions of the errors of measuring instruments," Metrologiya (Supplement to Izmeritel'naya Tekhnika), No. 7, 8–15 (1977).
51. S. G. Rabinovich, *Measurement Errors* [in Russian], (Energiya, Leningrad, 1978).
52. L. Sachs, *Applied Statistics: A Handbook of Techniques*, (Springer, New York, 1984).
53. T. N. Siraya, "Methods for checking the statistical homogeneity of groups of observations," in *General Questions in Metrology*, Proceedings of the Metrological Institutes of the USSR [in Russian], D. I. Mendeleev All-Union Scientific-Research Institute of Metrology No. 200 (260) (Energiya, Leningrad, 1977), pp. 20–29.
54. T. N. Siraya, "Methods of confluent analysis for construction of linear dependences," in *Methods for Processing the Results of Measurements, Proceedings of the Metrological Institutes of the USSR [in Russian]*, Scientific-Industrial Corporation "D. I. Mendeleev All-Union Scientific-Research Institute of Metrology" No. 242 (302) (Leningrad, 1979), pp. 33–44.
55. K. P. Shirokov, V. O. Arutyunov *et al.*, "Basic concepts of the theory of dynamic measurements," Izmeritel'naya Tekhnika, No. 12, 9–13 (1975).
56. K. P. Shirokov, "Theoretical questions regarding the formation of derivative units," in *General Questions in Metrology*, Proceedings of the Metrological Institutes of the USSR [in Russian], D. I. Mendeleev All-Union Scientific-Research Institute of Metrology No. 200 (260) (Energiya, Leningrad, 1977), pp. 12–19.
57. V. M. Sviridenko, "Logical-gnoseological aspect of the problem of the accuracy of measurements," Izmeritel'naya Tekhnika, No. 5, 6–8 (1971).
58. B. L. van der Waerden, Mathematical Statistics (Springer, New York 1969).
59. A. M. Vasilyev, *An Introduction to Statistical Physics*, (Translated from Russian) (MIR, Moscow, 1983).
60. R. L. Winkler and W. L. Hays, *Statistics: Probability, Inference and Decision*, 2nd ed. (Holt, Rinehart and Winston, New York, 1975).
61. T. L. Yakovleva, "About statistical homogeneity of the samples of measuring instrument errors," Metrologiya (Supplement to Izmeritel'naya Tekhnika), No. 2, 19–23 (1979).

Index

T

U

V

W